AF759961

Nano-Biomedical Engineering 2009

Proceedings of the Tohoku University
Global Centre of Excellence Programme

Global Nano-Biomedical Engineering
Education and Research Network Centre

Nano-Biomedical Engineering 2009

Proceedings of the Tohoku University Global Centre of Excellence Programme

Global Nano-Biomedical Engineering Education and Research Network Centre

Sendai International Centre,
Sendai, Japan 27 – 28 March 2009

Editor

Takami YAMAGUCHI, MD PhD

School of Biomedical Engineering, Tohoku University

ICP

Imperial College Press

Published by

Imperial College Press
57 Shelton Street
Covent Garden
London WC2H 9HE

Distributed by

World Scientific Publishing Co. Pte. Ltd.
5 Toh Tuck Link, Singapore 596224
USA office: 27 Warren Street, Suite 401-402, Hackensack, NJ 07601
UK office: 57 Shelton Street, Covent Garden, London WC2H 9HE

British Library Cataloguing-in-Publication Data
A catalogue record for this book is available from the British Library.

NANO-BIOMEDICAL ENGINEERING 2009
Proceedings of the Tohoku University Global Centre of Excellence Programme

Copyright © 2009 by Imperial College Press
All rights reserved. This book, or parts thereof, may not be reproduced in any form or by any means, electronic or mechanical, including photocopying, recording or any information storage and retrieval system now known or to be invented, without written permission from the Publisher.

For photocopying of material in this volume, please pay a copying fee through the Copyright Clearance Center, Inc., 222 Rosewood Drive, Danvers, MA 01923, USA. In this case permission to photocopy is not required from the publisher.

ISBN-13 978-1-84816-352-2
ISBN-10 1-84816-352-5

Printed in Singapore.

FOREWORD

It is my greatest pleasure to publish this book on the nano-biomedical engineering. Since we started the Tohoku University Global COE Programme "Global Nano-Biomedical Engineering Education and Research Network Centre", numerous excellent studies have been conducted by the members of the programme. It should particularly be noted that young investigators including PhD students and post-doctoral fellows supported by the programme have carried out prominent researches. This book is edited and published to allow international researchers of this field of nano-biomedical engineering understand and utilize those results of studies.

Nanotechnology will undoubtedly become one of the most important fields of science and technology of the 21st century. Amongst its many possible applications, biomedical engineering is one of the most promising. The advancement of biology, in combination with information technology, has opened a new era in medicine that will be the basis of sophisticated diagnostic and revolutionary therapeutic products. Nanotechnology, combined with biomedical engineering, will further reinforce this trend.

Development of biomedical engineering is particularly important in the East Asian Pacific Rim region due to the rapid changes in the composition of its population. The most rapidly developing countries of the 20th century in this area are all now confronting serious problems with ageing societies that require innovations in health care and medicine. This is not, however, simply a burden to society as is often suggested; the challenge of dealing with ageing should be looked upon as having the potential to build a new type of economy and industry.

Although it is frequently misunderstood that engineering and biology are very different fields of science, they are historically close and share a common basis. Even in previous times when biology was part of natural history, its advancement was supported by cutting-edge technologies of the time, such as chemistry and use of the microscope. Moreover, engineering has developed under a strong influence from biological fields, such as plant biology and medicine.

Particularly since the explosive advance in molecular biology following Watson and Crick's establishment of DNA's crystalline structure in 1953, it has been very clear that life has a chemical foundation, governed by the rules of physics and chemistry. In this sense, the history of biology since 1953 can be regarded as a process in which biology has transformed itself more and more into a quantitative science, sharing the same basic principles as other physical sciences.

We can now declare that we are in the age when life can be examined in the light of physical and mechanical principles, which leads to a conviction that any aspect of the life sciences can and should also be subject to engineering applications. We believe that the role of engineering, based on quantitative sciences such as physics, is a sine qua non in improving medicine and surgery, both of which are practical applications of biology toward human beings. We need to bear in mind, however, that engineering does not simply offer convenient means to assist medical practitioners.

A profound understanding of life through physical and chemical principles is required to invent and develop truly novel, innovative technologies in the field of medicine. Although we are sometimes dazzled by the success of novel devices and forget the fundamental research that finally led to them, as engineers, we know that what is most important is not just a new product, but the fundamental research and innovation at a concept level. The biomedical engineering we are pursuing in the current program provides such a balance between the fundamental research and industrial innovation.

In the Tohoku University Global COE Programme, the “Global Nano-Biomedical Engineering Network Centre”, we seek to integrate nano-biomedical engineering activities within East Asian and Pacific Rim countries. It is not restricted to those areas, but our intention is to start to organise the most active institutions in this region in the hope that worldwide collaborations will be implemented. To consolidate the infrastructure and growth of our and partner countries, we need to encourage our young scientists and engineers to collaborate in a global environment. I hope this book become a foundation of these collaborations.

December 2008

TAKAMI YAMAGUCHI
Global COE Leader
Department of Biomedical Engineering,
Graduate School of Biomedical Engineering, Tohoku University

CONTENTS

Section 1

Nano-Biomechanics

MEASUREMENT INTEGRATED SIMULATION TO PROVIDE ACCURATE INFORMATION OF BLOOD FLOWS

TOSHIYUKI HAYASE *

Institute of Fluid Science, Tohoku University,
2-1-1 Katahira, Aoba-ku, Sendai 980-8577, Japan

KENICHI FUNAMOTO

Institute of Fluid Science, Tohoku University,
2-1-1 Katahira, Aoba-ku, Sendai 980-8577, Japan

In order to make an accurate diagnosis in cases of serious cardiovascular diseases such as aortic or cerebral aneurysms, a method to obtain detailed information on the blood flow velocity and pressure is essential, measurement and simulation being the basic tools for this. In this article, a general formulation of measurement integrated simulation is first explained and then the results of numerical experiments for ultrasonic measurement integrated simulation of an aneurysmal aorta and MR measurement integrated simulation of a cerebral artery are presented.

Keywords: numerical realization; flow observer; blood flow; aneurysmal aorta; ultrasonic measurement; cerebral artery; MR measurement; flow simulation.

1. Introduction

In order to make an accurate diagnosis in cases of serious cardiovascular diseases such as aortic or cerebral aneurysms, a method to obtain detailed information on the blood flow velocity and pressure is essential, measurement and simulation being the basic tools for this.

Among a number of measurement methods, ultrasound color Doppler imaging is widely used since it provides a real-time image of the blood flow structure and vessel configuration non-invasively with relatively compact equipment [1]. However, the velocity measurement with this apparatus provides only the velocity component along the ultrasound beam, and, therefore, it is difficult to comprehend the exact three-dimensional blood velocity field.

As a counterpart of measurement, numerical simulation of the blood flow has been extensively studied. Realistic solutions of the blood velocity and pressure are obtained by solving the fundamental equations of the flow for real

---

* Toshiyuki Hayase is a Tohoku University Global COE Member.

vessel geometries obtained by visualizing methods such as X-ray computer tomography (CT) or magnetic resonance imaging (MRI) [2,3]. However, the simulation has an inherent problem of difficulty in specifying boundary conditions and/or initial conditions, and the calculated blood flow is similar but not exactly identical to the real one.

In order to overcome shortcomings of measurement and simulation, many studies have been carried out to combine them so as to develop a new methodology. Assimilation is a method commonly used in numerical weather prediction [4]. In a numerical simulation to predict future weather conditions, the initial condition is repeatedly updated at certain time intervals using the latest computational results and the measurement data around the computational grid points. A Similar concept, namely, interactive computational-experimental methodology (ICEME) was proposed by Humphrey et al [5] for application to engineering problems. Zeldin and Meade applied the Tikhonov regularization method, which is common in inverse problems, to obtain an optimum solution to estimate the real flow from the numerical and measurement results [6]. Studies have been made to apply CFD (computational fluid dynamics) schemes to modify PIV (particle imaging velocimetry) measurements to satisfy physical constraints such as the continuity equation [7]. The observer and Kalman filter, which are fundamental tools in modern control theory, have been applied to flow problems [8]. The authors have proposed a concept of numerical realization [9], which is defined as a numerical simulation with a mechanism to include information on real phenomena appropriately. Figure 1 explains the numerical realization based on the concept of the observer in control theory [10]. A finite number of measurable output signals are defined in the real flow as well as in the simulation, and a feedback signal proportional to the difference between the output signals of the measurement and the simulation is added to the simulation to converge to the real flow. This methodology, measurement-integrated (MI) simulation, has been successfully applied to obtain field information for a turbulent flow in a square duct [11] and in a Karman vortex street behind a square cylinder [12].

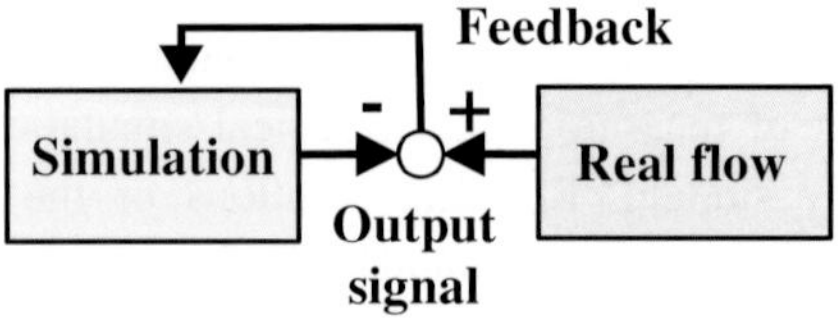

Figure 1. Numerical realization based on flow observer.

For medical application of MI simulation, the authors have proposed ultrasonic-measurement-integrated (UMI) simulation of the blood flow by integrating the ultrasound blood velocity measurement and the numerical simulation based on the flow observer. Since existing theory of observers is not applicable to complex flow problems, design and evaluation of the feedback schemes were made using numerical experiment in various conditions: two-dimensional unsteady problems [13,14] or three-dimensional steady [15] and unsteady problems [16]. Measurement integrated simulation was also applied to magnetic resonance (MR) measurement. A numerical experiment revealed that blood flow in a cerebral artery was successfully reproduced by MR measurement integrated simulation [17].

In this article, a general formulation of the MI simulation is first explained and then the results of numerical experiments for UMI simulation in an aneurysmal aorta and MR measurement integrated simulation in a cerebral artery are presented.

2. General Formulation

The dynamic behavior of an incompressible and viscous flow field is governed by the Navier-Stokes equation and a pressure equation derived from the equation of continuity with appropriate initial and the boundary conditions. A discrete model is described as follows:

$$\begin{cases} \dfrac{d\mathbf{u}_N}{dt} = g_N(\mathbf{u}_N) + h_N(\mathbf{p}_N) + \mathbf{f}_N \\ \Delta_N \mathbf{p}_N = q_N(\mathbf{u}_N) + \nabla_N \mathbf{f}_N \end{cases}, \qquad (1)$$

where $\mathbf{u}_N$ and $\mathbf{p}_N$ are the $3N$-dimensional vector consisting of velocity components at N grid points and the N-dimensional vector consisting of pressure divided by the density at N grid points, respectively. $\mathbf{f}_N$ is external force which is used as the feedback signal in MI simulation. ∇_N and Δ_N are matrices to express the discrete model of ∇ and Δ, respectively. g_N, h_N, and q_N are given as follows.

$$\begin{aligned} g_N(\mathbf{u}_N) &= -(\mathbf{u}_N \cdot \nabla_N)\mathbf{u}_N + \nu \cdot \Delta_N \mathbf{u}_N \\ h_N(\mathbf{p}_N) &= -\nabla_N \mathbf{p}_N \\ q_N(\mathbf{u}_N) &= -\nabla_N\left((\mathbf{u}_N \cdot \nabla_N)\mathbf{u}_N\right) \end{aligned}, \qquad (2)$$

The real flow without external force is described by the following equation:

$$\begin{cases} \frac{d}{dt} D_N(\mathbf{u}) = D_N\left(g(\mathbf{u})\right) + D_N\left(h(p)\right) \\ D_N(\Delta p) = D_N\left(q(\mathbf{u})\right) \end{cases}, \tag{3}$$

where D_N is the operator to extract the information of the real flow at the computational grid points.

In MI simulation, on the other hand, the external force is applied as the feedback signal denoted by a function of real flow and numerical simulation. In the present study, we consider the case of feedback signal $\mathbf{f}_N$ denoted by a linear function of the difference of velocity and pressure between real flow and numerical simulation:

$$\begin{aligned} \mathbf{f}_N \equiv & -\mathbf{K}_\mathbf{u}\mathbf{C}_\mathbf{u}\left\{\mathbf{u}_N - D_N\left(\mathbf{u} + \boldsymbol{\varepsilon}_\mathbf{u}\right)\right\} \\ & -\mathbf{K}_\mathbf{p}\mathbf{C}_\mathbf{p}\left\{\mathbf{p}_N - D_N\left(\mathbf{p} + \boldsymbol{\varepsilon}_\mathbf{p}\right)\right\}, \end{aligned} \tag{4}$$

where $\mathbf{K_u}$ and $\mathbf{K_p}$ denote the $3N \times 3N$ and $N \times N$ feedback gain matrices of velocity and pressure, respectively, $\mathbf{C_u}$ and $\mathbf{C_p}$ denote the $3N \times 3N$ and $N \times N$ diagonal matrices to identify the measurable velocity and pressure data, respectively, and $\boldsymbol{\varepsilon}_\mathbf{u}$ and $\boldsymbol{\varepsilon}_\mathbf{p}$ denote the $3N$ and N-dimensional vector of velocity and pressure measurement error, respectively.

3. Numerical Experiment of Measurement Integrated Simulation

Results of the numerical experiment are shown in the following for cases: (1) UMI simulation in an aneurysmal aorta [16] and (2) MR measurement integrated simulation in a cerebral artery [17]. Calculation was performed with SGI Prism in the AFI Research Center, Institute of Fluid Science, Tohoku University.

3.1. *Ultrasonic Measurement Integrated Simulation in Aneurysmal Aorta*

Based on the concept of the flow observer, we have developed Ultrasonic-Measurement-Integrated (UMI) simulation by integrating ultrasonic color Doppler imaging and numerical simulation to reproduce the real blood flow numerically with the aid of feedback [16]. Figure 2 shows a schematic diagram of the UMI simulation. In the UMI simulation, at a number of grid points selected as feedback points in the feedback domain, feedback signals proportional to the optimal estimation of the difference between the velocity vectors determined by measured and computed Doppler velocities are generated and added to the numerical simulation as artificial force during the computational process.

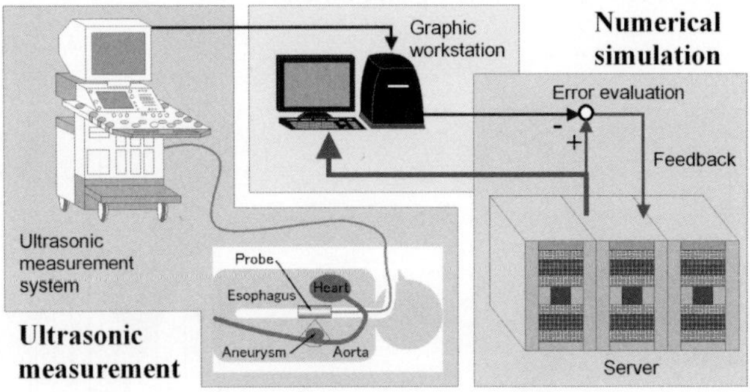

Figure 2. Schematic diagram of UMI simulation system.

In this section, a numerical experiment of UMI simulation of a three-dimensional unsteady blood flow field in an aneurysm was performed to evaluate the effectiveness of UMI simulation for real blood flow. A numerical solution obtained by assuming realistic boundary conditions was first defined as a model of real blood flow (a standard solution), and then UMI simulation was investigated with a focus on the reproduction of the standard solution. In the UMI simulation, a simplified boundary condition that was different from that of the standard solution was applied, and feedback signals were generated by comparing projected velocity information of simulated measurement and computation. In the projection of velocity vectors, a transesophageal ultrasonic measurement of blood flow was assumed in the three-dimensional aneurysmal domain using one probe, which measures only a one-directional velocity component along the ultrasonic beam (Doppler velocity).

The objective flow in this study was the blood flow in an aneurysmal aorta. A 76-year-old female patient, who had no significant complications, with a chronic aortic aneurysm in her descending aorta participated in this study. The cardiac output was 7.0×10^{-5} m^3s^{-1} and the heart rate was 1.0 Hz during the measurement. Figure 3(a) shows the full blood vessel configuration reconstructed from the sliced images acquired by X-ray CT (AquilionTM 16, Toshiba, Tokyo, Japan) by means of commercial three dimensional (3D) reconstruction software (Mimics 7.3, Materialise, Leuven, Belgium). The UMI simulation in this study dealt with the blood flow in a partial domain in the vicinity of the aneurysm as shown in Fig. 3(b). Table 1 summarizes the parameters used in the 3D unsteady blood flow analysis. An orthogonal equidistant computational grid was generated by introducing a staggered grid system with $N_x \times N_y \times N_z = 43\times30\times91$ grid points.

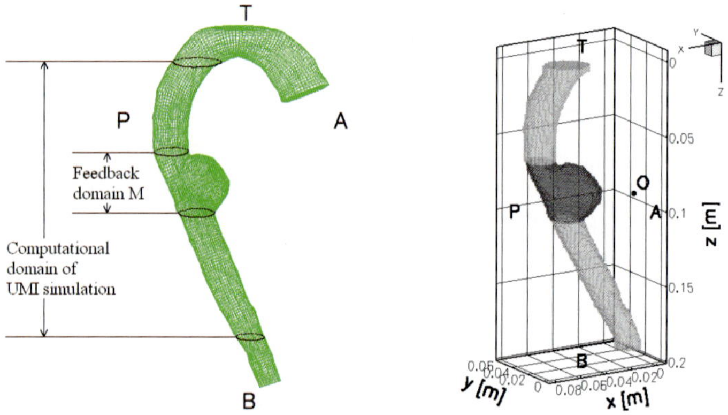

(a) (b)

Figure 3. Computational domains of (a) whole aorta with an aneurysm for FLUENT and of the descending aorta with feedback domain M (dark gray zone) and origin O of ultrasonic beam or ultrasound probe position for UMI simulation.

Table 1. Computational conditions for three-dimensional unsteady blood flow analysis.

Heart rate	1.02 Hz
Cardiac cycle T	0.98 s
Cardiac output	7.00×10^{-5} m^3/s
Entrance flow	4.90×10^{-5} m^3/s
Maximum mean velocity u'_{max} (U)	0.37 m/s
Entrance vessel diameter D (L)	29.25×10^{-3} m
Kinematic viscosity ν	4.0×10^{-6} m^2/s
Characteristic time D/u'_{max}	0.080 s
Womersley number $D\sqrt{2\pi/T\nu}$	37
Maximum Reynolds number $u'_{max}\cdot D/\nu$	2700

The space-averaged error norms, $\bar{e}_M(\mathbf{u},t)$ and $\bar{e}_M(p,t)$, of the velocity vector and the pressure in the aneurysm (feedback domain M) were calculated at each moment as shown in Fig. 4. The ordinary simulation has a relatively large error

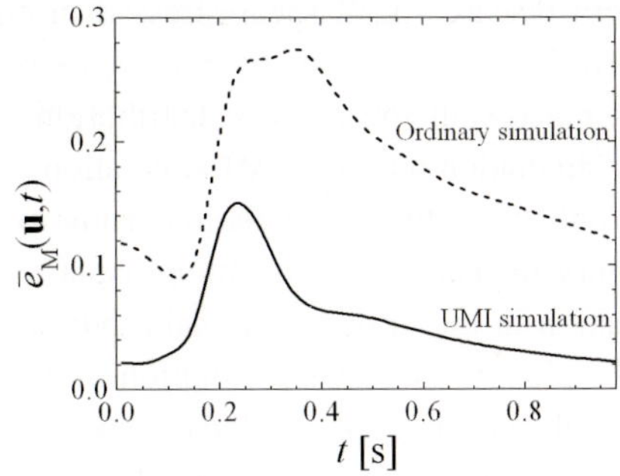

Figure 4. Time-variations of space-averaged error norms of (a) velocity vector and (b) pressure in the aneurysm of the ordinary simulation and in the UMI simulation with ideal feedback at Kv* = 5.

in the deceleration phase (0.13 s $\leq t \leq$ 0.33 s). The UMI simulation resulted in smaller error than the ordinary simulation at all phases. The calculation of space-time-averaged error norms, $\bar{e}_{MT}(\mathbf{u})$ and $\bar{e}_{MT}(p)$, of UMI simulation indicates that the application of feedback in feedback domain M using one probe reduced the error in the aneurysm by a factor of 0.31 for the velocity vector in one cardiac cycle.

In order to investigate the effectiveness of UMI simulation for providing information on hemodynamic stress on a blood vessel for advanced medical

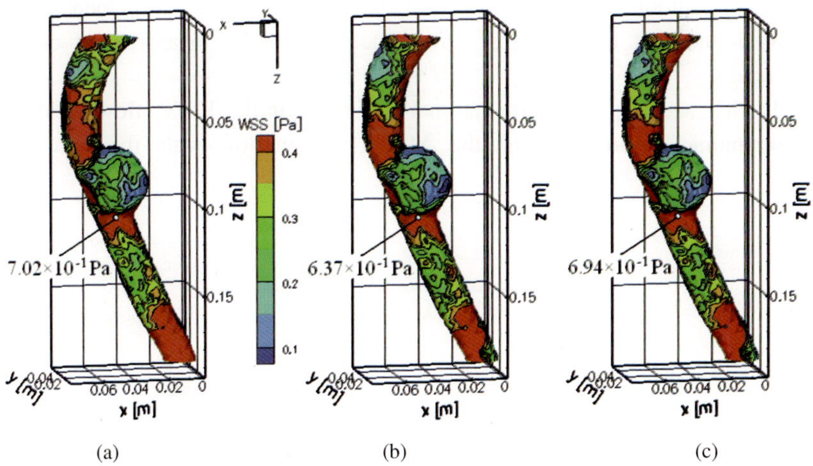

Figure 5. Comparison of time-averaged wall shear stress distribution on the blood vessel between (a) the standard solution, (b) the ordinary simulation, and (c) the UMI simulation with ideal feedback at Kv* = 5. Each white dot implies the local maximum value of wall shear stress at the aneurysmal neck.

diagnosis of circulatory diseases, wall shear stress distribution was calculated by first-order numerical differentiation of the velocity vectors. Figure 5 compares the time-averaged wall shear stress distribution between the standard solution, the ordinary simulation and the UMI simulation. All simulations show similar distributions, but the ordinary simulation tends to estimate lower wall shear stress in the aneurysm. The UMI simulation displays a similar distribution as the ordinary simulation in the upstream domain, but accurately provides the wall shear stress distribution of the standard solution in the aneurysm. In Fig. 5, a white dot represents the position of the local maximum value of the time-averaged wall shear stress at the aneurysmal neck.

3.2. *MR Measurement Integrated Simulation in Cerebral Artery*

We have previously proposed magnetic resonance (MR)-measurement-integrated (MR-MI) simulation by integrating MR measurement and numerical simulation [17]. Figure 6 shows a schematic diagram of the system of MR-MI simulation of blood flow. In such simulation, artificial body forces, which are generated by comparing 3D velocity vectors obtained by PC MRI measurement and the corresponding computational results, are added to the numerical simulation. This feedback process is expected to enable reproduction of the real blood flow field computationally by convergence of computational results with those of PC MRI measurement. With this methodology, it is likely that a numerical solution whose resolution is better than the MR measurement can be acquired and that the effect of errors included in the measurement can be eliminated.

In this section, the computational accuracy of the blood flow field and hemodynamic information of the MR-MI simulation with the simplified

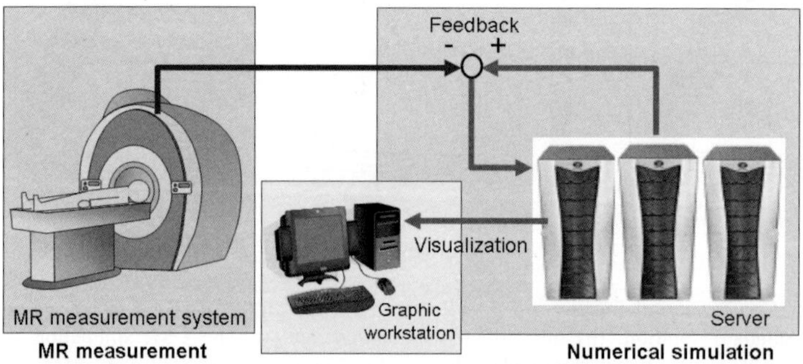

Figure 6. Schematic diagram of MR-measurement-integrated (MR-MI) simulation system.

boundary conditions were investigated by numerical experiments dealing with the reproduction of steady blood flow field in an aneurysm which developed at a bifurcation of a cerebral artery.

We first obtained a standard solution as a model of real blood flow with velocity vectors obtained by PC MRI measurement at the boundaries. MR-MI simulation was then performed using the data of the standard solution, and its ability to reproduce the standard solution was examined. The objective was the blood flow in a cerebral aneurysm at the bifurcation between basilar and superior cerebellar arteries in a 70-year-old female patient. This study was approved by the institutional review board of Hamamatsu University School of Medicine, and informed consent was obtained from the patient.

MR imaging was performed by a 1.5 T MR scanner with a head coil. The PC MRI data, which contain information on the blood vessel configuration and the velocity vector of blood flow, were obtained at 20 phases in one cardiac cycle with a sampling time of Δt = 0.046 s. The configuration including the cerebral aneurysm and the parent artery, was reconstructed by means of image data processing software, Flova (Renaissance of Technology Corp., Hamamatsu, Japan), from the PC MRI data. We assumed that the blood vessel was rigid because the deformation was subtle, and we generated several computational grid systems with different resolutions. Compromising between the reproducibility of the blood vessel shape and the computational load, a staggered equidistant grid system with $N_x \times N_y \times N_z = 30 \times 27 \times 26$ ($\Delta x = \Delta y = \Delta z = 0.5\times10^{-3}$ m) shown in Fig. 7 was chosen. With the same software, one inlet, A, and two outlets, B and C, were determined (see Fig. 7), and the velocity information on the boundaries was exported by linear interpolation.

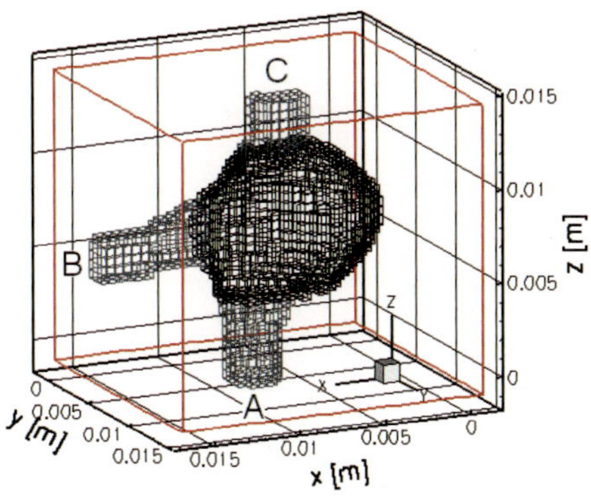

Figure 7. Computational grid (30×27×26) with the feedback domain (black grid lines).

Table 2. Computational conditions.

Entrance flow	2.68×10^{-6} m^3/s
Mean velocity u'_1	0.23 m/s
Maximum mean velocity u'_{max} (U)	0.42 m/s
Entrance diameter $D\,(L)$	4.0×10^{-3} m
Density ρ	1.0×10^{3} kg/m^3
Viscosity μ	4.0×10^{-3} Pa·s
Maximum Reynolds number Re_{max}	423

For the numerical experiment concerning the reproduction of the steady flow, the first phase in the MR measurement (the initial phase in the systole) was assumed to be steady flow, and a convergent solution obtained with the velocity profiles obtained by PC MRI measurement at boundaries was defined as the standard solution. The mean velocity u'_1 upstream is noted in Table 2.

In the MR-MI simulation, we applied a uniform velocity profile at the upstream boundary and free flow as well as pressure zero conditions at the two downstream boundaries, assuming that the complicated velocity vectors at the boundaries were unknown since the application of the exact velocity profiles is difficult. The computational time increment was set at Δt = 0.01 s for the numerical experiment of the reproduction of steady flow.

Figure 8 shows the relationship between $\bar{e}_M(\mathbf{u}_c, \mathbf{u}_s, t_\infty)$ at $t = t_\infty$ and the feedback gain. The result of the MR-MI simulation becomes different from that of the ordinary simulation, and the error with respect to the standard solution becomes small due to the feedback. The error monotonically decreases to the minimum value. In this study, the computational accuracy was best improved by setting the feedback gain at $K_v^* = 3.5$, which is the approximate value $K_v^* = 4$ at which the computation diverged. In this case, the error in the velocity field in the blood flow in the cerebral aneurysm was reduced to 13% of that in the ordinary simulation ($K_v^* = 0$).

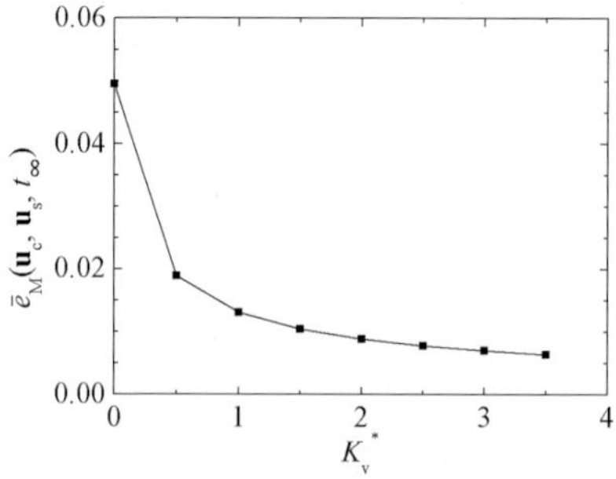

Figure 8. Steady space-averaged error norm of the velocity vector in the aneurysm with the feedback gain.

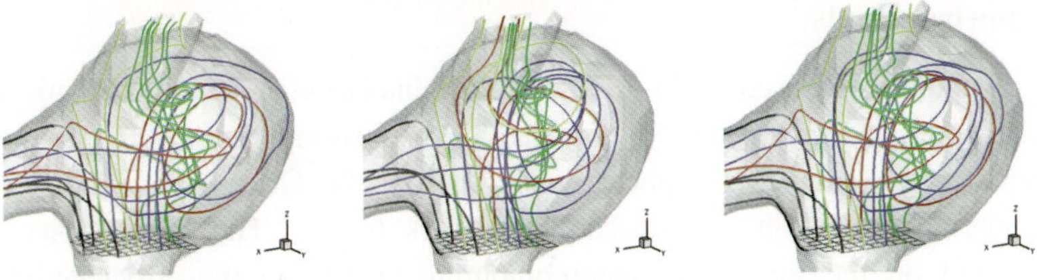

(a) (b) (c)

Figure 9. Streamlines departing from the upstream surface of the feedback domain of (a) standard solution, (b) ordinary simulation ($K_v{}^* = 0$) and (c) MR-MI simulation with $K_v{}^* = 3.5$.

Figure 9 compares the vortex structures in the cerebral aneurysm visualized by the streamlines which depart from multiple points on the first cross section of the feedback domain between the standard solution, the ordinary simulation ($K_v^* = 0$), and the MR-MI simulation with $K_v^* = 3.5$. Those different colored streamlines are categorized by their patterns. All results have a z-directional helical structure (see green streamlines in Fig. 9), which flows from the upstream boundary A to the downstream boundary C, and a swirl in the cerebral aneurysm (see red and blue streamlines). Here, since this result was obtained under the assumption of steady flow, it does not represent the real blood flow field. Though the ordinary simulation (Fig. 9(b)) shows a flow structure similar to the standard solution (Fig. 9(a)), especially for the helical structure, the shapes and positions of the streamlines are different. Several streamlines that swirl in the cerebral aneurysm exit from the opposite outlet compared with those in the standard solution (see red streamlines). In Fig. 9(c), owing to the improvement of the computational accuracy in the cerebral aneurysm by the MR-MI simulation, we can understand the blood flow structure in the cerebral aneurysm of the standard solution more accurately than in the case of the ordinary simulation.

4. Conclusions

In order to accurately diagnose serious cardiovascular diseases such as an aortic or a cerebral aneurysm, a method to obtain detailed information on the blood flow velocity and pressure is essential, measurement and simulation being the basic tools for this. Herein, a general formulation of the measurement integrated simulation was first explained and then the results of numerical experiments for ultrasonic measurement integrated simulation in an aneurysmal aorta and MR measurement integrated simulation in a cerebral artery were presented.

Acknowledgments

The computer code used in this study was originally developed at the University of California at Berkeley by T. Hayase, J. A. C. Humphrey and R. Greif. The authors acknowledge the support received from the Global Nano-Biomedical Engineering Education and Research Network Centre. The calculations were performed on the integrated supercomputing system in the Advanced Fluid Information Research Center, Institute of Fluid Science, Tohoku University.

References

1. *Biomedical Engineering Handbook, Second Edition, Volume 1*, edited by J. D. Bronzino, CRC Press (2000).
2. T. W. Taylor and T. Yamaguchi, Three-dimensional simulation of blood flow in an abdominal aortic aneurysm — steady and unsteady flow cases. *J Biomech Eng* **116**, 89–97 (1994).
3. E. S. Di Martino, G. Guadagni, A. Fumero, G. Ballerini, R. Spirito, P. Biglioli and A. Redaelli, Fluid-structure interaction within realistic three-dimensional models of the aneurysmatic aorta as a guidance to assess the risk of rupture of the aneurysm. *Med Eng Phys* **23**, 647–655 (2001).
4. S. G. Benjamin, D. Dévényi, S. S. Weygandt, K. J. Brundage, J. M. Brown, G. A. Grell, D. Kim, B. E. Schwartz, T. G. Smirnova, T. L. Smith and G. S. Manikin, An hourly assimilation-forecast cycle. *The RUC Monthly Weather Review* **132**, 495–518 (2004).
5. J. A. C. Humphrey, R. Devarakonda and N. Queipo, Interactive computational-experimental methodologies (ICEME) for thermofluids research: application to the optimized packaging of heated electronic components. *Computers and Computing in Heat Transfer Science and Engineering*, edited by K. T. Yang and W. Nakayama, CRC Press and Begell House, New York, 293–317 (1993).
6. B. A. Zeldin and A. J. Meade, Jr., Integrating experimental data and mathematical models in simulation of physical systems. *AIAA J* **35**, 1787–1790 (1997).
7. T. Ido, Y. Murai and F. Yamamoto, Postprocessing algorithm for particle-tracking velocity based on ellipsoidal equations. *Experiments in Fluids* **32**, 326–336 (2002).
8. M. Uchiyama and K. Hakomori, Measurement of instantaneous flow rate through estimation of velocity profiles. *IEEE Trans Automat Contr* **AC-28**, 380–388 (1983).
9. T. Hayase, K. Nisugi and A. Shirai, Numerical realization of flow field by integrating computation and measurement. *Proceedings of Fifth World Congress on Computational Mechanics, WCCM V*, Paper-ID: 81524, 1–12 (2002).

10. R. E. Skelton, *Dynamic Systems Control*, John Wiley & sons (1988).
11. T. Hayase and S. Hayashi, State estimator of flow as an integrated computational method with feedback of online experimental measurement. *J Fluids Eng Trans ASME* **119**, 814–822 (1997).
12. K. Nisugi, T. Hayase and A. Shirai, Fundamental study of hybrid wind tunnel integrating numerical simulation and experiment in analysis of flow field. *JSME Int J, Ser. B* **47**, 593–604 (2004).
13. K. Fumamoto, T. Hayase, A. Shirai, Y. Saijo and T. Yambe, Fundamental study of ultrasonic-measurement-integrated simulation of real blood flow in the aorta. *Annals Biomed Eng* **33**, 413–426 (2005).
14. K. Funamoto, T. Hayase, Y. Saijo and T. Yambe, Detection and correction of aliasing in ultrasonic measurement of blood flows with ultrasonic-measurement-integrated simulation. *Tech Health Care* **13**, 331–344 (2005).
15. K. Funamoto, T. Hayase, Y. Saijo and T. Yambe, Numerical experiment of transient and steady characteristics of ultrasonic-measurement-integrated simulation in three-dimensional blood flow analysis. *Annals Biomed Eng* (submitted).
16. K. Funamoto, T. Hayase, Y. Saijo and T. Yambe, Numerical experiment for ultrasonic-measurement-integrated simulation of three-dimensional unsteady blood flow. *Annals Biomed Eng* **36**, 1383–1397 (2008).
17. K. Funamoto, Y. Suzuki, T. Hayase, T. Kosugi and H. Isoda, Numerical validation of MR-measurement-integrated simulation of blood flow in a cerebral aneurysm. *Annals Biomed Eng* (submitted).

MICROELASTIC MAPPING BY AFM OF ENDOTHELIAL CELLS EXPOSED TO SHEAR STRESS

MASAAKI SATO [1,2)*], KENICHI SUZUKI [2)], YOSUKE UEKI [2)†], TOSHIRO OHASHI [2)]

1) Department of Biomedical Engineering, Graduate School of Biomedical Engineering,
2) Department of Bioengineering and Robotics, Graduate School of Engineering,
Tohoku University, Aramaki-Aoba 6-6-01, Aoba, Sendai, Japan 980-8579

The surface topography and local elastic moduli of endothelial cells exposed to shear stress were measured using atomic force microscopy. Bovine aortic endothelial cells were exposed to shear stress of 2 Pa for 6 h, 12 h or 24 h. In addition, a confocal laser-scanning microscope equipped with an atomic force microscope was used to observe the actin filament structure of these endothelial cells to elucidate the relationship between mechanical properties and the cytoskeletal structure. The elastic modulus, calculated using the Hertz model, was measured at 50 x 50 points at 1 μm intervals within 40 min. For endothelial cells sheared for 6 h and 12 h, the elastic modulus in the upstream region was found to be higher than that the downstream region. For endothelial cells sheared for 24 h, the elastic modulus in both the upstream and downstream regions increased. Fluorescent images showed thick, elongated actin filaments oriented in the direction of flow at the ventral surface of the cells. In the middle plane of the cells, actin filaments developed around the nucleus, while in the upper plane, short, thick actin filaments were observed but thick stress fibers were not present. A high elastic modulus resulted from the stress fibers. These results indicate that the higher elastic modulus observed in the upstream and downstream regions of sheared endothelial cells was mainly due to the development of stress fibers at the ventral surface and middle plane of the cell.

1. Introduction

Endothelial cells are known to change their morphology in response to fluid shear stress. Kataoka and Sato [1] have previously reported that endothelial cells were elongated and oriented parallel to the flow direction under shear stress of 2 Pa, accompanied by drastic changes in actin microstructure. Stress fibers started to form 40 min after the onset of flow and were aligned in the direction of flow after approximately 3 h. These observations have suggested that actin filaments play an important role in changing cell shape under varying shear flow conditions. To understand the mechanism underlying the dynamic changes in

* Masaaki Sato is a Tohoku University Global COE Member.
† Yosuke Ueki is a Tohoku University Global COE Research Assistant.

endothelial cell shape in response to mechanical stimuli, it is important to study the changes in cytoskeletal structure and location of focal adhesions.

Using a pipette aspiration technique, Sato *et al.* [2] first noted that the mechanical stiffness of endothelial cells increased after exposure to shear stress. It was considered that the development of actin filaments in response to shear stress contributed to these mechanical properties. Using atomic force microscopy, Miyazaki *et al.* [3] and Mathur *et al.* [4] indicated that the mechanical properties in the central region of the cells above the nucleus was higher than that in peripheral regions in rabbit aortic endothelial cells and human umbilical vein endothelial cells, respectively. Sato *et al.* [5] also used an atomic force microscope to measure the local mechanical properties in cultured bovine aortic endothelial cells exposed to shear stress. The results showed that the elastic modulus in the peripheral region, especially in the upstream region, was higher than that in the central region above the nucleus in sheared endothelial cells. This correlates with increased actin filament structure in the peripheral region.

Since the cytoskeletal structure of actin filaments changes spatially and temporally within the cell, it is necessary to assess the local mechanical properties throughout an endothelial cell exposed to shear stress in relation to the three-dimensional (3-D) distribution of actin filaments. Elastic mapping by use of atomic force microscopy is a suitable technique and has previously been reported for endothelial cells [6], kidney cells [7] and fibroblasts [8].

In this study, the surface topography and microelastic mapping of bovine aortic endothelial cells exposed to shear stress were obtained by atomic force microscopy. The endothelial cells were exposed to shear stress of 2 Pa for 6 h, 12 h or 24 h in a flow chamber. The actin filament structure of the same endothelial cells was observed simultaneously using a confocal laser-scanning microscope equipped with an atomic force microscope in order to elucidate the relationship between mechanical properties and cytoskeletal structure.

2. Materials and Methods

2.1. *Endothelial Cell Culture*

Bovine aortic endothelial cells were harvested from thoracic and abdominal aortas by means of culture procedures developed by Shasby and Shasby [9]. Bovine aortic endothelial cells were grown to confluence in 25 cm^2 tissue culture flasks in Dulbecco's modified Eagle medium (Invitrogen, USA) containing 10% (v/v) heat-inactivated fetal bovine serum (JRH Bioscience,

USA) and penicillin-streptomycin (Invitrogen, USA). Confluent endothelial cells were detached from substrates using 0.05% trypsin-EDTA (Invitrogen, USA), cultured at a 1:4 split ratio and used between the third and tenth passages. For exposure to shear stress, endothelial cells were plated in a 35-mm-diameter cell culture dish (Asahi Techno Glass, Japan) and cultured until reaching confluence. Leibovitz's L-15 Medium (Invitrogen, USA) was used to maintain a constant pH during the experiment.

2.2. *Flow Experiment*

A flow system consisting of a damping chamber, a flow chamber, a reservoir, and a roller pump (Master Flex, USA) was used to apply steady shear stress to the cells, as reported previously [5]. Briefly, the flow chamber, which consists of an I/O unit, a silicone gasket and the cell culture dish, has a flow section measuring 22 x 14 x 0.5 mm^3. Endothelial cells plated in this flow section are exposed to steady shear stress of 2 Pa, the pulsatile part of fluid flow (due to the roller pump) being eliminated by the damping chamber. During the experiment, the temperature of culture medium was maintained at 37 ± 0.5°C by soaking the damping chamber and reservoir in a thermostatic chamber, and the pH was maintained by introducing mixed gas (5% CO_2 and 95% air) into the reservoir.

2.3. *Measurement of Cellular Topography and Mechanical Properties Using Atomic Force Microscopy*

An atomic force microscope system, developed in our laboratory [5], was used to measure the topography and mechanical properties of living endothelial cells. This atomic force microscope system was combined with an inverted confocal laser-scanning microscope (Olympus LSM-GB200, Tokyo, Japan) to visualize cytoskeletal structure simultaneously. The system, particularly the dish-holding stage (XYZ-PZT stage), has been modified and is shown schematically in Fig. 1. It incorporates a triangular silicon nitride cantilever (Olympus, Tokyo, Japan) with a spring constant k of 0.02 N/m. The deflection of the cantilever due to the force of interaction with the sample was detected by measuring the motion of a laser beam (780 nm) reflected off the gold-coated cantilever with a photo detector (Hamamatsu Photonics, Hamamatsu, Japan). The cell culture dish containing the endothelial cells was placed on the XYZ-PZT stage equipped with a piezo actuator (NEC Tokin, Japan) with a maximum XY scan range of 56 μm and a Z range of 16 μm. An image of both the tip of the atomic force microscope cantilever and endothelial cells under conventional light microscopy was monitored by a CCD camera.

Three-dimensional topography was measured by recording the feedback due to Z displacement while keeping the cantilever deflection constant. The Z displacement was measured at each pixel point in a scan area consisting of 100 pixels x 100 lines at 500 nm intervals within 5 min. For measurement of mechanical properties, the relationship between the external force, F, and the indentation depth, δ, was obtained for different locations within a cell. The force-indentation response was modeled using the Hertz equation (1):

$$F = \delta^2 \frac{\pi}{2} \frac{E}{(1-\nu^2)} \tan\alpha \tag{1}$$

where E is the elastic modulus, ν is the Poisson's ratio and α is the opening angle of the cantilever. In this study, ν was assumed to be 0.49, considering the incompressibility of cells, and α was 70.6°. The elastic modulus was measured at 50 x 50 points at 1-μm intervals within 40 min. Statically cultured control endothelial cells were compared with sheared endothelial cells, the elastic modulus being estimated at three different indentation depths of 200 nm, 300 nm and 500 nm.

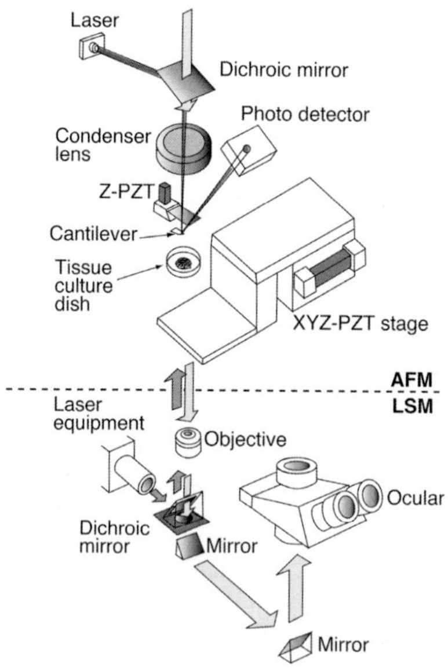

Figure 1. Schematic diagram of an atomic force microscope (AFM) combined with an inverted laser scanning confocal microscope system (LSM). Reprinted from [15] with permission from Elsevier.

2.4. *Fluorescent Staining and Observation of Actin Filaments*

Following force measurement by the atomic force microscope, the endothelial cells were rinsed in PBS (-) and then fixed with 10% formaldehyde for 5 min. The cells were rinsed again in PBS and soaked in Triton X for 5 min. The endothelial cells were then stained with 150 nM rhodamine-phalloidin for 20 min. The confocal laser-scanning microscope was used to observe distribution of actin filaments of endothelial cells. The culture dish was firmly fixed by screws to the holder on the stage of the microscope. The image of endothelial cells during measurement with the atomic force microscope by conventional light microscopy was recorded. It was confirmed by recording the same field of the microscope after the treatment for actin filament observation that the same cells were targeted for both the measurement of cell elasticity and the observation of actin filament structure. Fluorescent actin filaments were imaged at every 200-nm interval in the Z-direction from the ventral side to the upper side. The ventral surface was firstly defined by its abundant actin filaments at the nearest plane to the culture glass. The ventral surface was confirmed, in another experiment by near-field microscopy, to correspond approximately to the plane 200 nm above the glass. Typical images are shown in this paper at three planes with different heights: the ventral surface, the middle plane and the upper plane of the cell. The middle and upper planes are located 2000 nm and 4000 nm above from the ventral surface.

2.5. *Image Analysis*

A 3-D image of actin filaments was reconstructed using the images obtained in the Z-direction by deconvolution software (Auto Deblur/Auto Visualize, AutoQuant Imaging, USA). The correlation between the elastic modulus and the fluorescent intensities of the actin filaments were analyzed by freeware of ImageJ 1.34S (by W. Rasband, NIH, USA) and plugin software Colocalization Finder 1.1 (by C. Laummonerie and J. Mutterer, France). The parts of colocalization of the elastic modulus above 600 Pa and the fluorescent intensity above 45 in 256 gray-scale are colored white in the superimposed images of the distribution of the elastic modulus and the actin filaments.

3. Results

Representative microelastic mapping for statically cultured endothelial cells is shown in Fig. 2 (a), (b) and (c) which show indentation depths of 200 nm, 300 nm and 500 nm, respectively. Broken white lines indicate the contours measured by atomic force microscopy. The height values shown in the figures were measured from an arbitrary plane. It can be clearly seen that the endothelial cells are confluent and that their shape is polygonal. Values for the elastic modulus depend on the indentation depth. The deeper indentation depth at the peripheral region indicates increased stiffness in this area. However, in the central region of the cell, including the upper portion of nucleus, the opposite phenomenon is observed: namely a lower elastic modulus in the case of a 500-nm indentation depth. The mechanical properties of a cell are considered to have a strong correlation with its microstructure, especially the distribution of actin filaments. Therefore, the actin filaments of the same cells were stained by rhodamine-phalloidin and observed by confocal laser scanning microscopy. The images taken at the three different Z planes, namely, the ventral surface, the middle plane (ventral surface + 2000 nm) and the upper plane (ventral surface + 4000 nm) are shown in Fig. 3. Many actin filaments, especially thick stress fibers (bundles of actin filaments), are observed to be present on the ventral surface. In the middle plane, both thick stress fibers and short actin filaments exist. The nuclei, N, are clearly visible. In the upper plane, actin filaments are short and small in mass. When we compare the distribution of elastic modulus in Fig. 2 with the structure of actin filaments in Fig. 3, the portions showing a higher elastic modulus correspond well with the locations of thick stress fibers at the ventral surface and middle plane. The superimposed images of distribution of elastic modulus (indentation depth = 300 nm) and the fluorescence intensities of actin filaments are shown in Fig. 4 at three different planes by the contour map of the cell. Colocalization of a higher elastic modulus and a higher fluorescent intensity of actin filaments is shown in white. This colocalization is found around the nuclei in the middle plane and at the nuclei in the upper plane. Since higher colocalization was found in the case of an indentation depth of 500 nm in the middle plane (data not shown), the indentation depth of 300 nm was too small to detect the effect of stress fibers where colocalization was not found.

Microelastic mapping (indentation depth = 300 nm) of cells exposed to a shear stress of 2 Pa for exposure times of 6 h, 12 h, and 24 h is shown in Fig. 5. The cell shape contour is indicated by broken lines in each figure.

In comparison with the control cells, changes in the shape of cells exposed to shear stress are observed. After exposure to fluid flow, cells elongate and are oriented in the direction of flow, and this increases with the exposure time. The elastic modulus in peripheral regions is also higher than in central regions for all three exposure times. At the exposure times of 6 h and 12 h, there is a characteristic change towards a heterogeneous distribution of the elastic modulus. Values of the elastic modulus at the upstream side of the cell are found to be higher than at the downstream side. However, after exposure times of 24 h, the elastic moduli at both the upstream and downstream sides seem to be equal, as shown in Fig. 5 (c).

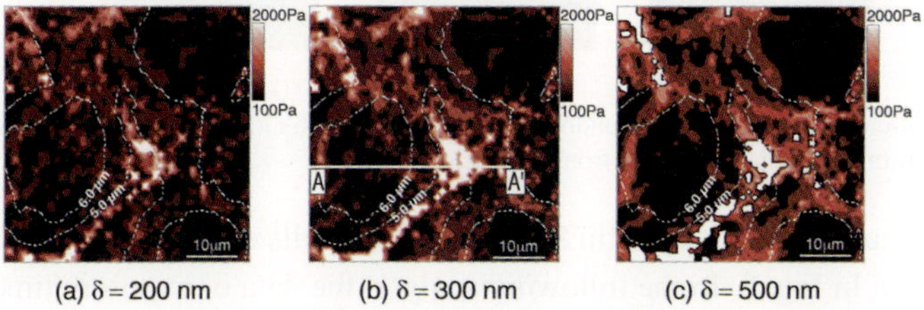

Figure 2. Microelastic mapping for statically cultured endothelial cells at three different indentation depths, δ: (a) δ = 200 nm, (b) δ = 300 nm, (c) δ = 500 nm. Reprinted from [15] with permission from Elsevier.

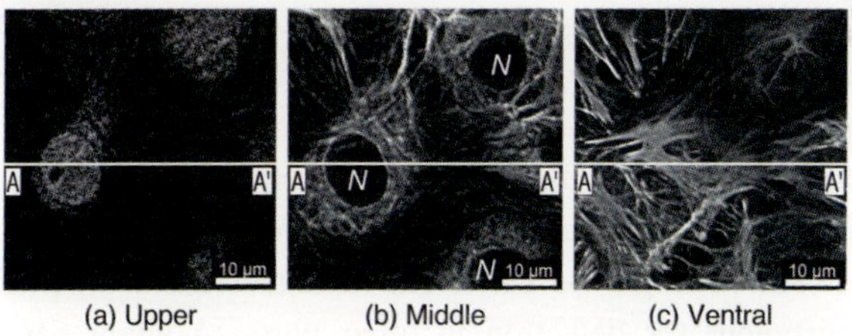

Figure 3. Images of actin filaments stained by rhodamine-phalloidin in three different planes of statically cultured endothelial cells: (a) ventral surface, (b) middle plane (ventral surface + 2000 nm), (c) upper plane (ventral surface + 4000 nm). Reprinted from [15] with permission from Elsevier.

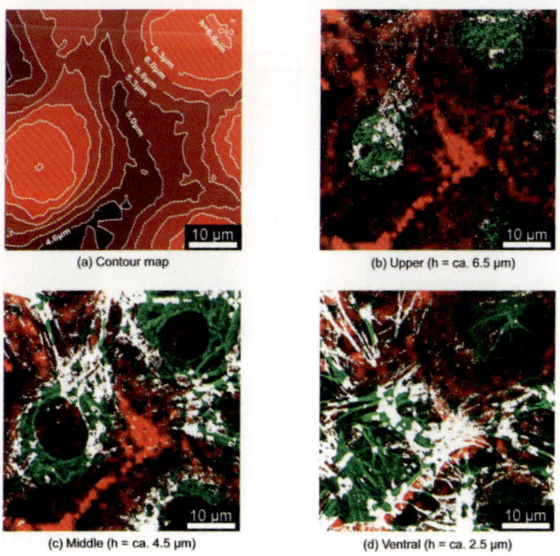

Figure 4. (a) Contour map of statically cultured endothelial cells. Colocalization (white color) of higher elastic modulus (>380 Pa, red color) and higher fluorescent intensity of actin filaments (>45 in 256 gray-scale, green color) in the (b) upper plane, (c) middle plane, and (d) on the ventral surface. Reprinted from [15] with permission from Elsevier.

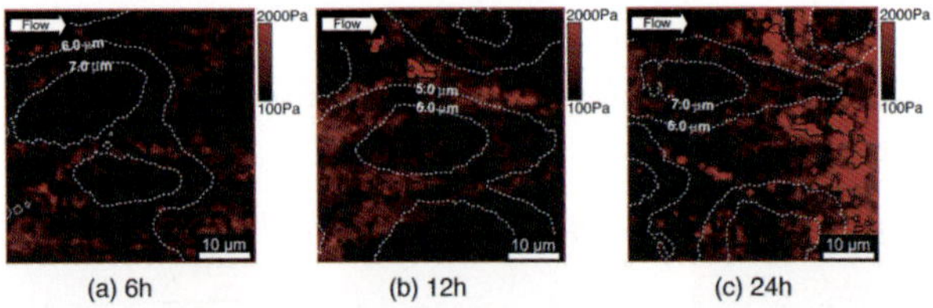

Figure 5. Microelastic mapping of endothelial cells sheared for (a) 6 h, (b) 12 h, and (c) 24 h. Reprinted from [15] with permission from Elsevier.

Distribution images of actin filaments in the cells exposed to shear stress are summarized in Fig. 6. In the following analysis the data of exposure time of 12 h are shown as a typical example. From the images of Fig. 6, actin filaments or stress fibers were found to develop relatively well at the upstream side of the ventral surface and middle plane. Colocalization of higher elastic modulus and higher fluorescent intensity of actin filaments is shown in white in Fig. 7. The colocalization is clearly seen at the upstream side and in part of the side-peripheral regions of the ventral plane in Fig. 7. In the middle and upper planes the colocalization is found in part of the peripheral region. Although regions of high actin filament fluorescence are found in every plane, colocalization is not observed. Indentation might be sufficient to detect the actin filament in some parts, but insufficient for detection in other parts. A slight parallel shift, say 1-2 μm, of the both images was observed in a certain area. Although the reason for this discrepancy is not yet clear, artifacts due to slight movement of the dish and/or cellular shrinkage during fixation may have occurred.

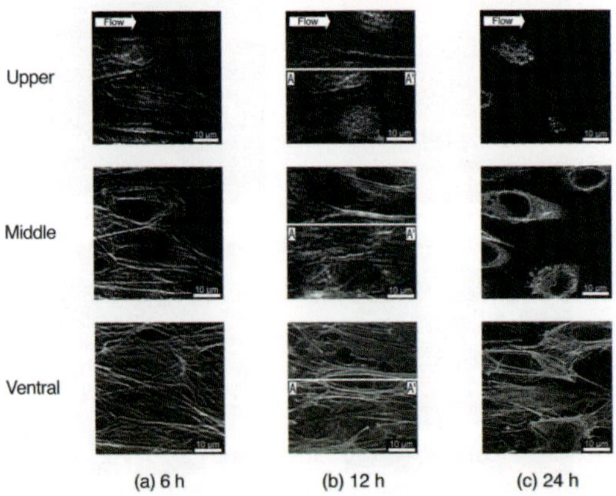

Figure 6. Images of actin filaments of endothelial cells sheared for (a) 6 h, (b) 12 h, and (c) 24 h. Images at three different planes (ventral, middle and upper) are shown at each exposure time. Reprinted from [15] with permission from Elsevier.

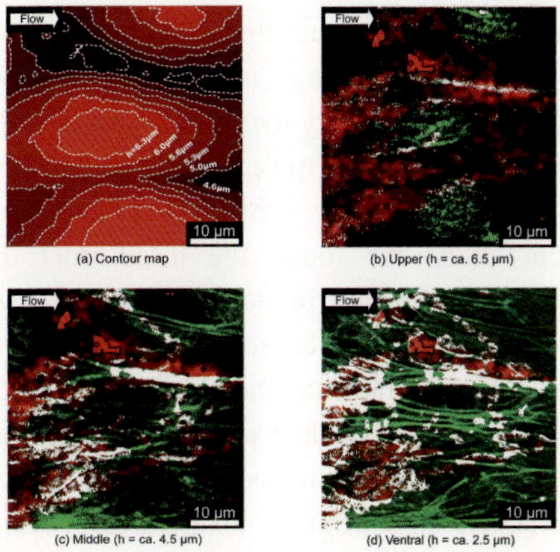

Figure 7. (a) Contour map of endothelial cells sheared for 12 h. Colocalization (white color) of higher elastic modulus (>600 Pa, red color) and higher fluorescent intensity of actin filaments (>45 in 256 gray-scale, green color) in the (b) upper plane, (c) middle plane, and (d) on the ventral surface. Reprinted from [15] with permission from Elsevier.

4. Discussion

Mahaffy *et al.* [10] studied the Hertz model using NIH3T3 fibroblasts as typical eukaryotic cells. They reported that the extended ranges of Hertz model application were remarkable, despite the fact that cells were only 2-4 μm thick and very soft. No significant deviations from a constant were observed for indentations of up to 25% of the total depth. Therefore, we adopted indentation depths of 200 nm, 300 nm and 500 nm. These depths have no effect on the measured elastic modulus value for cell peripheral regions with a thickness of 2 μm. Many thick actin filaments were observed on the ventral surface and in the middle plane; however, only thin, short actin filaments were distributed in the upper planes of the cells. From this finding, the higher elastic modulus values measured at the peripheral regions can be considered to be related to the thick actin filaments distributed on the ventral surface and in the middle plane.

The technique of AFM has a limitation in addressing the deep internal structure of the cytoskeleton, because the indentation depth is too small to detect the actin filaments in the central area at the ventral surface and in the middle plane. Cell height is approximately 2 μm at the periphery and 5 μm at the top. The indentation depth adopted in this study was 200-500 nm. Therefore, we can detect the actin structure located close to the cell surface; that is, around the periphery of the ventral surface, around the mesogastrium in the middle plane, and around the top of the upper plane.

The present AFM study showed that the maximum elastic modulus is approximately 2 kPa, this value perhaps arising from stress fibers or the bundle of actin filaments. We have previously measured mechanical properties of stress fibers isolated from cultured smooth muscle cells and reported the value of the elastic modulus to be approximately 2 MPa [11]. In contrast to the previous study, in the present study, we found the three-order difference in elastic modulus by the application of measuring techniques. When we applied the AFM technique to living cells, we measured some collective mechanical properties, including all the cytoskeletal filaments and ground substance in cytoplasm. Hu *et al.* [12] have investigated mechanical properties of elongated cells with and without stress fibers using a three-dimensional magnetic twisting device. They pointed out that mechanical anisotropy of adherent cells originates from intrinsic cytoskeletal tension within the stress fibers. Deformation patterns of the cytoskeleton were sensitive to loading direction, suggesting anisotropic mechanical signaling. Our present results also suggest that mechanical loading such as shear stress has the same effect on the formation of cytoskeletal patterns and the local elastic modulus of cells. How the local elastic modulus of cells is related to the local numbers of stress fibers quantitatively is a very important problem. This question should be examined on the structural basis of mechanotransduction to elucidate the mechanisms in the near future.

In a previous paper, Sato *et al.* [5] reported that the stiffness of endothelial cells exposed to shear stress of 2 Pa increased with the duration of exposure. After 6-h exposure, the stiffness was higher at the upstream side of the cell than at the downstream side. However, after 24-h exposure, the stiffness was similar on both sides of the cell. A similar phenomenon was observed in the present study. After 6-h and 12-h exposures, the elastic moduli increased at the upstream side of the cell. However, the elastic moduli in the central regions were still lower than the peripheral regions after 24-h exposure. These changes in elastic moduli corresponded well with the distribution of actin filaments. In another previous paper, we analyzed the stress distribution in endothelial cells exposed to shear stress by finite element analysis [13]. The results indicated the internal stress first becomes higher at the upstream side. This might depend upon the cell topography. The development of actin filaments might be related to the stress distribution.

In a related study, Haga *et al.* [8] measured the spatial distribution of elastic moduli of live mouse fibroblasts (NIH3T3) under physiological conditions. Their indentation depth was between 50 nm and 200 nm. They reported that the nuclear portion of the cellular surface was about 10 times softer than the surroundings. This result is similar to that of the control cells in our experiment.

Haga *et al.* also compared the elasticity of cells with the distribution of three types of cytoskeletal filaments, namely, actin filaments, microtubules and intermediate filaments, observed by confocal laser scanning microscopy. They reported that the elasticity of the cells was not only related to the distribution of the actin network, but also to intermediate filament organization, whereas microtubules only had a minimal effect on the measured elasticity. In our study, we did not examine the effect of microtubules or intermediate filaments, and so we cannot refer to their contribution to our experimental results. However, we can stress that the higher elastic modulus originated from stress fibers.

5. Conclusion

The surface topography and microelastic mapping of living endothelial cells exposed to shear stress were obtained using atomic force microscopy. The actin filament structure of the same endothelial cells was also observed using a confocal laser-scanning microscope equipped with an atomic force microscope. Our conclusions can be summarized as follows:

1. The elastic modulus in the peripheral region is higher than that in the central region, the above nucleus, in control endothelial cells.
2. Endothelial cells sheared for 6 h or 12 h had higher elastic modulus values in upstream regions compared with downstream regions. However, after 24-h exposure, the elastic modulus in both the upstream and the downstream regions increased.
3. The high elastic modulus originated from the stress fibers.

Acknowledgement

This work was supported in part by the Global COE Program "Global Nano-Biomedical Engineering Education and Research Network Centre" and Grants-in-Aid for Scientific Research from the Ministry of Education, Culture, Sports, Science and Technology (MEXT) in Japan (Nos. 14208100, 15086203).

References

1. N. Kataoka and M. Sato, The change of F-actin distribution and morphology of cultured bovine aortic endothelial cells in the early stage of fluid shear stress exposure. *Trans Jap Soc Mech Eng Ser B* **64**, 1801–1808 (1998).
2. M. Sato, N. Ohshima and R. M. Nerem, Viscoelastic properties of cultured porcine aortic endothelial cells exposed to shear stress. *J Biomech* **29**, 461–467 (1996).

3. H. Miyazaki and K. Hayashi, Atomic force microscopic measurement of the mechanical properties of intact endothelial cells in fresh arteries. *Med Biol Eng Comp* **37**, 530–536 (1999).
4. A. B. Mathur, A. M. Collinsworth, W. M. Reichert, W. E. Kraus and G. A. Truskey, Endothelial, cardiac muscle and skeletal muscle exhibit different viscous and elastic properties as determined by atomic force microscopy. *J Biomech* **34**, 1545–1553 (2001).
5. M. Sato, K. Nagayama, N. Kataoka, M. Sasaki and K. Hane, Local mechanical properties measured by atomic force microscopy for cultured bovine endothelial cells exposed to shear stress. *J Biomech* **33**, 127–135 (2000).
6. N. Almqvist, R. Bhatia, G. Primbs, N. Desai, S. Banerjee and R. Lal, Elasticity and adhesion force mapping reveals real-time clustering of growth factor receptors and associated changes in local cellular rheological properties. *Biophys J* **86**, 1753–1762 (2004).
7. E. A-Hassan, W. F. Heinz, M. D. Antonik, N. P. D'Costa, S. Nageswaran, C-A. Schoenenberger and J. H. Hoh, Relative micro elastic mapping of living cells by atomic force microscopy. *Biophys J* **74**, 1564–1578 (1998).
8. H. Haga, S. Sasaki, K. Kawabata, E. Ito, T. Ushiki and T. Sambongi, Elasticity mapping of living fibroblasts by AFM and immunofluorescence observation of the cytoskeleton. *Ultramicroscopy* **82**, 253–358 (2000).
9. D. M. Shasby and S. S. Shasby, Effects of calcium on transendothelial albumin transfer and electrical resistance. *J Appl Physiol* **60**, 71–79 (1986).
10. R. E. Mahaffy, C. K. Shih, F. C. MacKintosh and J. Kas, Scanning probe-based frequency-dependent microrheology of polymer gels and biological cells. *Phys Rev Let* **85**, 880–883 (2000).
11. S. Deguchi, T. Ohashi and M. Sato, Tensile properties of single stress fibers isolated from cultured vascular smooth muscle cells. *J Biomech* **39**, 2603–2610 (2006).
12. S. Hu, L. Eberhard, J. Chen, J. C. Love, J. P. Butler, J. J. Fredberg, G. M. Whitesides and N. Wang, Mechanical anisotropy of adherent cells probed by a three-dimensional magnetic twisting device. *Am J Physiol Cell Physiol* **287**, C1184–C1191 (2004).
13. T. Ohashi, H. Sugawara, T. Matsumoto and M. Sato, Surface topography measurement and intracellular stress analysis of cultured endothelial cells exposed to fluid shear stress. *JSME Int J Ser C* **43**, 780–786 (2000).
14. M. Sato, K. Suziki, Y. Ueki and T. Ohashi, Microelastic mapping of living endothelial cells exposed to shear stress in relation to three-dimensional distribution of actin filaments. *Acta Biomat* **3**, 331–339 (2007).

INNER EAR BIOMECHANICS

HIROSHI WADA [*], MICHIO MURAKOSHI, KOJI IIDA, SHUN KUMANO, KENJI ISHIHARA

Department of Bioengineering and Robotics, Tohoku University, 6-6-01 Aoba-yama, Sendai 980-8579, Japan

The elongation and contraction of outer hair cells (OHCs) greatly contribute to the high sensitivity and frequency selectivity of the mammalian inner ear. The motility of OHCs is believed to be based on the conformational changes of the motor protein prestin in the plasma membrane of OHCs. In the present study, first, to observe individual prestin molecules, prestin in the plasma membrane of Chinese hamster ovary (CHO) cells which were modified to express prestin was observed using an experimental approach combining atomic force microscopy (AFM) with quantum dots (Qdots), used as topographic surface markers. As a result, ring-like structures, each with four peaks and one valley at its center, were recognized in the vicinity of the Qdots, suggesting that these structures are prestin expressed in the plasma membranes of the CHO cells. Second, purification of prestin from prestin-expressing CHO cells was attempted for further clarification of the structure of prestin. Results indicate that 120 μg of purified prestin can be obtained from 2×10^9 prestin-expressing CHO cells.

1. Introduction

The OHCs in the mammalian cochlea exhibit elongation and contraction in response to acoustical stimulation [1]. This OHC motility results in cochlear amplification, leading to the high sensitivity and sharp frequency selectivity of the mammalian cochlea (Fig. 1). The source of such motility is believed to be conformational changes of the motor protein prestin, densely embedded in the plasma membrane of OHCs.

Previous morphological studies of OHCs using an electron microscope have shown the lateral membrane of the OHCs to be densely covered with particles about 10 nm in diameter, these particles being believed to be a motor protein [2]. The lateral membrane has also been observed by AFM and the existence of many particles with diameters of about 10 nm has also been reported [3]. Imaging by AFM of prestin-expressing CHO cells has revealed that particle-like structures 8–12 nm in diameter are possibly prestin [4]. However, since there are many kinds of intrinsic membrane proteins other than

[*] Hiroshi Wada is a Tohoku University Global COE Member.

prestin in the plasma membranes of OHCs and CHO cells, it was impossible to clarify which structures observed in such membranes were prestin.

In the present study, two lines of research for the clarification of the structure of prestin were conducted in parallel. First, an experimental approach combining AFM with Qdots, used as topographic surface markers, was employed. Inside-out plasma membranes were isolated from prestin-expressing CHO cells. Such membranes were then incubated with anti-prestin primary antibodies and Qdot-conjugated secondary antibodies. The plasma membranes of both types of CHO cells were subsequently observed by fluorescence microscopy and then scanned by AFM.

Second, an attempt was made to purify prestin from CHO cells stably expressing prestin. Purification of prestin would allow further analysis, *e.g.*, 2-D crystal structure analysis by AFM, to obtain knowledge about the structure of prestin at an angstrom level. Since it is unclear which detergents are suitable for solubilization of prestin, the best detergent for this was selected from 8 kinds of detergent commonly used for membrane protein isolation. Using the selected detergent, purification of prestin by anti-FLAG affinity chromatography was then performed.

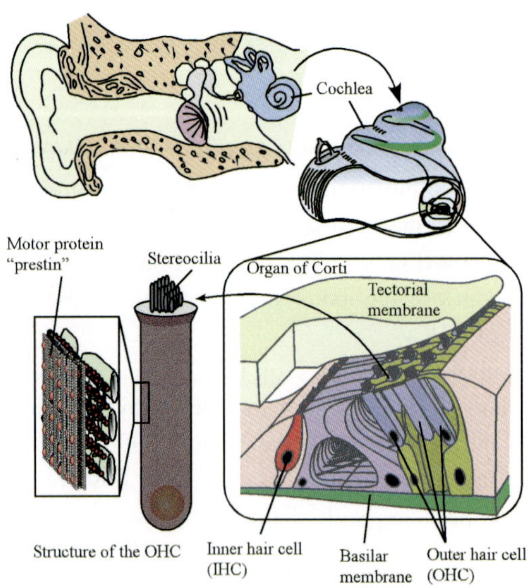

Figure 1. Human auditory system. Sensory cells, such as OHCs, inner hair cells and various kinds of the other cells, sit on the basilar membrane. OHCs subject the membrane to force, leading to cochlear amplification, resulting in the high sensitivity of mammalian hearing. Prestin is thought to be the origin of the motility of OHCs.

2. Immune Atomic Force Microscopy

2.1. *Materials and Methods*

2.1.1. *Sample preparation*

In this study, CHO cells stably expressing C-terminal FLAG-tagged prestin were used [5]. The inside-out plasma membranes were isolated from the prestin-expressing CHO cells. The cultured cells were plated on a glass-bottomed dish. After overnight culture, the cells were washed twice with an external solution (145 mM NaCl, 5.8 mM KCl, 1.3 mM $CaCl_2$, 0.9 mM $MgCl_2$, 10 mM HEPES, 0.7 mM Na_2HPO_4 and 5.6 mM glucose; pH 7.3) for removal of unwanted materials attached to the surfaces of the cells and substrate. The dish was immersed for 3 min in a hypotonic buffer (10 mM piperazine-*N*,*N*′-bis(2-ethanesulfonic acid), 10 mM $MgCl_2$ and 0.5 mM ethylene glycol bis(2-aminoethyl ether)-*N*,*N*,*N*′,*N*′-tetraacetic acid; pH 7.2). The cells were then sheared open by gentle exposure to a stream of the hypotonic buffer, resulting in the isolation of the inside-out basal plasma membranes. The isolated plasma membranes were incubated with a high-salt buffer (2 M NaCl, 2.7 mM KCl, 1.5 mM KH_2PO_4 and 1 mM Na_2HPO_4; pH 7.2) for 30 min at room temperature to remove the cytoskeletal materials and the peripheral proteins. The isolated membranes were then incubated with 0.05% trypsin for 1 min at room temperature to remove the remaining materials.

The isolated plasma membranes were fixed with 4% paraformaldehyde in PBS for 30 min at room temperature. After fixation, the membranes were rinsed three times with PBS and incubated with Block Ace (Dainippon Pharmaceutical, Osaka, Japan) for 30 min at 37°C. The membranes were incubated with goat antiprestin N terminus primary antibody (Santa Cruz Biotechnology, Santa Cruz, CA, USA) in PBS overnight at 4°C. The membranes were then incubated with Qdot® 655-conjugated rabbit antigoat IgG secondary antibody (Invitrogen, Carlsbad, CA, USA) in PBS for 60 min at 37°C. Finally, the membranes were washed with Hanks' balanced salt solution (HBSS; 5.33 mM KCl, 0.44 mM KH_2PO_4, 137.93 mM NaCl, 0.34 mM Na_2HPO_4 and 0.56 mM glucose; pH 7.3).

2.1.2. *Fluorescence microscopy*

Fluorescence images were obtained from the isolated membranes of the prestin-expressing CHO cells by an inverted fluorescence microscope, which comprised a part of the AFM used in this study. This system enables simultaneous recording of fluorescence and AFM images of the same sample.

2.1.3. *Atomic force microscopy*

An AFM (NVB100, Olympus), in which the AFM unit is mounted on an inverted fluorescence microscope (IX70, Olympus), was used. To minimize sample damage during scanning, AFM images were obtained using the oscillation imaging mode (Tapping mode™, Digital Instruments). The original AFM images were flattened by use of a software program (NanoScope v4.23, Digital Instruments) to eliminate background slopes and to correct dispersions of individual scanning lines. In the present study, Qdots were used as topographic surface markers. As the prestin molecules were labeled with primary and secondary antibodies, the distance between a Qdot and a prestin molecule was estimated to be approximately 50 nm [6].

2.2. *Results and Discussion*

The cytoplasmic surfaces of the isolated plasma membranes of the prestin-expressing CHO cells were observed by the tapping mode of AFM. Figure 2 is a fluorescence image of prestin labeled with Qdots in a prestin-transfected CHO cell (Fig. 2a) and the corresponding AFM images at low- (Fig. 2b) and medium-magnifications (Fig. 2c). The inset in each image is the digital zoom of the boxed area. The medium-magnification image (Fig. 2c) was obtained from the boxed area in Fig. 2b. As shown in Fig. 2a, the distribution of prestin labeling with Qdots in the plasma membrane of the prestin-expressing CHO cell was seen as a pattern of bright patches, as indicated by the arrowheads in its inset. The fluorescence of Qdots in the area surrounding the patches but not in the substrate area (asterisk, Fig. 2a) was also observed although its intensity was weak, indicating that prestin is nonuniformly distributed all over the plasma membrane of the CHO cells. In the low-magnification AFM image (Fig. 2b), the smooth surface of the cytoplasmic face of the prestin-transfected CHO cell was confirmed. However, as shown in its digital-zoomed image (inset, Fig. 2b), bulges about 7 nm in height were observed, as shown by the arrowheads.

Figure 3 shows high-magnification 3-D AFM images of a prestin-expressing CHO cell. Qdots 8 nm in height were clearly observed on the cytoplasmic face of the isolated plasma membrane of the prestin-transfected CHO cell, as indicated by black arrowheads. As shown in the magnification in the right of Fig. 3, a ring-like structure about 10 nm in diameter with four peaks and one valley at its center, possibly corresponding to a prestin molecule, was observed in the vicinity of the Qdot (a, arrow). In the vicinity of the other Qdots, similar structures were observed at the locations indicated by arrows (a–d), the magnifications of which are shown in the bottom panels. The observed configuration of prestin, *i.e.*, four peaks at its periphery, suggests that prestin has the tetrameric structure in the plasma membranes of prestin-transfected CHO cells.

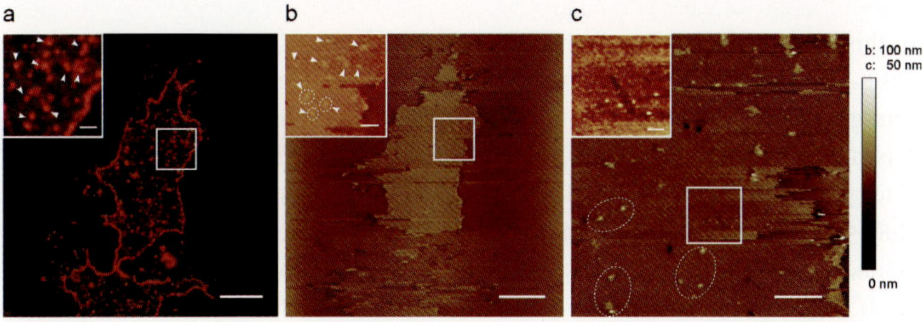

Figure 2. Fluorescence image of prestin labeled with Qdots and the corresponding AFM images of the prestin-transfected CHO cell. a. Fluorescence image of prestin labeled with Qdots. The inset shows a digital-zoomed image of the boxed area. b. AFM image at low magnification. The inset is the digital zoom of the boxed area. c. Medium-magnification AFM image obtained from the boxed area, shown in b. The inset shows the digital zoom of the boxed area. Scale bars of a, b and c are 10, 10 and 2 μm, respectively. Scale bars of the insets of a, b and c are 2 μm, 2 μm and 500 nm, respectively. Reprinted from [7] with the kind permission of Springer Science+Business Media.

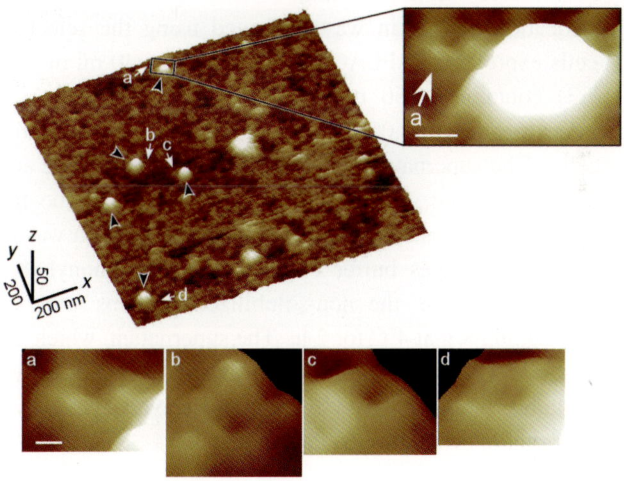

Figure 3. Membrane topology of prestin on the cytoplasmic face of the plasma membrane of the prestin-transfected CHO cell. a. High-magnification 3-D AFM images of the plasma membrane of the prestin-transfected CHO cell. Qdots 8 nm in height were clearly observed, as indicated by black arrowheads. As shown in the magnification at the right, a ring-like structure with four peaks and one valley at its center, possibly corresponding to a prestin molecule, was observed in the vicinity of the Qdot (a, arrow). Similar structures were observed at the locations indicated by arrows (a–d). Reprinted from [7] with the kind permission of Springer Science+Business Media.

3. Purification of the Motor Protein Presitn

3.1. *Materials and Methods*

In this study, CHO cells which had been modified to stably express C-terminal 3×FLAG-tagged prestin were used [5]. Generally, membrane proteins require detergents for solubilization. Although there are many kinds of detergents, it is unclear which are suitable to solubilize prestin. In this study, 8 different detergents, commonly used for membrane protein solubilization, *i.e.*, octyl glucoside, dodecyl maltoside, nonyl thiomaltoside, sucrose monocholate, sucrose monocaprate, sucrose monolaurate, MEGA-9 and CHAPSO were tested. First, 5×10^6 cells were incubated on ice with 100 μl of 10 mM Hepes buffer (pH 7.3) containing 140 mM NaCl, 5 mM KCl, 1 mM $MgCl_2$, 1.5 mM $CaCl_2$ and the different detergents at various concentrations for 30 min. After incubation, the samples were centrifuged at 20,360 × g at 4°C for 2 h. Proteins contained in the supernatant and those in the precipitate were defined as solubilized proteins and non-solubilized proteins, respectively. The ratio of prestin in supernatants (solubilized prestin) to that in the precipitates (non-solubilized prestin) was analyzed by quantitative Western blotting.

Next, purification of prestin was attempted using the selected detergent. First, 2×10^9 cells expressing 3×FLAG-tagged prestin in 40 ml of 10 mM Hepes buffer (pH 7.3) containing 150 mM Na_2SO_4 were disrupted by sonication, followed by centrifugation at 1,000 × g for 7 min at 4°C to remove nuclei and undisrupted cells. The supernatant was centrifuged at 20,360 × g at 4°C for 2 h to separate the cytoplasmic soluble fraction and the membrane fraction. To solubilize membrane proteins, the obtained membrane fraction was resuspended in 40 ml of the same Hepes buffer containing 10 mM nonyl thiomaltoside. After 2-h incubation on ice, the non-solubilized proteins were removed by centrifugation at 20,360 × g at 4°C for 2 h. The supernatant, which contained the solubilized prestin, was applied to a column of anti-FLAG affinity gel. After washing the column, bound proteins were competitively eluted with a buffer containing 250 μg/ml of 3×FLAG peptide.

3.2. *Results and Discussion*

The relationships between the detergent concentration and the solubilization efficiency using the 8 kinds of detergents are shown in Fig. 4. When the concentrations of detergent are higher than their critical micelle concentrations (cmc), the detergents form micelles. In such cases, membrane proteins are reported to be solubilized by detergents [8]. In the present study, when octyl glucoside, nonyl thiomaltoside, sucrose monocholate or MEGA-9 was used, solubilization efficiency sharply increased around each cmc. Since this tendency agrees with the above-mentioned knowledge about membrane protein solubilization and with findings of the previous reports in which target membrane proteins were solubilized in active form [9-11], these 4 kinds of detergents may uniformly solubilize prestin without denaturation. On the other hand, when dodecyl maltoside, sucrose monocaprate, sucrose monolaurate or CHAPSO was used, solubilization efficiency gradually increased with increasing concentrations of detergents. This tendency is contrary to the above-mentioned knowledge about membrane protein solubilization. Furthermore, the results showing a gradual increase of solubilization efficiency indicate that a part of prestin was solubilized at low detergent concentrations and that another part of prestin was solubilized at high detergent concentrations, suggesting that the condition of prestin solubilized at low detergent concentrations was different from that of prestin solubilized at high detergent concentrations. Hence, there is a possibility that these 4 kinds of detergents nonuniformly solubilized prestin, perhaps with denaturation. Therefore, in the present study, the candidates for use in prestin solubilization were limited to the former four detergents.

It has been reported that the use of nonionic detergents containing a sugar group, including octyl glucoside and nonyl thiomaltoside, are expected to increase the possibility of solubilization of membrane proteins without denaturation and that nonyl thiomaltoside is more stable than octyl glucoside [12]. In the present study, therefore, nonyl thiomaltoside was used for solubilization of prestin.

Purification of prestin was performed. To visualize all proteins in the sample and to detect prestin, 1/10,000 volume of each sample obtained at each step of the purification procedure was subjected to SDS-PAGE with Coomassie brilliant blue staining and Western blotting with probing by anti-FLAG antibody. Representative data are shown in Fig. 5. From the cells (lane 1), nuclei and undisrupted cells (lane 2) were removed by sonication followed by centrifugation at 1,000 × *g*. The obtained supernatant (lane 3) containing prestin was separated into a cytoplasmic soluble fraction (lane 4) and a membrane fraction (lane 5) by centrifugation at 20,360 × *g*. Prestin was detected in the isolated membrane fraction (lane 5). Membrane proteins were then solubilized from the membrane fraction with 10 mM nonyl thiomaltoside. Non-solubilized proteins (lane 6) and solubilized proteins (lane 7) were separated by centrifugation at 20,360 × *g*. The solubilized proteins were then applied to an anti-FLAG affinity column. Some of the prestin did not bind to

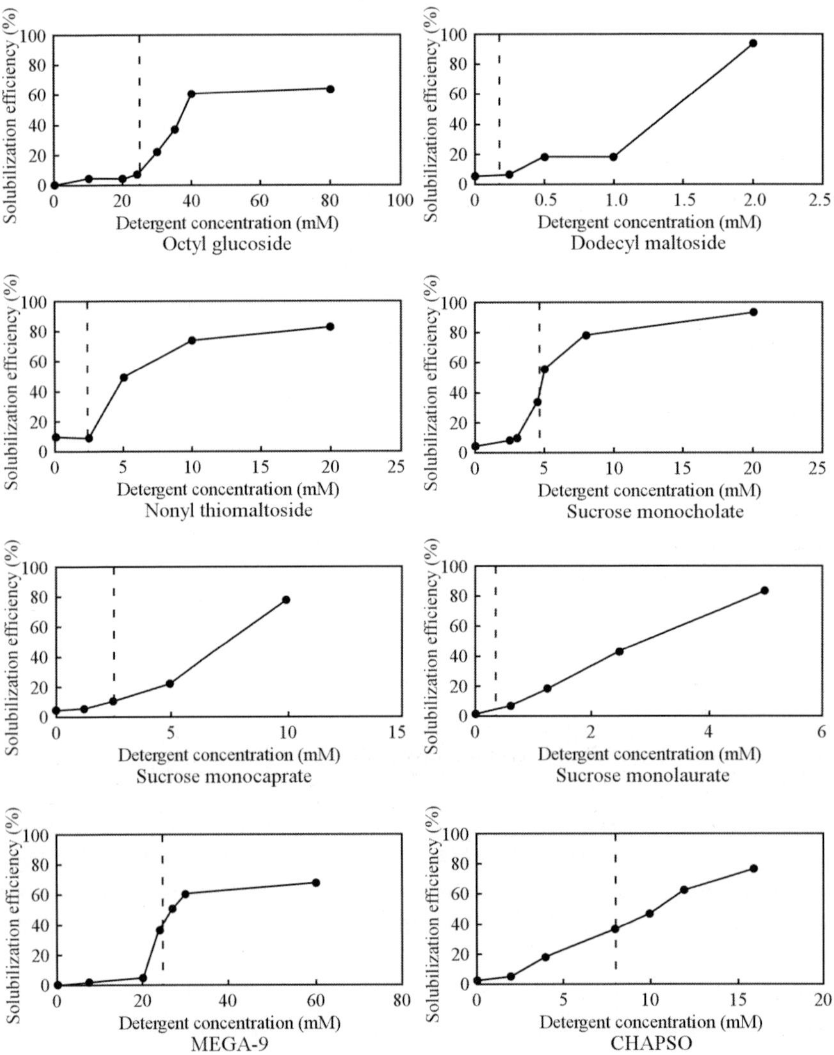

Figure 4. Relationships between the detergent concentration and the solubilization efficiency [13]. The filled circles, connected by solid lines, indicate the solubilization efficiency obtained by quantitative Western blotting. Dashed lines show each critical micelle concentration (cmc). In the case of octyl glucoside, nonyl thiomaltoside, sucrose monocholate and MEGA-9, the solubilization efficiency of prestin sharply increased around the cmc, while when the other detergents were used, solubilization efficiency gradually increased with increasing detergent concentration.

the column and was found in the flow-through fraction (lane 8). To remove nonspecifically retained contaminants, the column was washed with the buffer (lane 9). The prestin was then competitively eluted from the column using 3×FLAG peptide (lane 10).

Since purified prestin was not detected by Coomassie brilliant blue staining due to its low concentration (Fig. 5(a), lane 10), 1/200 volume of the obtained purified prestin and the same volume of the sample obtained from the untransfected cells (negative control) were loaded onto another SDS-PAGE gel. The gel was stained with Silver Stain Kit (Bio-Rad Laboratories), which has high sensitivity. The result is shown in Fig. 6. In lane 1, which shows the obtained purified prestin, a major band around 100 kDa (arrowhead) indicating prestin and a weak band around 40 kDa were detected. In lane 2, which shows the sample obtained from the untransfected cells, a band around 40 kDa was also detected. Since 40-kDa protein was detected in lane 2, this protein was probably derived from CHO cells.

The amount of obtained purified prestin was determined by quantitative Western blotting. As a result, it was estimated that about 120 μg of purified prestin was obtained from 2×10^9 3×FLAG-tagged prestin-expressing CHO cells.

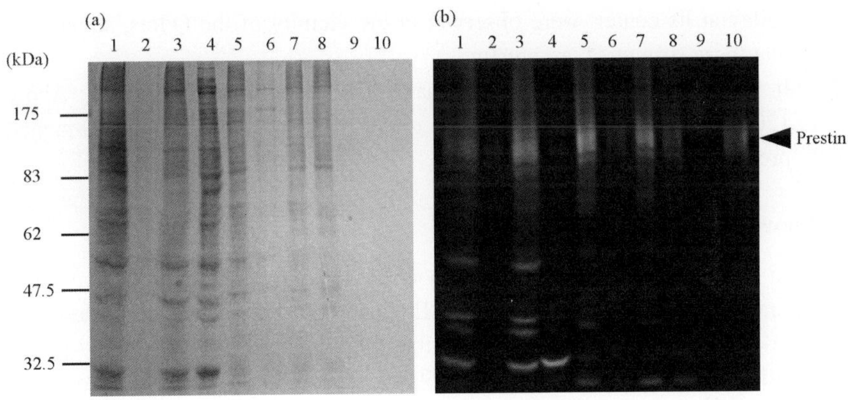

Figure 5. SDS-PAGE and Western blot analysis of prestin purification [13]. (a) Coomassie brilliant blue-stained gel and (b) corresponding immunoblot. SDS-PAGE was performed on a 10% polyacrylamide gel. The arrowhead shows prestin. Lane 1: total proteins of FLAG-tagged prestin-expressing CHO cells. Lane 2: nuclei and undisrupted cells. Lane 3: supernatant after sonication and centrifugation at 1,000 × g. Lane 4: cytoplasmic soluble fraction. Lane 5: membrane fraction. Lane 6: non-solubilized membrane proteins. Lane 7: solubilized membrane proteins. Lane 8: flow-through fraction. Lane 9: wash fraction. Lane 10: purified prestin eluted from the column using 3×FLAG peptide.

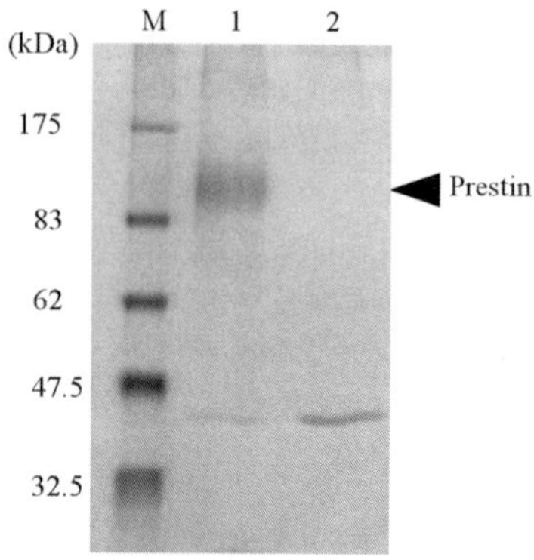

Figure 6. Silver-stained SDS-PAGE gel loaded with purified prestin and with the sample obtained from the untransfected CHO cells [13]. SDS-PAGE was performed on a 10% polyacrylamide gel. The arrowhead shows prestin. Lane M: molecular weight marker. Lane 1: purified prestin (Fig. 5, sample 10). Lane 2: sample obtained from untransfected cells (negative control).

4. Conclusions

1. Prestin molecules expressed in the plasma membranes of prestin-expressing CHO cells were labeled with Qdots about 8 nm in height, which were clearly imaged by AFM. Ring-like structures, each with four peaks and one valley at its center, were observed in the vicinity of the Qdots, suggesting that these structures are prestin.
2. It was clarified that nonyl thiomaltoside can solubilize prestin. About 120 μg of purified prestin was obtained from $2{\times}10^9$ 3×FLAG-tagged prestin-expressing CHO cells.

Acknowledgments

Hiroshi Wada acknowledges the support of Tohoku University Global COE Program "Global Nano-Biomedical Engineering Education and Research Network Centre," as well as that of the following grants: Gtrant-in-Aid for Scientific Research on Priority Areas 15086202 from the Ministry of Education, Culture, Sports, Science and Technology of Japan, Grant-in-Aid for Scientific Research (B) 18390455 from the Japan Society for the Promotion of Science, Grant-in-Aid for Exploratory Research 18659495 from the Ministry of Education, Culture, Sports, Science and Technology of Japan, the Human Frontier Science Program, a Health and Labour Science Research Grant from the Ministry of Health, Labour and Welfare of Japan, a grant from the Iketani Science and Technology Foundation and a grant from the Daiwa Securities Health Foundation.

References

1. W. E. Brownell, C. R. Bader, D. Bertrand and Y. de Ribaupierre, *Science* **227**, 194 (1985).
2. A. Forge, *Cell Tissue Res.* **265**, 473 (1991).
3. C. Le Grimellec, M.C. Giocondi, M. Lenoir, M. Vater, G. Sposito and R. Pujol, *J. Comp. Neurol.* **451**, 62 (2002).
4. M. Murakoshi, T. Gomi, K. Iida, S. Kumano, K. Tsumoto, I. Kumagai, K. Ikeda, T. Kobayashi and H. Wada, *J. Assoc. Res. Otolaryngol.* **7**, 267 (2006).
5. K. Iida, K. Tsumoto, K. Ikeda, I. Kumagai, T. Kobayashi and H. Wada, *Hear. Res.* **205**, 262 (2005).
6. W. K. Hartmann, N. Saptharishi, X. Y. Yang, G. Mitra and G. Soman, *Anal. Biochem.* **325**, 227 (2004).
7. M. Murakoshi, K. Iida, S. Kumano and H. Wada, *Pflüger Archiv.* (in press).
8. A. Helenius and K. Simons, *Biochem. Biophys. Acta.* **415**, 29 (1975).
9. V. De Pinto, R. Benz and F. Palmieri, *Eur. J. Biochem.* **183**, 179 (1989).
10. S. Izawa, Y. Sakai-Tomita, K. Kinomura, S. Kitazawa, M. Tsuda and T. Tsuchiya, *J. Biochem.* **113**, 573 (1993).
11. T. Shimamoto, S. Saito and T. Tsuchiya, *J. Biochem.* **97**, 1807 (1985).
12. T. Tsuchiya and S. Saito, *J. Biochem.* **96**, 1593 (1984).
13. K. Iida, M. Murakoshi, S. Kumano, K. Tsumoto, K. Ikeda, T. Kobayashi, I. Kumagai and H. Wada, *JBSE* **3**, 221 (2008).

COMPUTATIONAL BIOMECHANICS FOR INVESTIGATING CARDIOVASCULAR DISEASES

TAKAMI YAMAGUCHI [1)*], HITOSHI KONDO [2)], YUJI SHIMOGONYA [2)], YOHSUKE IMAI [2)], NORIAKI MATSUKI [1)], TAKUJI ISHIKAWA [2)]

1) Department of Biomedical Engineering, Tohoku University
6-6-01, Aoba, Aramaki, Aoba-ku, Sendai 980-8579, Japan

2) Department of Bioengineering and Robotics, Tohoku University
6-6-01, Aoba, Aramaki, Aoba-ku, Sendai 980-8579, Japan

Due to the recent rapid growth in computational power, computational biomechanics is now thought of as the most rapidly growing subfield in biomedical engineering. We have investigated cardiovascular diseases on micro to macro levels using conjugated computational mechanics to analyze fluid, solid, and biochemical interactions. In this report, we introduce our recent studies on cerebral aneurysm growth based on a hemodynamic hypothesis and malaria-infected red blood cell mechanics using a particle method.

1. Introduction

Because of the rapid growth in computational power, computational biomechanics is now the most rapidly growing subfield in biomedical engineering. We have been involved in a large project, facilitated by The Ministry of Education, Culture, Sports, Science and Technology, named the "Next-Generation Supercomputer Project of Research and Development of the Next-Generation Integrated Simulation of Living Matter." This national project seeks to utilize massive computer power to open a new scientific field, "computational biomedicine." Since some disease processes might be clarified and treatment may be supported using computational biomechanics, the study of cardiovascular diseases in terms of computational biomechanics is an important area of scientific research.

The human cardiovascular system is under the integrated nervous and humoral control of the entire body, *i.e.*, homeostasis. Multiple feedback mechanisms with mutual interactions among systems, organs, and even tissues provide integrated control of the whole body. These control mechanisms have different spatial coverages, from the micro- to macroscale and different time

[*] Takami Yamaguchi is a Tohoku University Global COE Leader.

constants, from nanoseconds to decades. We believe that these variations in spatial and temporal scales should be taken into account in discussing phenomena of the cardiovascular system.

Based on this background, we investigated the cardiovascular system from the micro to macro levels using conjugated computational mechanics to analyze fluid, solid, and biochemical mechanics. In this report, we introduce our recent work on cerebral aneurysm growth based on a hemodynamic hypothesis and malaria-infected red blood cell mechanics using a particle method.

2. Simulation of Initiation and Growth of a Cerebral Aneurysm

A cerebral aneurysm is an extremely important disorder in clinical medicine, since the rupture of aneurysms can lead to serious pathologic conditions, such as subarachnoid hemorrhages. However, the mechanism of aneurysm growth is not yet understood. A cerebral aneurysm can be characterized by a saccular expansion of the arterial wall. Moreover, the strength degradation of the arterial wall may not be sufficient to explain the saccular expansion [1]. To understand this phenomenon, considering biological reactions of the arterial wall is important. We have focused on the increase in the volume of the extracellular matrix, or in the number of cells in the arterial wall, as candidates for the key factor on cerebral aneurysm growth [2].

2.1. *Modeling and Methods*

We investigated how an aneurysm is formed in a model cerebral artery with curvature. Figure 1 shows the initial geometry of the artery employed in this study, which was modeled on a part of an internal carotid artery. The artery model was 30 mm in length along the central axis, with a diameter of 3 mm, curvature radius of 3.6 mm, and torsion of 15°. These values are not very different from clinical observations.

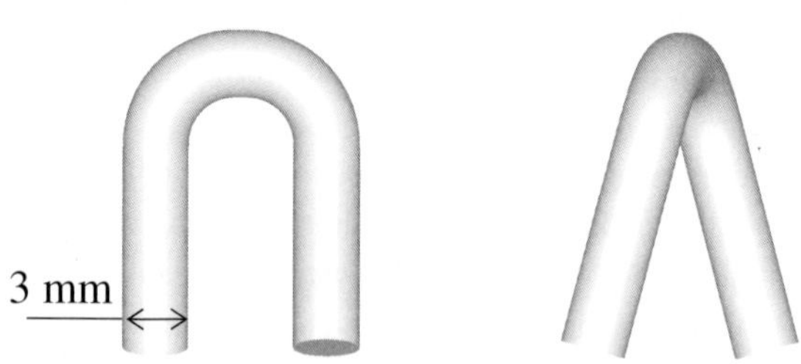

(a) Front view (b) Side view

Figure 1. Geometry of the artery model.

For the calculation of blood flow, blood was assumed to be an incompressible and Newtonian fluid with a density $\rho = 1.05 \times 10^3$ kg/m^3 and viscosity $\mu = 3.5 \times 10^{-3}$ Pa·s. The governing equations for such a blood flow are the equation of continuity and the Navier–Stokes equations. We solved the steady flow at the averaged Reynolds number in the internal carotid artery (Re = 200). The change in the diameter of typical cerebral arteries during one pulsation was small and its effect on the wall shear stress (WSS) is not very significant. Moreover, as we solved the steady flow in this study, we neglected the wall deformation due to pulsation. Boundary conditions were a parabolic velocity profile at the inlet, zero pressure at the outlet, and the no-slip condition on the wall. Blood flow calculation was accomplished through an in-house three-dimensional flow solver based on a MAC algorithm. The total number of grid points was 52,065.

The arterial wall was discretized by triangle elements. The computational grid generated on the arterial wall is shown in Fig. 2(a), where 16,384 triangle elements (8,256 nodal points) were generated in total. The spring network model was used to mechanically model the arterial wall. In this model, the mechanical behavior of the arterial wall was expressed by two types of spring, stretch/compression and bending, as depicted in Fig. 2(b). The stretch/compression spring, which corresponded to a side of a triangle element, expressed the resistance to stretch/compression of the membrane. The other spring expressed the bending resistance of the membrane. Thus, the effect of wall thickness was approximated by this spring.

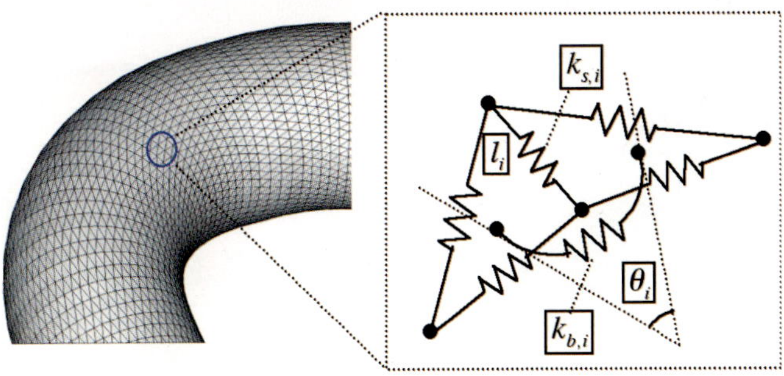

(a) Computational grid (b) Mechanical modeling of the arterial wall

Figure 2. Modeling of an arterial wall.

The arterial wall expansion in the hypothesis can be expressed by the natural length elongation of a stretch/compression spring. In this study, we formulated the degree of the elongation as follows:

$$\frac{l_i - l_i^0}{l_i^0} = \alpha(\tau_i - \tau_{th}) \text{ if } \tau_i > \tau_{th} \quad (1)$$

where l_i^0 and l_i are the natural length of the stretch/compression spring element i before and after biological reactions, respectively; τ_i is the WSS due to blood flow on the element i; τ_{th} is the threshold of the WSS; and α is a parameter for the degree of the biological reactions. This equation was applied only to the stretch/compression element, in which $\tau_i > \tau_{th}$. We employed a simple linear equation because we wanted to start from a simple model as a first step, since the detail of the relationship is unclear.

To solve the deformation of the arterial wall due to the change in the natural length of springs, we performed the following process: (a) initially change the natural length of springs without deformation, *i.e.*, in the same geometry with Fig. 1; (b) calculate spring forces acting on each node; (c) move each node during a small time step by adding artificial viscous drag force at each node; and (d) continue (b) and (c) until convergence criteria are satisfied. The details of the numerical methods can be found elsewhere [3].

2.2. *Results and Discussion*

First, we performed the calculation of the steady blood flow of Re = 200 for the artery model. Figure 3 illustrates the distribution of the WSS due to blood flow. The WSS value was relatively high on the curve, and an especially high WSS region was concentrated on one side of the artery due to the arterial torsion.

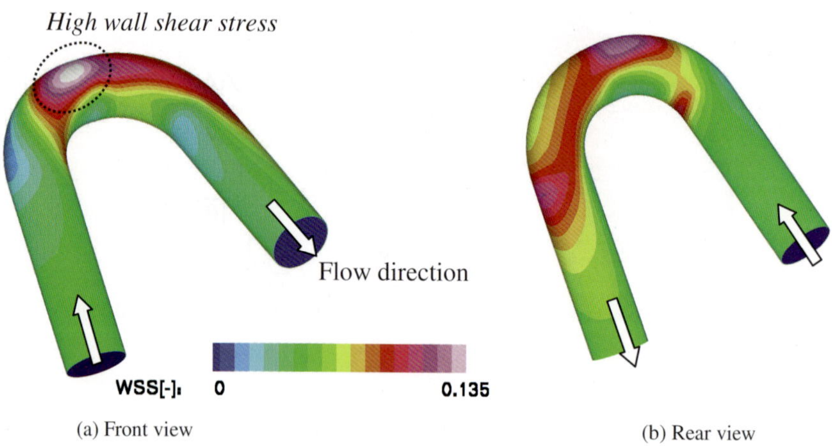

Figure 3. Wall shear stress distribution.

We then performed the growth simulation of cerebral aneurysms under the following three situations: (a) only the strength degradation of the wall, (b) only the biological reaction (surface area expansion), and (c) both the strength degradation and the biological reaction. The threshold value τ_{th} was assumed to be 0.12, which was equivalent to 90% of the maximum WSS value in this computation.

Figure 4 shows the results of the growth simulation of cerebral aneurysms in the case when only the strength degradations of the wall occur at the site of $\tau > \tau_{th}$. In Fig. 4(a), stretch/compression spring constants at the site of $\tau > \tau_{th}$ decreased uniformly by 30%. Figure 4(b) and (c) show the case of 60% and 90%, respectively. Young's modulus degradation of the cerebral aneurysmal wall has been reported to be about 30% [4]. Thus, the 90% degradation in Fig. 4(c) represented an extremely low-strength condition. However, the resultant shape was not saccular. Note that the WSS and shape change calculations were performed only once in each case and the feedback of shape change on the WSS distribution was not considered. The same holds for the following results.

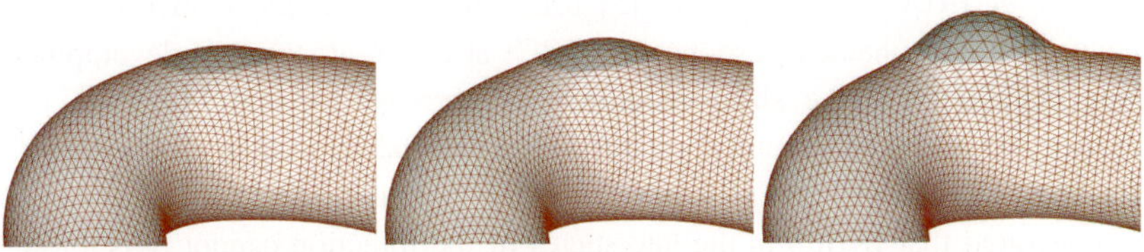

Figure 4. Wall deformation due to strength degradation.

Next, we applied the present model to the WSS distribution. Figure 5 shows the results of the growth simulation of cerebral aneurysms in the case when only the biological reactions (surface area expansion) occur at the site of $\tau > \tau_{th}$. In Fig. 5(a), the parameter α expressing the degree of the biological reaction

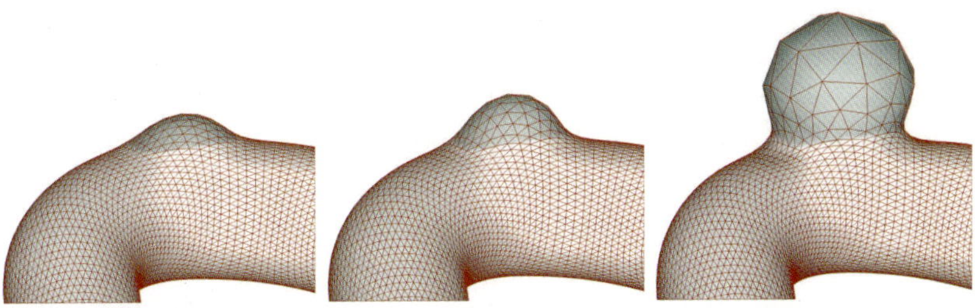

Figure 5. Wall deformation due to biological reaction.

(see Eq. (1)) was assumed to be $\alpha = 30$. Figure 5(b) and (c) represent the cases of $\alpha = 50$ and $\alpha = 100$, respectively. The resultant shape in the case of $\alpha = 100$ was saccular and consistent with former clinical observations. Although the saccular expansion can be formed by considering the biological reaction, in the case of strength degradation, the saccular expansion of the wall was not formed, even with an unrealistically large degradation in wall strength. Thus, considering the biological reaction in addition to purely mechanical processes is necessary to understand the growth of a cerebral aneurysm. We believe that the present computational model is a powerful tool for understanding the phenomenon.

3. Modeling the Hemodynamics Arising from Malarial Infection

Malaria is one of the most serious infectious diseases on earth, affecting several hundred million patients, with several million deaths arising from malarial infection. When a parasite invades and matures inside a red blood cell (RBC), the infected RBC (IRBC) becomes stiffer and cytoadherent. Experimental techniques for cell mechanics have been applied to study the mechanical property of IRBCs. Suresh *et al.* [5] used optical tweezers to investigate the change in the mechanical response of IRBCs at different stages of development. They clarified that the shear modules can increase by tenfold in the final schizont stage. More recently, microfluidics has been used to investigate the effect of the IRBC stiffness on capillary obstruction. Shelby *et al.* [6] demonstrated that IRBCs in the late stages of the infection cannot pass through microchannels having a diameter smaller than that of the IRBCs. The cytoadherent property of IRBCs has also been investigated experimentally. Cooke *et al.* [7] identified some ligand–receptor pairs for the adhesion. They also reported that few receptor interactions show stable binding under flow conditions. The IRBCs roll on intercellular adhesion molecule 1, vascular cell adhesion molecule 1, and P-selectin, and rolling IRBCs are arrested at the site of CD36.

While these experimental results imply that the stiffness and cytoadherence of IRBCs result in the microvascular occlusion, these experiments were limited to the effect of a single infected cell. The effect of malarial infection on hemodynamics has not been well studied, primarily due to limitations of the current experimental techniques. Observing RBC interactions with other cells is still difficult, even with the recent confocal microscopic techniques. Instead, numerical modeling can be a strong tool for furthering our understanding of the pathology of malaria. Numerical modeling was performed for the stretching of an IRBC using optical tweezers [5,8]. We have also developed a numerical model of hemodynamics involving adhesive interactions [9]. In the following section, we discuss the effect of the adhesive property of IRBCs on the microvascular occlusion.

3.1. *Modeling and Methods*

Our model was based on a particle (Lagrangian and free-mesh) method. All components of blood were represented by particles. In two-dimensional problems, the membrane of IRBCs was represented by a one-dimensional network consisting of a finite number of particles. A membrane particle was connected to two neighboring particles with a spring, providing a stretching force. We also considered the resistance to bending, as shown in Fig. 6(a). The adhesive property of IRBCs was also modeled by springs (Fig. 6(b)). A connection between two particles was represented a cluster of many ligand–receptor bindings. If the distance between a particle of IRBC membrane and a particle of endothelial cells or a particle of neighboring RBC membrane was less than a threshold value, the two particles were connected by a stretch/compression spring. The connection was maintained until the distance again became larger than the threshold value. The details of the modeling method can be found in reference [9].

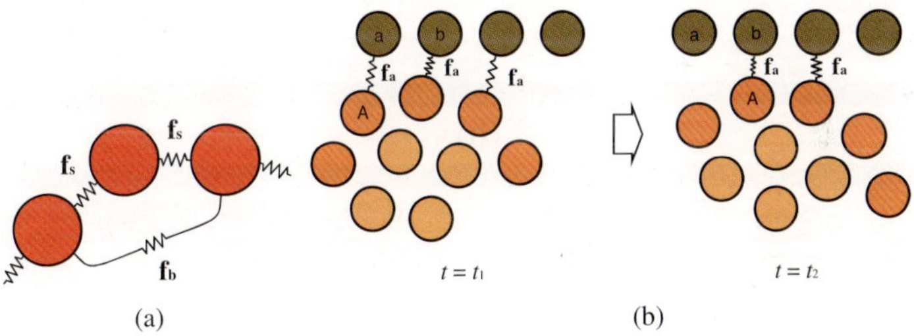

Figure 6. Spring model of the property of IRBCs: (a) membrane model, (b) adhesion model.

3.2. *Results and Discussion*

We simulated the flow in parallel plates set 12 μm apart. The IRBC was assumed to be at the schizont stage. The given boundary conditions were constant velocity μ = 4.0 [mm/s] at the inlet, constant zero pressure at the outlet, and the no-slip condition at the wall surface. The flow condition was similar to the arteriolar flow. Figure 7 shows snapshots of the numerical results, when the blue particles adhered to the other particles. The IRBC moved downstream rolling slowly on the wall surface. Such a rolling motion was observed experimentally [7,10] due to synergistic interactions with multiple adhesion molecules on endothelial cells [10]. Since the velocity of the IRBC that had adhesive interactions with endothelial cells was lower than that of healthy RBCs, the following healthy RBC, indicated by the character A, caught the IRBC, and then these two cells formed a cluster. This phenomenon is known as rosetting, which refers to the formation of a cluster of cells involving an IRBC and more than two healthy RBCs. According to an *in vivo* study [11], stable rosettes formed mainly in capillary and postcapillary venules, and the rosettes then produced a sufficient degree of aggregation to block the microcirculation. In the arteriolar flow examined here, however, the shear stress was sufficiently high to preclude rosetting. The rosette could be disrupted in a short time period, and the other RBCs deformed and passed to the side of the IRBC.

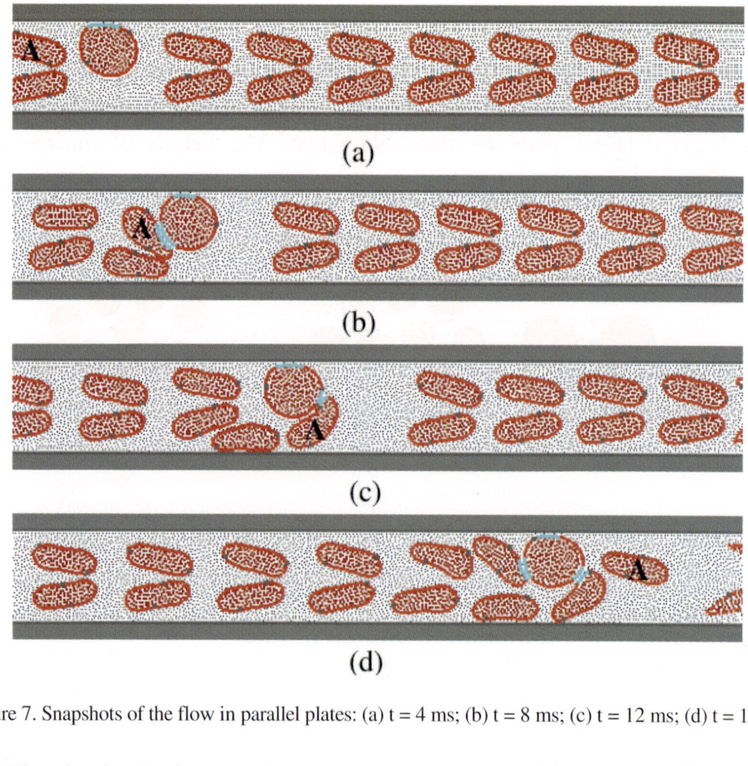

Figure 7. Snapshots of the flow in parallel plates: (a) t = 4 ms; (b) t = 8 ms; (c) t = 12 ms; (d) t = 16 ms.

The slowly moving IRBC on endothelial cells increased the flow resistance significantly. The difference in the velocity between the IRBC and the healthy RBCs caused complex interactions with the healthy RBCs. These complex interactions resulted in a further pressure drop, as shown in Fig. 8. In our previous study, we also confirmed that the increase/decrease of adhesive interactions increased/decreased the flow resistance [9]. These results suggest that the adhesive interactions with endothelial cells and sequential interactions with healthy RBCs have a critical role in the microvascular obstruction, particularly in the late stage of malarial infection. In the late stage, many knobs develop at the surface of the IRBC membrane. The adhesive interactions are promoted by the development of the knobs because the level of the PfEMP1 parasite protein increases. The microvascular obstruction is therefore accelerated in the late stage of the infection.

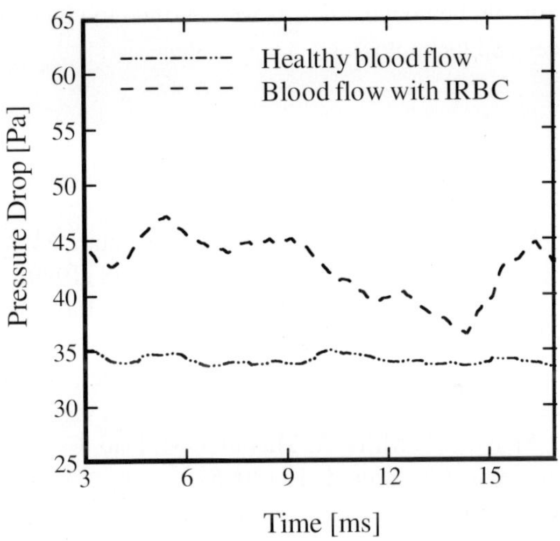

Figure 8. Comparison of the pressure drop between the healthy blood flow and the blood flow with an IRBC.

4. Conclusions

In this report, we reviewed our recent studies on computational mechanics for arterial diseases. In clinical applications, however, we must consider biological complexities in the analysis of blood flow, especially with respect to disease processes. A disease is not just a failure of a machine. It is an outcome of complex interactions among multilayered systems and subsystems. They interact mutually across the layers in a strongly nonlinear and multivariable manner. Note also that a living system, either as a whole or as a subsystem, such as the cardiovascular system, is always under the integrated nervous and humoral control of the whole body, *i.e.*, homeostasis. Multiple feedback mechanisms with mutual interactions among systems, organs, and even tissues provide integrated control of the entire body. These control mechanisms have different spatial coverage, from the micro- to macroscale, and different time constants, from nanoseconds to decades. Although it has not been fully acknowledged, much longer timescale phenomena, such as evolution and differentiation of living systems, must also be considered if we are to understand the living system *per se*. Therefore, in future analyses, these biological phenomena should be included in discussing physiological, as well as pathological (*i.e.*, disease) processes. We expect this to be accomplished in the future by integrating new understandings of macro- and microscale hemodynamics, if we continue to coincide with advances of related sciences and technologies.

References

1. Y. Feng, S. Wada, T. Ishikawa, K. Tsubota and T. Yamaguchi, *J. Biomech. Sci. Eng.* **3**, 124 (2008).
2. H. Masuda, Y. J. Zhuang, T. M. Singh, K. Kawamura, M. Murakami, C. K. Zarins and S. Glagov, *Thromb. Vasc. Biol.* **19**, 2298 (1999).
3. Y. Shimogonya, T. Ishikawa, Y. Imai, D. Mori, N. Matsuki and T. Yamaguchi, *J. Biomech. Sci. Eng.* **3**, 431 (2008).
4. H. J. Steiger, R. Aaslid, S. Keller and H.-J. Reulen, *Heart Vessels* **5**, 41 (1989).
5. S. Suresh, J. Spatz, J. P. Mills, A. Micoulet, M. Dao, C. T. Lim, M. Beil and T. Seufferlein, *Acta Biomater.* **1**, 15 (2005).
6. J. P. Shelby, J. White, K. Ganesan, P. K. Rathod and D. T. Chiu, *PNAS* **100**, 14618 (2003).
7. B. M. Cooke, N. Mohandas and R. L. Coppel, *Semin. Hematol.* **41**, 173 (2004).
8. M. M. Dupin, I. Halliday, C. M. Care and L. L. Munn, *Int. J. CFD* **22**, 481 (2008).
9. H. Kondo, Y. Imai, T. Ishikawa, K. Tsubota and T. Yamaguchi, *Ann. Biomed. Eng.* (submitted).
10. B. G. Yipp, S. Anand, T. Schollaardt, K. D. Patel, S. Looareesuwan and M. Ho, *Blood* **96**, 2292 (2000).
11. K. Silamut, N. H. Phu, C. Whitty, G. D. H. Turner, K. Louwrier, N. T. H. Mai, J. A. Simpson, T. T. Hien and N. J. White, *Am. J. Pathol.* **155**, 395 (1999).

CHANGE IN MECHANICAL PROPERTIES AND ACTIVITIES OF MATRIX METALLOPROTEINASES IN RAT AORTAS STIMULATED WITH CYTOKINES

WENJING HUANG [1)*], NAOYA SAKAMOTO [2)],
TOSHIRO OHASHI [2)], MASAAKI SATO [1)†]

1) Graduate School of Biomedical Engineering, Tohoku University,
6-6-01 Aramakiaza aoba, Aobaku, Sendai, 980-8579, Japan

2) Graduate School of Engineering, Tohoku University,
6-6-01 Aramakiaza aoba, Aobaku, Sendai, 980-8579, Japan

Aortic aneurysms are characterized by inflammation and excessive degradation of extracellular matrix, which is mainly composed of elastin and collagen. These processes are thought to be induced by inflammatory cytokines, such as interleukin-1β (IL-1β) and tumor necrosis factor-α (TNF-α). However, the detailed role of cytokines in the formation of aortic aneurysms is unclear. In this study, we examined the relationship between the inflammatory cytokines and the mechanical properties of rat aortas, focusing on the contribution of matrix metalloproteinases (MMPs). Aortic walls extracted from rats were incubated in serum-free culture medium with or without IL-1α or TNF-α for 7 days and uniaxial tensile testing of the specimens was then performed. A modified form of the Michaelis-Menten equation was used to quantify the mechanical properties of the aorta wall. We determined for each specimen the initial slope E_i and the final slope of curve E_f. Activities of MMP-2 and MMP-9 in culture media used for cultivation of aortas were also determined by gelatin zymography. The stress-strain curve of aortas stimulated with cytokines was shifted to the left when compared to culture without cytokines (control). There were no significant differences in the final slope E_f between aortas stimulated with cytokines and control. However, the E_i of rat aorta stimulated with cytokines had an upward tendency and was significantly higher for aortas stimulated with TNF-α when compared with controls. These results indicated that degradation of elastic fibers was induced by cytokines. The activities of MMP-2 and MMP-9 for aortas stimulated with cytokines also tended to increase when compared to controls. In addition, we found that the activities of MMP-2 were significantly correlated with the initial slope E_i. These results support the hypothesis that inflammatory cytokines play an important role in the formation of aneurysms by increasing the activities of MMP-2 and MMP-9, leading to degradation of elastin.

---

* Wenjing Huang is a Tohoku University Global COE Research Assistant.

† Masaaki Sato is a Tohoku University Global COE Member.

1. Introduction

An aortic aneurysm is an inflammatory disorder characterized by localized degradation of extracellular matrix including elastin and collagen. The loss of elastin, which is normally responsible for elasticity of the aorta under physiological conditions, is thought to lead to initiation of aortic aneurysms, and the mechanical properties of aortic aneurysm tissue become less compliant [1,2]. The degradation of extracellular matrix may be induced by an increase in inflammatory cytokines such as interleukin-1β (IL-1β), tumor necrosis factor-α (TNF-α) in aortic aneurysms [3], which may be responsible for the increased expression of matrix-degrading enzymes and matrix metalloproteinases (MMPs) in aortic aneurysms [4,5]; however, to date, the relationship between the inflammatory cytokines and the mechanical properties of aortic aneurysms remain uncertain. In this study, we cultured rat aortas with inflammatory cytokines and investigated the mechanical properties in order to determine their relationship and to elucidate the mechanisms behind aneurysm formation.

2. Materials and Methods

2.1. *Sample Preparation*

Aortas were extracted from male Wistar rats weighing 230 to 260g. After the aortas were cut into 5.5-mm (length) rings, specimens were incubated in serum-free Dulbecco's Modified Eagle's Medium (Invitrogen, UK) for 7 days with or without 10 ng/ml IL-1β or 20 ng/ml TNF-α in a 5% CO_2 atmosphere at 37°C. Culture medium was changed every 2 days, and the ring specimens were used for uniaxial tensile testing in order to evaluate mechanical properties, after which culture media was collected for zymography to evaluate the MMP activity. In this study, rat aortas cultured without cytokines were used as a control.

2.2. *Biomechanical Analysis*

A tensile testing machine (Shimadzu, Japan) was used to measure the mechanical properties of aortas. The ring specimens mounted between two pins and connected to the tensile testing machine via jigs were stretched at a speed of 3 mm/min. The change in the distance between the two pins was measured with a perceptor scope (Hamamatsu, Japan), and the tensile force was measured with a load cell. The force and the distance were simultaneously recorded with a X-Y recorder (Yokogawa, Japan). True stress was calculated assuming specimen incompressibility.

In this study, the mathematical model reported by Raghavan *et al.* [6] was used to characterize the stress-strain curves of the aorta specimens as follows:

$$\varepsilon = \left(K + \frac{A}{B+\sigma} \right) \sigma \tag{1}$$

where, σ is the stress and ε is the strain. K, A, and B are model parameters, which can be obtained by curve fitting using the free software ORIGIN (OriginLab Corporation, USA). From Eq. (1), we obtained the initial slope of curve E_i, which was computed by 1/(K + A/B) and the final slope E_f by 1/K – E_E.

2.3. *Gelatin Zymography*

Gelatinolytic activity was determined by zymography. Culture medium was collected and concentrated by ultrafiltering with the Centricon® centrifugal filter device (Millipore, USA) at 2500 × g for 50 min. Concentrated media was prepared in non-reducing sample buffer and was then subjected to 7.5% SDS-polyacrylamide gel electrophoresis containing 1 mg/ml gelatin (Sigma, USA). After electrophoresis, the gel was washed three times in 2.5% Tween 20 (Pharmacia Biotech, Sweden) at room temperature to remove SDS, and was then incubated for 24 hours at 37°C in buffer containing 50 mM Tris-HCl, pH 7.5, 5 mM $CaCl_2$, 100 mM NaCl, and 1% Tween 20. Gels were then stained in Coomassie blue R-250 (Bio Rad, USA) and destained in a destaining solution (10% acetic acid and 40% methanol). Gelatinase activity was quantified from scanned gel images using ImageJ (NIH, USA), on which the intensity of each band was integrated over length to produce an arbitrary measure of light intensity.

3. Results

Figure 1(a) shows stress-strain curves for rat aortas. The stress-strain curves of aorta rings stimulated by cytokines were shifted to the left when compared to the control. E_i of aorta rings stimulated with cytokines tended to be higher when compared to control and statistical significance was obtained for TNF-α (Fig. 1(b)). Changes in the final slope E_f for the aorta rings stimulated with cytokines were not seen (Fig. 1(c)).

A representative result for gelatin zymography is shown in Fig. 2(a). The activities of the pro-form and activated form of MMP-2 and MMP-9 were detected in this study. Band intensities of activated MMP-2 and MMP-9 were expressed in relative terms against the activities in the controls (Fig. 2(b) and (c)). Activities of both MMP-2 and MMP-9 for ring specimens stimulated by cytokines tended to be higher when compared with controls.

In order to determine the roles of MMP activities on the mechanical properties of the aorta wall, the relationship between activated MMPs and E_i was studied. As a result, a positive correlation between MMP-2, but not activated MMP-9, was observed (data not shown). On the other hand, we did not find a significant correlation between MMP-2 or MMP-9 and the final slope of the curve E_f (data not shown).

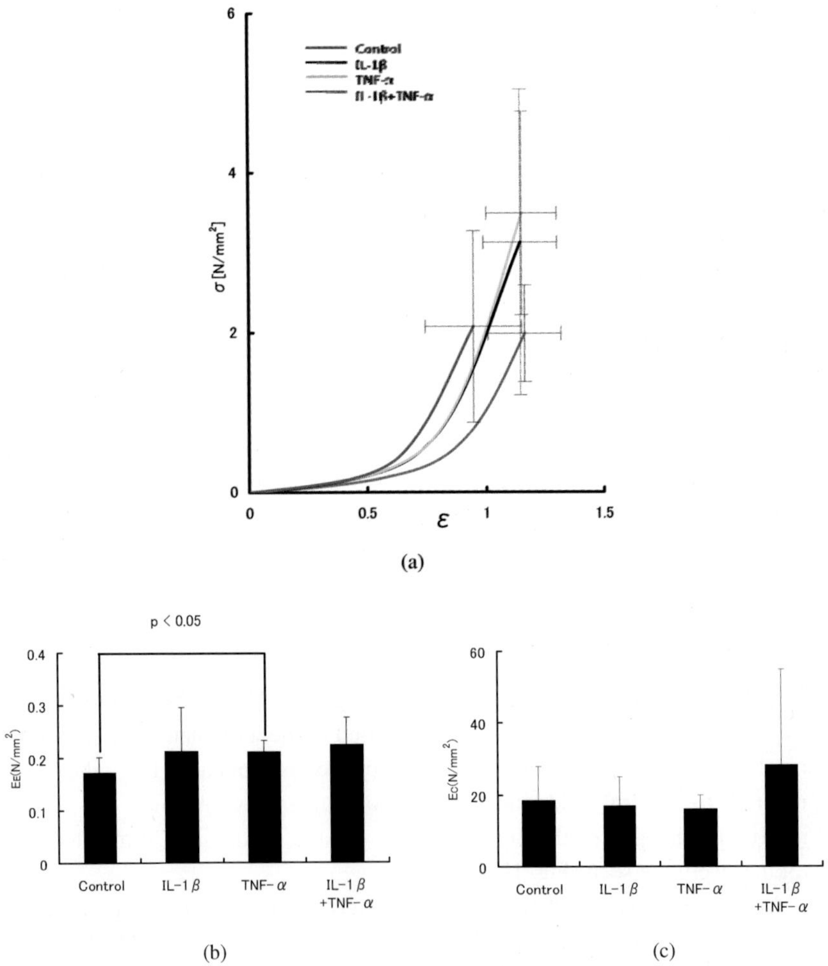

Figure 1. Results of tensile testing of rat aortas incubated without cytokines. (a) Stress-strain curves of rat aortas; (b) initial slope of curve and (c) final slope of curve for rat aortas.

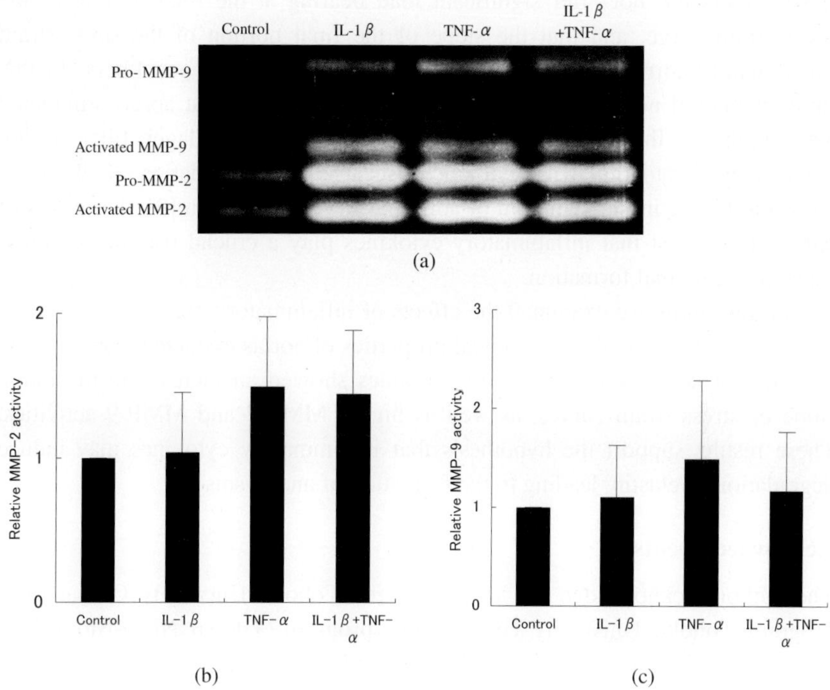

Figure 2. Results of gelatin zymography (a) Typical result of gelatin zymography; (b) Normalized activities of MMP-2; (c) Normalized activities of MMP-9.

4. Discussion

The mechanical properties of arterial walls are determined by the composition of elastin and collagen. In this study, we found that the stress-strain curves for ring specimens stimulated by cytokines were shifted to the left compared when to controls. In addition, the initial slope of curve E_i was higher. It was reported that rat aortas incubated with elastase, which degrades elastin, showed a leftward shift in the stress-strain curve and a higher initial slope when compared to controls, which is similar to what we obtained in this study [7,8]. In addition, we obtained a significant positive correlation between MMP-2 and the initial slope E_E. Our results suggest that after stimulation with cytokines, the mechanical properties of rat aortas were altered by the degradation of elastin.

At the final point of the stress-strain curve, all of the collagen fibers are recruited into load bearing. For this portion, the slope is the total modulus of the elastin and collagen fibers. However, as the elastic modulus of elastin (0.6 MPa) is much smaller than that of collagen fibers (10^3 MPa), we can assume that the elastin fibers did not offer significant load bearing at the final portion of the stress-strain curve, and that the slope of the final portion of the stress-strain curve mainly corresponds with the elastic modulus of collagen fibers. In this study, we could not find any significant changes in E_f for rat aortas stimulated with cytokines, thus suggesting that the degradation of collagen fibers is not induced by stimulation with cytokines. Because the degradation of elastin is known to be the initial symptom of aortic aneurysm formation [9], the results of our study suggest that inflammatory cytokines play a crucial role in the initial stage of aneurismal formation.

In this study, we examined the effects of inflammatory cytokines, including TNF-α and IL-1β on the mechanical properties of aortas extracted from rats. As a result, rat aortas stimulated with cytokines showed an increase in the initial slope of stress-strain curves, as well as higher MMP-2 and MMP-9 activities. These results support the hypothesis that inflammatory cytokines may induce degradation of elastin, leading to the formation of aneurysms.

Acknowledgments

The authors acknowledge the support of the Tohoku University Global COE program Tohoku University Global COE program "Global Nano-Biomedical Engineering Education and Research Network Centre".

References

1. R. W. Thompson and W. C. Parks, *Ann. N. Y. Acad. Sci.* **800**, 157–174 (1996).
2. D. A. Vorp, B. J. Schiro, M. P. Ehrlich, T. S. Juvonen, M. A. Ergin and B. P. Griffith, *Ann. Thorac. Surg.* **75**, 1210–1214 (2003).
3. K. M. Newmann, J. Jean-Claude, H. Li, W. G. Ramey and M. D. Tilson, *Circulation* **90**, 224–227 (1994).
4. P. B. Dobrin and R. Mrkvicka, *Cardiovasc. Surg.* **2**, 484–488 (1994).
5. M. M. Lalu, J. Cena, R. Chowdhury, A. Lam and R. Schulz, *Br. J. Pharmacol.* **149**, 31–42 (2006).
6. M. L. Raghavan, M. W. Webster and D. A. Vorp, *Ann. Biomed. Eng.* **24**, 573–582 (1996).
7. M. O'Connell, H. Kimura, E. Sho, M. Sho, R. L. Dalman and C. A. Taylor, *2003 Summer Bioengineering Conference*, June 25–29, Sonesta Beach Resort in Key Biscayne, FL, 871–872 (2003).
8. A. M. Briones, J. M. Gonzalez, B. Somoza, J. Giraldo, C. J. Daly, E. Vila, M. C. Gonzalez, J. C. McGrath and S. M. Arribas, *J. Physiol.* **552**, 185–195 (2003).
9. R. W. Thompson and W. C. Parks, *Ann. N. Y. Acad. Sci.* **800**, 157–174 (1996).

FORMULATION OF LINEARIZED ERROR DYNAMICS EQUATIONS OF MEASUREMENT-INTEGRATED SIMULATION

KENTARO IMAGAWA *†

Graduate School of Engineering, Tohoku University,
6-6-01 Aramaki-Aoba, Aoba-ku, Sendai, Miyagi 980-8579, Japan

TOSHIYUKI HAYASE ‡

Institute of Fluid Science, Tohoku University,
2-1-1 Katahira, Aoba-ku, Sendai, Miyagi 980-8577, Japan

To obtain accurate and detailed information on blood flow, Measurement-Integrated (MI) simulation, a kind of observer using a computational fluid dynamics (CFD) scheme, has been successfully applied to the blood flow in an aneurismal aorta with ultrasonic measurement and that in a cerebral aneurysm with Magnetic-Resonance (MR) measurement. However, a general theory for the design an MI simulation has not been established.

In this study, as a fundamental consideration to establish a theory for the design of MI simulation, we derive the basic equations of MI simulation and the linearized error dynamics which describes time-development of the difference between the MI simulation and the real flow.

1. Introduction

Detailed and accurate information on the blood flow field is essential for advanced diagnosis of circulatory diseases since hemodynamic stresses are closely related to the development and progress of such diseases. To obtain accurate and detailed information on real flows, Measurement-Integrated (MI) simulation has been proposed [1]. MI simulation has been successfully applied to reproduce the complex blood flow field in an aneurismal aorta by using Doppler velocity and numerical simulation [2,3] and that in a cerebral aneurysm with PC MRI (Phase-Contrast Magnetic Resonance Imaging) measurement [4].

---

* Kentaro Imagawa is a Tohoku University Global COE Research Assistant.

† Work partially supported by grant 2-4570.5 of the Swiss National Science Foundation.

‡ Toshiyuki Hayase is a Tohoku University Global COE Member.

MI simulation is a flow simulation scheme modified by adding a feedback signal proportional to the difference between the real flow and the simulation. The main feature of MI simulation, which distinguishes it from other existing flow observers, is usage of a large-dimensional nonlinear computational fluid dynamics (CFD) scheme as a mathematical model of the physical flow. Figure 1 shows a block diagram of MI simulation in the case of blood flow analysis combined with ultrasonic measurement. However, a general theory of MI simulation has not been established and, therefore, a feedback law is designed by a trial-and-error method based on physical considerations. Establishment of a general theory to design the feedback signal is indispensable for application of MI simulation to various problems.

In this study, as a fundamental consideration to establish a theory applicable to the design of MI simulation, we derive the basic equations of MI simulation and the linearized error dynamics which describes the time-development of the difference between the MI simulation and the real flow.

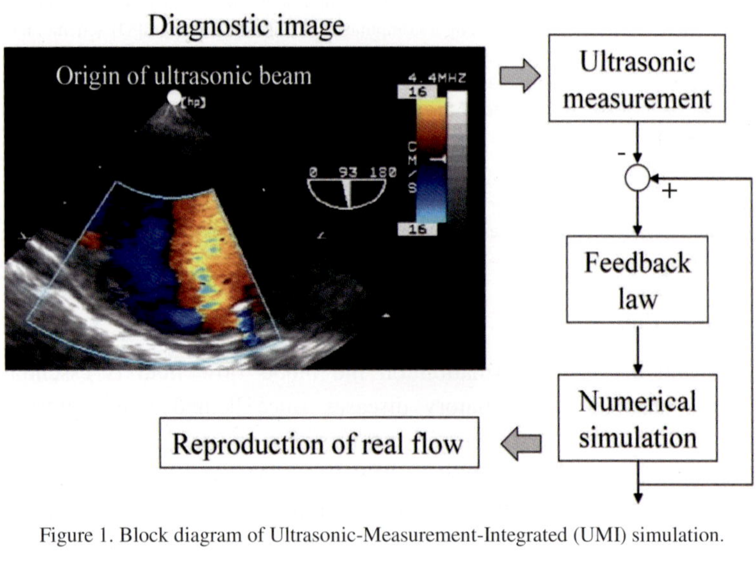

Figure 1. Block diagram of Ultrasonic-Measurement-Integrated (UMI) simulation.

2. Formulation of Linearized Error Dynamics

2.1. *Basic Equation*

The dynamic behavior of an incompressible and viscous flow field is governed by the Navier-Stokes equation and a pressure equation derived from the equation of continuity:

$$\begin{cases} \dfrac{\partial \mathbf{u}}{\partial t} = -(\mathbf{u}\cdot\nabla)\mathbf{u} + \nu\cdot\Delta\mathbf{u} - \nabla p + \mathbf{f} \\ \Delta p = -\mathrm{div}\{(\mathbf{u}\cdot\nabla)\mathbf{u}\} + \nabla\mathbf{f} \end{cases}, \tag{1}$$

with appropriate initial and the boundary conditions, where p is the pressure divided by the density and $\mathbf{f}$ is external force per unit mass, which is used as the feedback signal in MI simulation. In the following, we use a simplified notation for the governing equations:

$$\begin{cases} \dfrac{\partial \mathbf{u}}{\partial t} = g(\mathbf{u}) + h(p) + \mathbf{f} \\ \Delta p = q(\mathbf{u}) + \nabla\mathbf{f} \end{cases}. \tag{2}$$

A discrete model of Eqs. (2) is described as follows:

$$\begin{cases} \dfrac{d\mathbf{u}_N}{dt} = g_N(\mathbf{u}_N) + h_N(\mathbf{p}_N) + \mathbf{f}_N \\ \Delta_N\mathbf{p}_N = q_N(\mathbf{u}_N) + \nabla_N\mathbf{f}_N \end{cases}, \tag{3}$$

where $\mathbf{u}_N$ and $\mathbf{p}_N$ are the $3N$-dimensional vector consisting of velocity components at N grid points and the N-dimensional vector consisting of pressure at N grid points, respectively, and ∇_{N} and Δ_{N} are matrices to express the discrete model of ∇ and Δ, respectively.

On the other hand, the information of the real flow is described by the following equation:

$$\begin{cases} \dfrac{d}{dt}D_N(\mathbf{u}) = D_N(g(\mathbf{u})) + D_N(h(p)) \\ D_N(\Delta p) = D_N(q(\mathbf{u})) \end{cases}, \tag{4}$$

where D_N is the operator to take out the information of the real flow at the computational grid points. This equation is a model of real flow without external forces ($D_N(\mathbf{f})=0$).

In MI simulation, on the other hand, we apply the external force as the feedback signal denoted by a function of real flow and numerical simulation. In this study, we consider the case of feedback signal $\mathbf{f}_N$ denoted by a linear function of the difference of velocity and pressure between real flow and numerical simulation:

$$
\begin{aligned}
\mathbf{f}_N \equiv & -\mathbf{K}_\mathbf{u}\mathbf{C}_\mathbf{u}\left\{\mathbf{u}_N - \left(D_N(\mathbf{u}) + \boldsymbol{\varepsilon}_\mathbf{u}\right)\right\} \\
& -\mathbf{K}_\mathbf{p}\mathbf{C}_\mathbf{p}\left\{\mathbf{p}_N - \left(D_N(\mathbf{p}) + \boldsymbol{\varepsilon}_\mathbf{p}\right)\right\}
\end{aligned}, \qquad (5)
$$

where $\mathbf{K_u}$ and $\mathbf{K_p}$ denote the $3N \times 3N$ and $N \times N$ feedback gain matrices of velocity and pressure, respectively, $\mathbf{C_u}$ and $\mathbf{C_p}$ denote the $3N \times 3N$ and $N \times N$ diagonal matrices to identify the measurable velocity and pressure data, respectively, $\boldsymbol{\varepsilon}_\mathbf{u}$ and $\boldsymbol{\varepsilon}_\mathbf{p}$ denote the $3N$ and N-dimensional vector of velocity and pressure measurement error, respectively. By inserting Eq. (5) into Eq. (3), we obtain the basic equation of MI simulation:

$$
\begin{cases}
\dfrac{d\mathbf{u}_N}{dt} = g_N(\mathbf{u}_N) - \mathbf{K}_\mathbf{u}\mathbf{C}_\mathbf{u}\left\{\mathbf{u}_N - \left(D_N(\mathbf{u}) + \boldsymbol{\varepsilon}_\mathbf{u}\right)\right\} \\
\qquad + h_N(\mathbf{p}_N) - \mathbf{K}_\mathbf{p}\mathbf{C}_\mathbf{p}\left\{\mathbf{p}_N - \left(D_N(\mathbf{p}) + \boldsymbol{\varepsilon}_\mathbf{p}\right)\right\} \\
\Delta_N \mathbf{p}_N = q_N(\mathbf{u}_N) + \nabla_N \mathbf{K}_\mathbf{u}\mathbf{C}_\mathbf{u}\left\{\mathbf{u}_N - \left(D_N(\mathbf{u}) + \boldsymbol{\varepsilon}_\mathbf{u}\right)\right\} \\
\qquad - \nabla_N \mathbf{K}_\mathbf{p}\mathbf{C}_\mathbf{p}\left\{\mathbf{p}_N - \left(D_N(\mathbf{p}) + \boldsymbol{\varepsilon}_\mathbf{p}\right)\right\}
\end{cases}. \qquad (6)
$$

2.2. *Linearized Error Dynamics*

Next, we derive the linearized error dynamics equation. First, we subtract Eq. (4) from Eq. (6). Then by omitting the terms of the second order terms in the Taylor expansion with respect to the difference between the real flow and MI simulation for velocity $\mathbf{e_u}=\mathbf{u}_N - D_N(\mathbf{u})$ and pressure $\mathbf{e_p}=\mathbf{p}_N - D_N(p)$, we obtain the linearized error dynamics equations:

$$
\begin{aligned}
\frac{d\mathbf{e}_\mathbf{u}}{dt} = & \left(\left.\frac{dg_N}{d\mathbf{u}_N}\right|_{\mathbf{u}_N} - \mathbf{K}_\mathbf{u}\mathbf{C}_\mathbf{u}\right)\mathbf{e}_\mathbf{u} \\
& + \left(\left.\frac{dh_N}{d\mathbf{p}_N}\right|_{\mathbf{p}_N} - \mathbf{K}_\mathbf{p}\mathbf{C}_\mathbf{p}\right)\mathbf{e}_\mathbf{p} \\
& \underline{+\left\{g_N\left(D_N(\mathbf{u})\right) - D_N\left(g(\mathbf{u})\right)\right\}} \\
& \underline{+\left\{h_N\left(D_N(p)\right) - D_N\left(h(p)\right)\right\}} \\
& \underline{\underline{+\mathbf{K}_\mathbf{u}\boldsymbol{\varepsilon}_\mathbf{u} + \mathbf{K}_\mathbf{p}\boldsymbol{\varepsilon}_\mathbf{p}}}
\end{aligned}, \qquad (7)
$$

and

$$\mathbf{e_p} = \left(\Delta_N + \nabla_N \mathbf{K_p C_p}\right)^{-1} \left(\left.\frac{dq_N}{d\mathbf{u}_N}\right|_{\mathbf{u}_N} - \rho \nabla_N \mathbf{K_u C} \right) \mathbf{e_u}$$

$$\underline{+\left(\Delta_N + \nabla_N \mathbf{K_p C_p}\right)^{-1} \left\{ q_N\left(D_N(\mathbf{u})\right) - \Delta_N D_N(p) \right\}}$$

$$\underline{\underline{+\left(\Delta_N + \nabla_N \mathbf{K_p C_p}\right)^{-1} \nabla_N \left(\mathbf{K_u} \boldsymbol{\varepsilon}_\mathbf{u} + \mathbf{K_p} \boldsymbol{\varepsilon}_\mathbf{p}\right)}}, \quad (8)$$

where the underlined terms are caused by model error and the double-underlined ones are caused by measurement error.

3. Conclusion

In this study, as a fundamental consideration to establish a theory to design MI simulation, we derived basic equations for MI simulation and the linearized error dynamics which describes the time-development of the difference between the MI simulation and the real flow.

In future work, we intend to derive the formulation of eigenvalue analysis of the linearized error dynamics formulated in this report. The eigenvalue analysis will be applied to MI simulation of blood flows.

Acknowledgments

The authors acknowledge the support from the Tohoku University Global COE Program "Global Nano-Biomedical Engineering Education and Research Network Centre".

References

1. T. Hayase and S. Hayashi, State Estimator of Flow as an Integrated Computational Method with the Feedback of Online Experimental Measurement. *J. Fluids Eng.* **119**, 814–822 (1997).
2. K. Funamoto, T. Hayase, Y. Saijo and T. Yambe, Numerical Experiment for Ultrasonic-Measurement-Integrated Simulation of Three-Dimensional Unsteady Blood Flow. *Ann Biomed Eng* **36**, 1383–1397 (2008).
3. K. Funamoto, T. Hayase, A. Shirai, Y. Saijo and T. Yambe, Fundamental study of Ultrasonic-Measurement-Integrated Simulation of Real Blood Flow in the Aorta. *Ann Biomed Eng* **33**, 415–428 (2005).

4. K. Funamoto, Y. Suzuki, T. Hayase, T. Kosugi and H. Isoda, Numerical Experiment of MR-Measurement-Integrated Simulation of Steady Blood Flow in a Cerebral Aneurysm. *Proceedings of the ASME 2008 Summer Bioengineering Conference*, Marriott Resort, Marco Island, Florida, USA, (2008).

EVALUATION OF THE MECHANICAL PROPERTIES OF BLOOD VESSEL BIOMODELING WITH POLY (VINYL ALCOHOL) HYDROGEL

HIROYUKI KOSUKEGAWA [1)*], KEISUKE MAMADA [2)], KANJU KUROKI [3)], LEI LIU [1)†], KOSUKE INOUE [3)], TOSHIYUKI HAYASE [3)‡], MAKOTO OHTA [3)]

1) Graduate School of Engineering, Tohoku University, 6-6, Aramaki Aoba, Aoba-ku, Sendai, Miyagi, 980-8579, Japan

2) Graduate School of Medical Engineering, Tohoku University, 2-1, Seiryo-cho, Aoba-ku, Sendai, Miyagi, 980-8575, Japan

3) Institute of Fluid Science, 2-1-1, Katahira, Aoba-ku, Sendai, Miyagi, 980-8577, Japan

In vitro blood vessel biomodeing with realistic mechanical properties and geometrical structures is helpful for training in surgical procedures, especial those used in endovascular treatment. Poly (vinyl alcohol) hydrogel (PVA-H), which is made of poly (vinyl alcohol) (PVA) and water, may be useful as a material for blood vessel biomodeling due to its low surface friction resistance and good transparency. In order to simulate the mechanical properties of PVA-H were carried out with a dynamic mechanical analyzer, and the storage modulus (G') and loss modulus (G'') of PVA-H were obtained. PVA-H was prepared by the low-temperature crystallization method. They were made of PVA-H with various concentrations (C) and degrees of polymerization (DP). The G' and G'' of PVA-H increase, as the C or DP of PVA increase. These results indicate that it is possible to obtain PVA-H with desirable dynamic viscoelasticity. The dynamic viscoelasiticity of PVA-H obtained was similar to that of real artery measured in previous report. In conclusion, PVA-H is suggested to be useful as a material of blood vessel biomodeling.

1. Introduction

Cardiovascular diseases, such as ischemic heart disease and aneurysms, are serious diseases with a mortality rate as high as that of caner in many countries. Recently, endovascular intervention, a therapeutic approach for treatment of such diseases with various interventional devices such as guidewires or stents, is increasingly employed. Since it can reduce the invasiveness for patients, it can be contributed to improvement of the quality of life of patients. However, it

---

[*] Hiroyuki Kosukegawa was a Tohoku University Global COE Research Assistant (2008).
[†] Lei Liu is a Tohoku University Global COE Research Assistant.
[‡] Toshiyuki Hayase is a Tohoku University Global COE Member.

requires an advanced surgical skill and experience, and then training for the treatment is necessary.

In vitro blood vessel biomodeling with realistic mechanical properties and geometrical structures may be useful for the training of the intervention and the investigation of therapy. Blood vessel biomodeling also contributes for the reduction of animal experiments for the development of devices used in the intervention such as coil or stent. Ohta *et al.* developed a blood vessel biomodeling made of poly (vinyl alcohol) hydrogel (PVA-H) with low surface friction and good transparency [1]. They also performed the measurement of compressive moduli with various concentration of poly (vinyl alcohol) (PVA). However, other possibilities such as degree of polymerization have not been reported.

Since a living blood vessel is subjected to periodical change of stress by heart beat or respiratory movement, and the blood vessel is a viscoelastic material composed of collagen, smooth muscle, elastin, and so on [2], material for biomodeling may be required of having proper dynamic viscoelastic properties. Then a biomodeling with realistic mechanical properties has possibilities of simulating the motion of blood vessel. In this study, the measurements of dynamic viscoelasticities of PVA-H and the evaluation were carried out with various PVA-H.

2. Experimental

2.1. *Preparation of PVA-H*

PVA (JAPAN VAM & POVAL CO., LTD.) was dissolved in a mixed solvent of dimethyl sulfoxide (DMSO) (Toray Fine Chemicals Co., Ltd.) and distilled water (80/20, w/w) [3]. The PVA powder in the mixture solution was stirred for 2 hours at 100°C until dissolution, and then the solution was cast in a stainless steel mold with dimensions of 8 × 8 × 1 mm. The mold was maintained at -30°C for 24 hours to promote gelation of the PVA solution. After gelation, the resulting rectangular PVA gel was immersed in ethanol at room temperature for 24 hours and then immersed in distilled water for 4 days.

2.2. *PVA-H Samples*

PVA-H with various concentrations (C = 5, 10, 15, 18 wt%) of PVA solution were obtained by using PVA with degree of polymerization (DP) = 1700. In addition, PVA-H with various DP (DP = 1000, 1700, 3300) were obtained, and concentration of these samples is 1700.

2.3. *Dynamic Mechanical Analysis*

Dynamic mechanical analysis on PVA-H was carried out using a dynamic mechanical spectrometer DMS6100 (SII NanoTechnology Inc.). The dynamic viscoelasticities are separated into two kinds of moduli. The storage modulus, G', which is indicative of elasticity of material, and the loss modulus, G'', which is indicative of viscosity of material. These moduli are obtained by the following. First, the stress (S) and the strain (γ) generated on samples were measured by loading of them with small-amplitude sinusoidal oscillation, and then the complex modulus (G*) is given by Eq. (1).

$$G^* = S / \gamma \tag{1}$$

Secondly, G' and G'' are given by Eq. (2) and Eq. (3) by using G* and the phase angle (δ) between the stress and the strain.

$$G' = |G^*| \cos\delta \tag{2}$$

$$G' = |G^*| \sin\delta \tag{3}$$

After cooled to 0°C with liquid nitrogen, the PVA-H samples were measured at sinusoidal oscillatory frequency of 1 Hz as heated to 75°C at a rate of 2°C/min to evaluate the temperature influence.

3. Results

The G' and G'' of PVA-H with various C at 23°C are shown in Fig. 1 [4]. Both G' and G'' increased as the C of PVA increased. The G' and G'' of PVA-H with various DP of PVA at 23°C are shown in Fig. 2. Both G' and G'' increased as the DP of PVA increased. Dynamic viscoelasticity of PVA-H may also depend on the C and the DP of PVA. As described by Hyon [5], the density of the microcrystal formed by hydrogen bonds increases with increasing C or DP of PVA. Therefore, these dependences may be due to the increase of either C or DP. On the other hand, the G'' appeared to be slightly smaller than 1/10 of G' of all C values or of all DP values.

Figure 3 shows a comparison of the dynamic viscoelasticity at 23°C of PVA-H with that of dog artery [6]. According to this result, a broad range of dynamic viscoelasticity of PVA-H is possible by changing the PVA parameters (C and DP). Overall, the obtained PVA-H exhibited a similar ratio of G''/G' to the blood vessel. Part of the dynamic viscoelasticity of PVA-H obtained was similar to that of a dog artery.

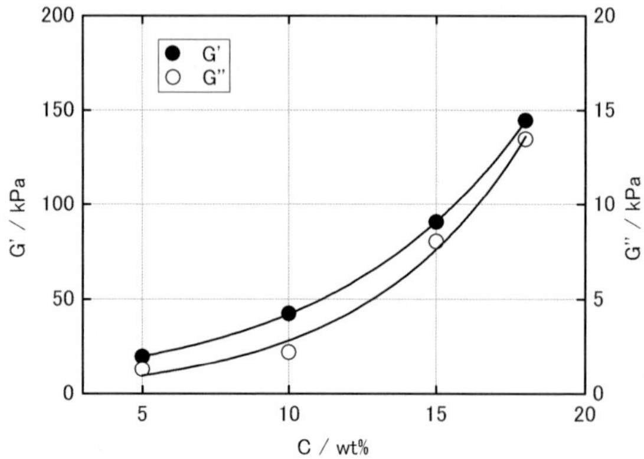

Figure 1. Dynamic viscoelasticity of PVA-H with various weight concentration of PVA

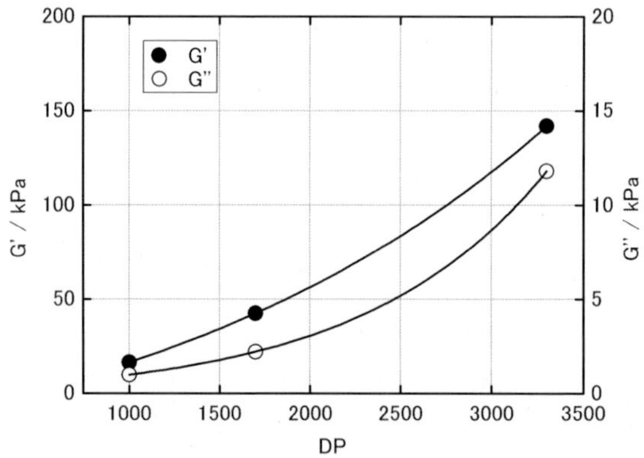

Figure 2. Dynamic viscoelasticity of PVA-H with various degree of polymerization of PVA

4. Discussions

This study examined the relationship between the dynamic viscoelasticity of PVA-H and the factors of PVA (C, DP), and the similarity between the dynamic viscoelasticity of PVA-H and that of a dog artery. Dynamic viscoelasticity of PVA-H can be flexibly changed by changing the C of PVA (Fig. 1). This result may be similar to the finding by Takigawa *et al.* [7] that the initial Young's modulus of PVA gel measured by stress-strain measurement depends on the concentration of PVA. This result indicates that the concentration of PVA may influence the elastic modulus of PVA-H in the case of large deformation as well as in small deformation. In addition, considering previously reported findings [8], PVA-H may exhibit dynamic viscoelasticity as expected from the scaling concept, which describes the elastic modulus of gel in a good solvent [9].

The dynamic viscoelasticity of PVA-H can be also changed by changing the DP of PVA. Cha *et al.* [10] reported a similar result for PVA-H, although the dynamic modulus was much higher than that obtained in the present study. The PVA-H prepared by Cha *et al.* included lower water content (20 wt%) than the PVA-H obtained in this study (80-90 wt%). These results indicate that the dynamic viscoelasticity of PVA-H can be changed by changing the DP of PVA without changing the C or the density of PVA-H.

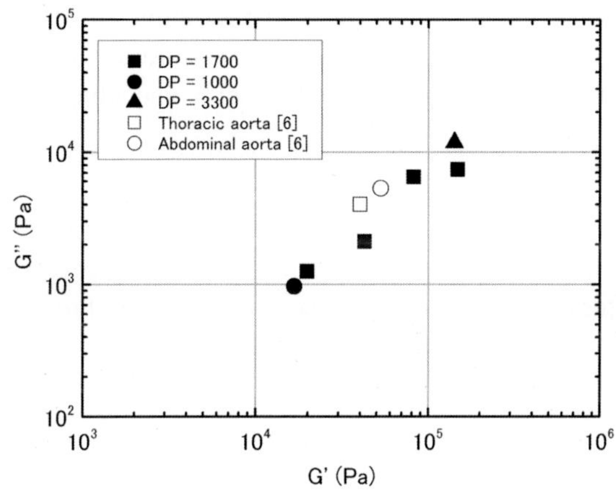

Figure 3. Comparison of the dynamic viscoelasticity of PVA-H with that of a dog artery

The results shown in Fig. 1 and Fig. 2 indicate that it is possible to flexibly control the dynamic viscoelasticity of PVA-H by adjusting the C or DP of PVA. Therefore, PVA-H with desirable dynamic viscoelasticity can be obtained by controlling the factors of PVA.

Figure 3 indicates that PVA-H may show dynamic viscoelasticity similar to that of a dog artery. Moreover, since the dynamic viscoelasticity of a human blood vessel is similar to that of a dog [11], PVA-H can represent the dynamic viscoelasticity of a human blood vessel. Of course, it is necessary to consider that each measurement was carried out under different conditions. In order to represent the dynamic viscoelasticity of a blood vessel more precisely, it would be necessary to equalize the experimental conditions: shape of test piece, inner pressure, and so on. In addition, an *in vivo* blood vessel is always subjected to longitudinal stretch equal to about 1.2 relative to the zero-stress state, and a blood vessel subjected to longitudinal stretch shows a higher elastic modulus than one without stretch [2]. According to the results shown in the present study, the mechanical properties of a blood vessel can possibly be simulated by PVA-H. However, in order to obtain a blood vessel biomodeling with realistic mechanical properties, it would be necessary to represent these properties with tubular-shaped PVA-H under the same conditions as those of an *in vivo* blood vessel.

4. Conclusion

PVA-H exhibits similar dynamic viscoelasticity to that of artery. By adjusting the concentration or the degree of polymerization of PVA, the dynamic viscoelasticity of PVA-H can be controlled. These results indicate that PVA-H has potential for use as material for blood vessel biomodeling.

Acknowledgments

This work is supported by Tohoku University Grobal COE Program "Global Nano-Biomedical Engineering Education and Research Network Centre" and by Japan Society for the Promotion of Science "Core to Core Project 20001".

References

1. M. Ohta, A. Handa, H. Iwata, D. A. Rüfenacht and S. Tsutsumi, Poly-vinyl alcohol hydrogel vascular models for *in vitro* aneurysm simulations: The key to low friction surfaces. *Technology and Health Care* **12**, 225–233 (2004).
2. Y. C. Fung, *Biomechanics 2nd edition*, Springer-Verlag, New-York, 321–391 (1997).
3. S. H. Hyon, W. I. Cha and Y. Ikada, Preparation of transparent poly (vinyl alcohol) hydrogel. *Polymer Bulletin* **22**, 119–122 (1989).
4. H. Kosukegawa, K. Mamada, K. Kuroki, L. Lei, K. Inoue, T. Hayase and M. Ohta, Measurements of Dynamic Viscoelasticity of Poly (vinyl alcohol) Hydrogel for the Development of Blood Vessel Biomodeling. *J Fluid Science and Technology* **3**, 533–543 (2008).
5. S. H. Hyon, PVA hydrogel with intermolecular force. *Kobunshi Kako* **39**, 44–50 (1990).

6. D. H. Bergel, The dynamic elastic properties of the arterial wall, *J Physiol.* **156**, 458–469 (1961).
7. T. Takigawa, H. Kashihara, K. Urayama and T. Masuda, Structures and mechanical properties of poly (vinyl alcohol) gels swollen by various solvents. *Polymer* **33**, 2334–2339 (1992).
8. T. Takigawa, H. Kashihara, K. Urayama and T. Masuda, Critical Behavior of Modulus of Poly (vinyl alcohol) Gels near the Gelation Point. *J The Physical Society of Japan* **59**, 2598–2599 (1990).
9. P. G. de Gennes, *Scalling Concept in Polymer Physics*, Cornell University Press, Ithaca and London, 128–164 (1979).
10. W. I. Cha, S. H. Hyon, M. Oka and Y. Ikada, Mechanical and Wear Properties of Poly (vinyl alcohol) Hydrogels. *Macromolecule Symposium* **109**, 115–126 (1996).
11. B. M. Learoyd and M. G. Taylor, Alterations with Age in the Viscoelastic Properties of Human Arterial Walls. *Circulation Research* **18**, 278–292 (1966).

NUMERICAL EXPERIMENT FOR ULTRASONIC-MEASUREMENT-INTEGRATED SIMULATION OF DEVELOPED LAMINAR PIPE FLOW USING AXISYMMETRIC MODEL

LEI LIU *

Graduate School of Engineering, Tohoku University, Sendai 980-8579, Japan

KENICHI FUNAMOTO and TOSHIYUKI HAYASE †

Institute of Fluid Science, Tohoku University, Sendai 980-8577, Japan

In the present study, the authors performed a fundamental numerical experiment in which UMI simulation was applied to a developed laminar pipe flow using an axisymmetric model in order to understand the effect of the feedback law on the accuracy of UMI simulation systematically. The effect of two types of ultrasonic probes, the linear scanning type and the sector scanning type, and the effect of irradiating angles of the ultrasonic beam in the linear probe were investigated. It was confirmed that the result of UMI simulation asymptotically approached the standard solution of developed laminar flow downstream of the feedback domain in all cases using the linear probe and the sector probe with axisymmetric feedback. Under the present conditions, a linear probe with a radiation angle of 70° was most effective, whereas there was not so much improvement in the accuracy in the case using the sector probe.

1. Introduction

Development and progression of arteriosclerosis and aneurysms are considered to be closely related to hemodynamics, and many experimental studies and numerical simulations have been performed.

Although there has been substantial improvement in medical measurement technology in recent years, it is still difficult to obtain accurate and detailed information on blood flow such as wall shear stress or pressure distribution in the living body with existing measurement techniques. On the other hand, computer simulation of blood flow has also been actively studied using realistic vessel geometry obtained with MRI or CT. Numerical simulation provides

* Lei Liu is a Tohoku University Global COE Research Assistant.
† Toshiyuki Hayase is a Tohoku University Global COE Member.

detailed blood flow information which is difficult to obtain in measurement. However, because it is generally difficult to set the initial and boundary conditions of blood flow as well as to accurately determine the mechanical properties and the geometry of blood vessels, computational results are sometimes insufficient for the exact reproduction of real blood flow.

With these backgrounds, the authors have proposed a new method called Ultrasonic-Measurement-Integrated (UMI) simulation, in which ultrasonic measurement and CFD are integrated to reproduce blood flow [1]. In previous studies, Funamoto *et al.* performed two-dimensional and three-dimensional numerical experiments of UMI simulation with realistic geometry of a thoracic aneurysm and confirmed the usability of UMI simulation [2-4]. However, the effect of the feedback on the UMI simulation is not understood in depth, although such understanding is essential if UMI simulation is to be widely applied other parts of the circulatory system.

The goal of the present study was to systematically advance our understanding of the effect of the feedback law on the accuracy of UMI simulation. We performed a numerical experiment of UMI simulation for a steady developed laminar pipe flow using an axisymmetric cylindrical model to investigate the effect of ultrasonic probes (linear and sector) and the irradiating angles ($\varphi = 70°, 110°$) of the linear probe.

2. Models and Methods

In the present study, we dealt with a steady developed laminar flow with Re = 1100 in a straight pipe (D = 10 mm, L = 80 mm). We defined the r axis as the radial direction and the z axis as the axial direction. Figure 1(a) shows the geometry of the calculation system. Feedback domain ($0 \leq r^* \leq 0.5$, $3 \leq z^* \leq 5$, "*" indicates a dimensionless value) is defined in the whole flow domain. We assumed the use of both linear and sector probes with the color Doppler imaging measurement mode. In the numerical experiment with the linear probe, the probe was placed parallel to the straight pipe and the irradiating angles of the ultrasonic beam φ were set at 70° or 110°. The sector probe was also placed parallel to the pipe and the position of ultrasonic origin was located at (r^*, z^*) = (3.25, 4). As shown in Fig. 1(b), a sector probe transmits the ultrasonic beam toward the pipe with the expanse angle of the ultrasonic beam ψ set at 40°, which corresponds to the setting of the linear probe (the angles between the z axis and the two ends of the ultrasonic beam are 70° and 110°, respectively). The inflow boundary condition was uniform flow, and the outflow was the free stream condition. We used the analytic solution for the steady developed laminar flow in a straight pipe (Poiseuille flow) as the standard solution. In UMI simulation, the feedback force is applied based on the difference between the standard solution and the numerical calculation. In the following, we term the simulation without feedback as "ordinary simulation" to distinguish it from "UMI simulation."

In the present study, we used all the grid points (820 points) in the feedback domain shown in Fig. 1(a) and (b) as the feedback points. The velocity vector obtained by UMI simulation is defined as $\vec{u}_c$ and that of the standard solution is $\vec{u}_s$. The purpose of the feedback is to force the velocity $\vec{u}_c$ to converge to $\vec{u}_s$.

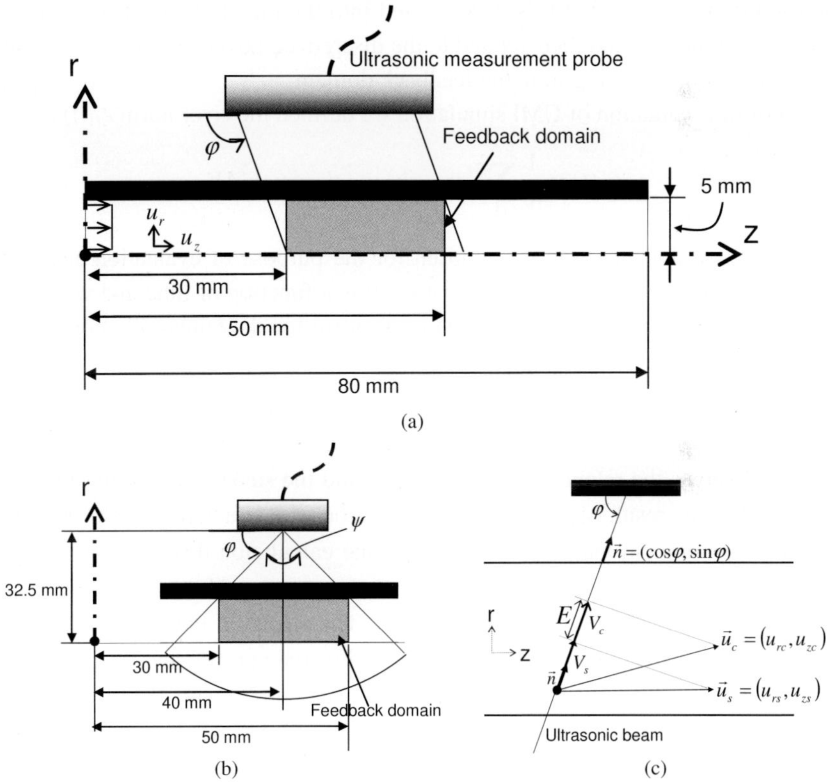

Figure 1. Overview of calculation system: (a) shows geometry and coordinate system for the case using the linear probe, (b) shows the sector probe and (c) is an explanation of the feedback law.

The Doppler velocities of the UMI simulation and the standard solution, V_c and V_s, are the projection of $\vec{u}_c$ and $\vec{u}_s$ in the ultrasonic beam direction, respectively. We defined the error E between V_s and V_c as

$$E = V_c - V_s, \tag{1}$$

The feedback force in the ultrasonic beam direction f_v applied to the control volume in the feedback domain to compensate the error E is

$$f_v = -K_v^* \cdot \frac{E}{U_{mean}} \cdot \frac{\Delta V}{D^3} \frac{\rho U_{mean}^2}{D}, \tag{2}$$

where K_v^* is the feedback gain (dimensionless value), which is an important factor concerning the convergence of UMI simulation. UMI simulation with K_v^* = 0 corresponds to an ordinary simulation. The feedback force in the ultrasonic beam direction f_v is decomposed into the radial component f_r and the axial component f_z, and then applied to the discretized governing equations at the control volume of each grid in the feedback domain.

For the evaluation of UMI simulation we defined the error norm $e_\Omega(t)$ as

$$e_\Omega(t) = \frac{1}{W} \sum_\Omega \left[\left| u_{zc} - u_{zs} \right| + \left| u_{rc} - u_{rs} \right| \right] \Delta V, \tag{3}$$

where Ω is a domain arbitrarily chosen for the purpose of evaluation and W is the volume of the domain Ω. Note that $e_\Omega(t)$ is a function of time and we define the steady-state error norm $e_{\Omega s}$ as the error norm for the temporally convergent solution.

3. Result

Figure 2 shows the relationship between K_v^* and the steady state error norm $e_{\Omega s}$ at t^*= 300. The result of $e_{\Omega s}$ using a linear probe with irradiating angle $\varphi = 70^\circ$ reached a minimum value at K_v^* = 44 and the calculation diverged for a larger gain. The linear probe with $\varphi = 110^\circ$ and the sector probe had the minimum value at K_v^* = 32 and K_v^* = 36, respectively. The results in Fig. 2 reveal that the accuracy of UMI simulation depends on the ultrasonic probes and the irradiating angles employed. Among these results, the linear probe with $\varphi = 70^\circ$ showed the greatest improvement in accuracy. The result of UMI simulation in each case is discussed in detail in the following section. In view of the stability of the calculation, we did not use the results with K_v^* corresponding to the minimum error norm but rather the results with slightly smaller gains: K_v^* = 40, 28, 32 for the linear probe with $\varphi = 70^\circ$, that with $\varphi = 110^\circ$ and the sector probe, respectively.

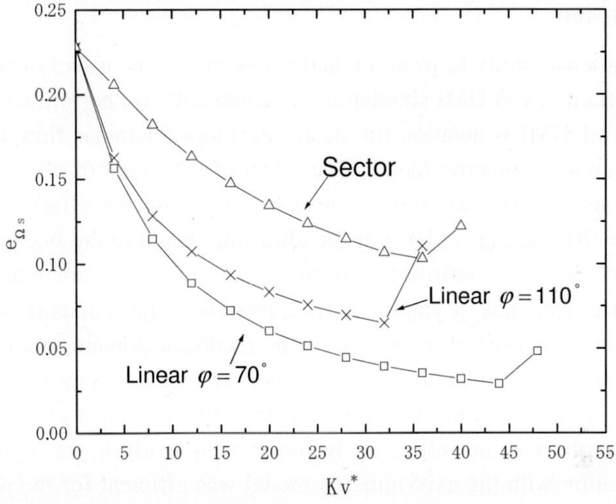

Figure 2. Change of steady error norm $e_{\Omega s}$ ($t^* = 300$) with feedback gain K_v^*.

Figure 3 shows the spatial change of the steady state error norm e_{zs} in each cross section in the axial direction. In the figure, the error norm of the ordinary simulation decreased gradually in the axial direction due to the development of the laminar boundary layer. On the other hand, the error decreased drastically in the feedback domain in the results of the UMI simulation. Especially in the case using linear probe with $\varphi = 70^\circ$, the error norm decreased to almost zero in the feedback domain.

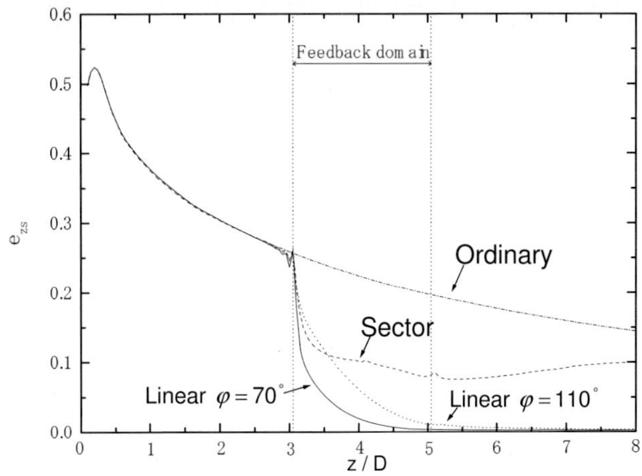

Figure 3. Change of error norm in each cross section e_{zs} in each case.

4. Conclusion

As a fundamental study to promote understanding of the effect of the feedback law on the accuracy of UMI simulation systematically, we performed a numerical experiment of UMI simulation for steady developed laminar flow in a straight pipe using an axisymmetric model. The effect of two types of ultrasonic probes, linear and sector (with expanse angle $\psi = 40^\circ$), and the effect of irradiating angles ($\varphi = 70^\circ$ and $\varphi = 110^\circ$) of the ultrasonic beam in the linear probe were investigated. It was confirmed that the result of UMI simulation with the axisymmetric feedback asymptotically approached the standard solution, *i.e.* developed laminar flow, downstream of the feedback domain in all cases using both the linear probe and the sector probe. Among conditions of the present numerical experiment, a linear probe with $\varphi = 70^\circ$ was most effective, whereas the sector probe was not efficient. In the present work, it was confirmed that UMI simulation with the axisymmetric model was efficient for the reproduction of axisymmetric pipe flow. The response of real three-dimensional flow to the feedback signal derived from one direction may be different from that in this study. In a future study, we intend to perform a numerical experiment of UMI simulation using a three-dimensional model to determine the effect of the feedback on the flow field.

Acknowledgments

The authors acknowledge the support received from the Global COE Program "2007 Tohoku University Global COE Program Global Nano-Biomedical Engineering Education and Research Network Centre."

References

1. K. Funamoto, T. Hayase, A. Shirai, Y. Saijo and T. Yambe, *Annals of Biomedical Engineering.* **33**, 415 (2005).
2. K. Funamoto, T. Hayase, Y. Saijo and T. Yambe, *Technology and Health Care.* **13**, 331 (2005).
3. K. Funamoto, T. Hayase, Y. Saijo and T. Yambe, *Proceedings of the ICBME.* CD-ROM (2005).
4. K. Funamoto, T. Hayase, Y. Saijo and T. Yambe, *JSME International Journal Series C.* **49**, 144 (2006).

THE EFFECT OF ATP ON FORCE RELAXATION OF ISOLATED ACTIN STRESS FIBERS EXPOSED TO STEP STRAIN

TSUBASA MATSUI [1)*], SHINJI DEGUCHI [2)], NAOYA SAKAMOTO [1)], TOSHIRO OHASHI [1)], MASAAKI SATO [1,2)†]

1) Department of Bioengineering and Robotics, Graduate School of Engineering, Tohoku University, 6-6-01 Aramaki-aza Aoba, Sendai, Miyagi 980-8579, Japan

2) Department of Biomedical Engineering, Graduate School of Biomedical Engineering, Tohoku University, 6-6-01 Aramaki-aza Aoba, Sendai, Miyagi 980-8579, Japan

Although cytoskeletons are thought to play important roles in viscoelastic properties of cells, their mechanical properties remain unclear. Recently, there has been increasing evidence that actin stress fibers (SFs) have pivotal roles in the regulation of mechanical properties of cells through regulation of ATP-dependent acto-myosin interaction. In this study, as a measure of such properties, the effect of ATP on force relaxation behavior of isolated SFs, was investigated. The results showed that the force relaxation behavior and relaxation time constants of SFs, which ranged widely from 4 sec to over 60 sec, were independent of ATP. The initial stiffness of SFs, defined as the force just after a step stretch divided by their elongation, in the absence of ATP was significantly greater than that in the presence of ATP, showing ATP-sensitive mechanical properties. These results suggest that the number of myosins that attach to actin may contribute to mechanical resistance to stretching.

Keywords: Actin stress fiber; Viscoelastic properties; Acto-myosin interaction; Force relaxation test.

1. Introduction

It is well known that cells constitute a dynamic system that sensitively reacts to the surrounding mechanical environment by changing their mechanical properties. For example, after exposure to shear stress, endothelial cells elongate in shape along with increases in elastic and viscous moduli [1]. These remodeling phenomena, in which changes in both distributions and mechanical properties of individual intracellular components occur, are thought to contribute to the maintenance of cell morphology and viscoelastic mechanical properties.

---

* Tsubasa Matsui is a Tohoku University Global COE Research Assistant.

† Masaaki Sato is a Tohoku University Global COE Member.

SFs, consisting of actin and myosin filaments, which are responsible for force generation within the cells, play important roles in celluar remodeling phenomena as mentioned above. It has been shown that local mechanical properties of the cells are affected by the distribution of SFs in the cells and that inhibition of acto-myosin interaction in SF induces reduction in the local stiffness of cells [2]. Therefore, it is believed that SFs play pivotal roles in the adjustment of mechanical properties of cells through regulation of the acto-myosin interaction.

Our group has previously reported the tensile properties of isolated SF [3]. However, the effect of acto-myosin interaction on mechanical properties of isolated SF still remains poorly understood. In the present study, we performed force relaxation tests as a measure of mechanical properties on isolated SFs with or without ATP, which is an essential factor for the activity of acto-myosin interaction.

2. Methods

2.1. *Cell Culture*

Bovine aortic smooth muscle cells (passage 5-9) (Cell Applications, USA) were maintained at 37°C in 5% CO_2/95% air on tissue culture dishes in a standard culture medium composed of Dulbecco's modified Eagle's medium (DMEM, Invitrogen, Netherlands) supplemented with 10% fetal bovine serum (JRH, USA), 3.7 g/l sodium hydrogen carbonate (Wako, Japan) and penicillin (100 U/ml)/streptomycin (100 μg/ml) (Invitrogen). For experiments, cells were trypsinized (0.05% Trypsin-EDTA, Invitrogen) and seeded onto glass bottom culture dishes, 60 mm in diameter (Sumilon, Japan).

2.2. *Isolation of Stress Fibers*

SFs were isolated from cells according to a previously reported method with modifications [4,5]. Briefly, cells were preincubated in potassium-HEPES buffer (125 mM potassium acetate, 12 mM glucose, 25 mM HEPES) supplemented with 1 mM $MnCl_2$ and 2.5 mM $MgCl_2$ for 3 min at room temprature. Cells were further incubated in a low ionic strength solution consisting of 2.5 mM triethanolamine (TEA) (Wako), 1 μg/ml leupeptin (Peptide Institute, Japan), and 1 μg/ml pepstatin (Peptide Institute) for 2 min at 4°C and were then exposed to a flow of low ionic strength solution using a Pasteur pipette. With this procedure, the apical plasma membrane and nucleus were removed. The extracted cells were treated for 5 min with 0.05 % Triton X-100 in PBS at 4°C. Extracted SFs were then washed with CSK stabilization buffer (50 mM KCl, 10 mM imidazole,

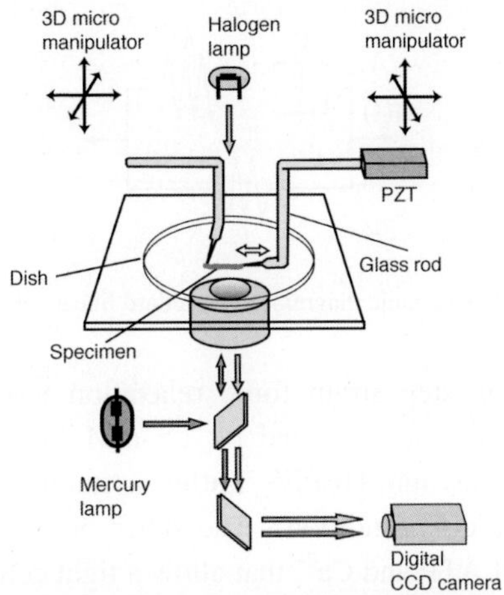

Figure 1. Schematic diagram of the micro-tensile tester (modified from [3]) [7].

1 mM EGTA, 1 mM $MgCl_2$, 10 mM PIPES, 10 mM dithiothreitol, 1 μg/ml leupeptin, and 1 μg/ml leupeptin pH 6.5) at 4°C. Finally, SFs were labeled using 1 nM Alexa 546-phalloidin (Molecular Probes, USA).

2.3. *Force Relaxation Test*

Force relaxation tests were performed using a micro-tensile tester (Fig. 1). Each end of a SF was attached to two glass microneedles which were pretreated with 3-aminopropyltriethoxysilane, dried, coated with 20% glutaraldehyde for 5 min and washed. One of the microneedles was driven using a piezoelectric ceramic actuator (NEC TOKIN, Japan) to stretch the specimen. Step input signals to the actuator were used to perform rapid stretching of the specimen, and the force applied to the specimen was determined by the deflection of the other microneedle. Images of the specimen were taken every 0.5 sec by a digital CCD camera (Hamamatsu Photonics, Japan) during the test. From sequential images, the deflection of the microneedle and the displacement of the specimen were analyzed using ImageJ software (NIH). The force applied to the specimen was then determined by the deflection of the microneedle. After each experiment, the spring constant of the microneedle was determined by a cross-calibration method using a standard cantilever with a known spring constant.

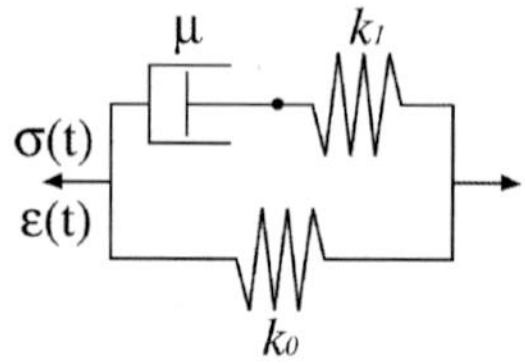

Figure 2. Schematic diagram of a standard linear solid model.

We performed the step strain force relaxation tests using two different chemical solutions. The one consisted of 2.5 mM $MgCl_2$, 0.5 mM ATP and 0.5 mM $CaCl_2$ in potassium-HEPES buffer assuming that the acto-myosin interaction is active (ATP condition). The other one consisted of potassium-HEPES buffer without ATP and Ca^{2+} that allow a tight connection between acto-myosin molecules (rigor condition).

We used a standard linear model to analyze the viscoelasticity analysis (Fig. 2). In the model, σ represents stress, ε is strain, k_0 and k_1 are the elastic components of the springs, and μ is the viscosity of the dashpot. Since force relaxation was realized at a constant strain, the following equation was used (1):

$$\sigma(t) = \varepsilon_0 (k_0 + k_1 e^{-\frac{t}{\tau}}), \tag{1}$$

where the time constant τ equals μ/k_1, and ε_0 is the strain right after rapid stretching (t = 0). We determined these model parameters by minimizing errors between the mathematical curves and data points for each specimen.

3. Results

Force relaxation of SF was observed after exposure to a single-step stretch with a displacement equivalent to 60% strain of the initial length. Figure 3 shows the force relaxation over time for seven specimens in the rigor condition (Fig. 3(a)) and five specimens in the ATP condition (Fig. 3(b)). During the first several tens of seconds, the force increased rapidly after initial stretching, followed by a steep decrease. From the estimation of the model parameters, time constants τ was 31.8 ± 29.7 sec (n = 7, mean ± SD) in the rigor condition and 28.5 ± 17.4 sec (n = 5) in the ATP condition. The initial stiffness was calculated based on the force acting on the specimen at t = 0 divided by the stretching displacement. The initial stiffness in the rigor and ATP conditions was 4.5 ± 2.0 nN/μm and 1.8 ± 1.2 nN/μm, respectively. Figure 4 shows a time-varying force normalized by the initial force obtained at t = 0. The force relaxation behavior in the rigor condition was similar in curve shape to that in the ATP condition.

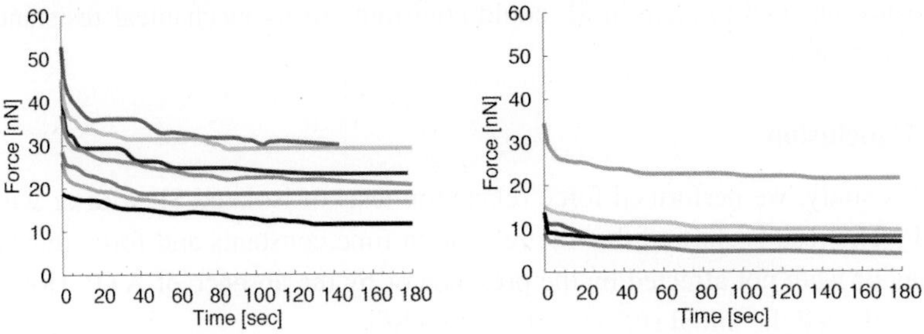

(a) Rigor Condition (b) ATP Condition

Figure 3. Force relaxation curves of stress fibers under (a) rigor and (b) ATP condition. [7]

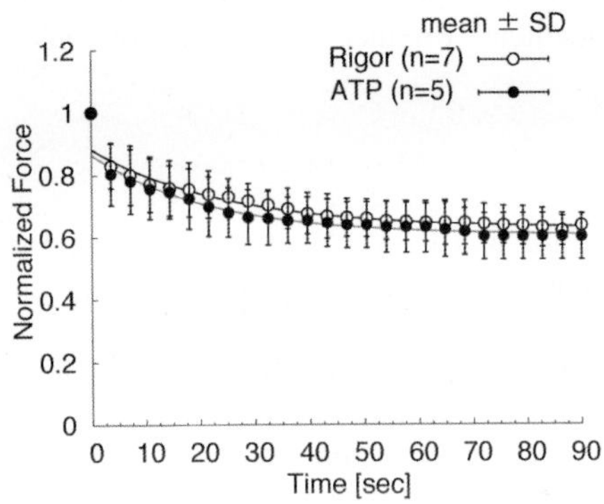

Figure 4. Normalized force relaxation curves of stress fibers under rigor and ATP conditions [7].

4. Discussion

In this study, we investigated the effect of acto-myosin interaction on force relaxation of isolated SFs. The force relaxation behavior and relaxation time constants of SF were not affected by chemical conditions. The time constants showed a wide range of values ranging from 4 sec to over 60 sec. The relaxation time constant of smooth muscle cells has been previously reported to be about 250 sec [6]. This difference between the relaxation time constants of SF and the cells suggests that SF may mainly act as an elastic element in the cells and that the other components, such as cytoskeletal meshwork or cytosol, may act as viscous elements. The initial stiffness of isolated SFs in the rigor condition was significantly larger than that in the ATP condition. In the ATP condition, acto-myosin interaction in SFs was thought to be in the binding-dissociation cycle. On the other hand, myosins in SFs in the rigor condition were thought to remain firmly attached to actin. It was consistent with the idea that the number of myosins attached to actin in SF could contribute to its mechanical resistance to step stretching.

5. Conclusion

In this study, we performed force relaxation tests of isolated SF with or without ATP. The results showed that the relaxation time constants and force relaxation behavior were not affected by the presence or by the absence of ATP. However, ATP affected the initial stiffness of isolated SFs.

Acknowledgments

This work was supported in part by Tohoku University Global COE Program "Global Nano-Biomedical Engineering Education and Research Network Centre."

References

1. M. Sato, N. Oshima and R. M. Nerem, Viscoelastic properties of cultured porcine aortic endothelial cells exposed to shear stress. *J Biomech* **29**, 461–467 (1996).
2. M. Nagayama, H. Haga, M. Takahashi, T. Saitoh and K. Kawabata, Contribution of cellular contractility to spatial and temporal variations in cellular stiffness. *Exp Cell Res* **300**, 396–405 (2004).
3. S. Deguchi, T. Ohashi and M. Sato, Tensile properties of single stress fibers isolated from cultured vascular smooth muscle cells. *J Biomech* **39,** 2603–2610 (2006).
4. K. Katoh, Y. Kano, M. Masuda, H. Onishi and K. Fujiwara, Isolation and contraction of the stress fiber. *Mol Biol Cell* **9**, 1919–1938 (1998).
5. H. Hirata, K. Ohki and H. Miyata, Dynamic change in the distribution of alpha5beta1 integrin on isolated ventral membrane: effect of divalent cation species. *Cell Motil Cytoskeleton* **59**, 131–140 (2004).
6. K. Nagayama, S. Yanagihara and T. Matsumoto, A novel micro tensile tester with feed-back control for viscoelastic analysis of single isolated smooth muscle cells. *Med Eng Phys* **29**, 620–628 (2007).
7. T. Matsui, S. Deguchi, N. Sakamoto, T. Ohashi and M. Sato, Effect of actin-myosin interaction on stress relaxation behavior of isolated stress fibers (in Japanese). *Proceedings of 20th Bioengineering Conference, 2007 Annual Meeting of BED/JSME,* Tokyo, No. 07-49, 105–106 (2008).

DEVELOPMENT OF AN *IN VITRO* TRACKING SYSTEM FOR CATHETER MOTION

MAKOTO OHTA [1], CHANG-HO YU [2]*, HIROYUKI KOSUKEGAWA [3]†, KEISUKE MAMADA [2], KANJU KUROKI [1], SHINZO OOTA [4], KAZUTO TAKASHIMA [5], KIYOSHI YOSHINAKA [6]

1) Institute of Fluid Science, 2-1-1, Katahira, Aoba-ku, Sendai, Miyagi 980-8577, Japan

2) Graduate School of Biomedical Engineering, Tohoku University, 6-6,Aramaki Aoba, Aoba-ku, Sendai, Miyagi 980-8579, Japan

3) Graduate School of Engineering, Tohoku University, 6-6, Aramaki Aoba, Aoba-ku, Sendai, Miyagi 980-8579, Japan

4) Brain attack center Oota memorial Hospital, 3-6-28, Okinogamicho Fukuyama, Hiroshima 720-0825, Japan

5) RIKEN-TRI Collaboration Center for Human-Interactive Robot Research, RIKEN, 2271-130, Anagahora, Shimoshidami, Moriyama-ku, Nagoya-shi, Aichi 463-0003, Japan

6) Department of Bioengineering, The University of Tokyo, 7-3-1, Hongo, Bunkyo-ku, Tokyo 113-8656, Japan

Endovascular treatments using catheters for cerebral aneurysms have been widely accepted as a less invasive way and a navigation system for catheters has been developed to support complicated treatments. Since the mechanical properties of a catheter play an important role in reaching the targeted disease, tracking of catheter motion during endovascular treatments may be useful for the increasing the confirmation of an operation. In this study, we developed an *in vitro* tracking system for catheter motion using Poly (vinyl alcohol) hydrogel (PVA-H) to mimic an arterial wall. The model made of PVA-H is transparent and the catheter motion can be observed clearly. This system will lead validation of computer-based navigation systems for operation assistance.

---

* Chang-Ho Yu was a Tohoku University Global COE Research Assistant (2008).

† Hiroyuki Kosukegawa was a Tohoku University Global COE Research Assistant (2008).

1. Introduction

Catheters are used for conveying medicines and implants for observation, diagnosis or treatments of infarction, aneurysm or stenosis in the endovascular fields. Recently, endovascular treatment has been proven to be useful as a reliable and less invasive treatment modality [1-3] and thus, the number of the cases so treated has increased. To reach and treat a diseased part with a catheter, medical doctors manipulate and control the catheter by using angiographic monitors. The techniques of treatments have been higher and then a catheter simulator or navigation has been developed [4,5]. Takashima *et al.* developed a computer-based navigation system for operation assistance [6,7]. Since their navigation system have been developed based on force and balance, the reconstruction of geometry and force field may be important. To validate such computer-based navigation systems, tracking the motion of a catheter is necessary and an *in vitro* model with transparent, mechanical properties and geometries mimicking those of a real artery, termed a biomodel, may be useful, Ohta *et al.* developed a biomodel using poly (vinyl alcohol) hydrogel (PVA-H) shown in Fig. 1 [8]. The mechanical properties such as Young's modulus of PVA-H are controllable with various techniques. For example, Kosukegawa *et al.* have described ways to elucidate the mechanical properties of biomodels using various concentrations of PVA solution, degrees of polymerization, saponification values and blending techniques [9]. Mamada *et al.* have reported that sensory evaluation of such procedures as touching, suturing or cutting of PVA-H mucosa model shown in Fig. 2 yields higher scores than those of a conventional material [10]. These results suggest that the force and balance field of an artery wall can be reconstructed by a PVA-H biomodel. In this paper, we describe the development of an *in vitro* tracking system for catheter motion by applying a PVA-H model and evaluation of the system with observation of video recordings of catheter motion.

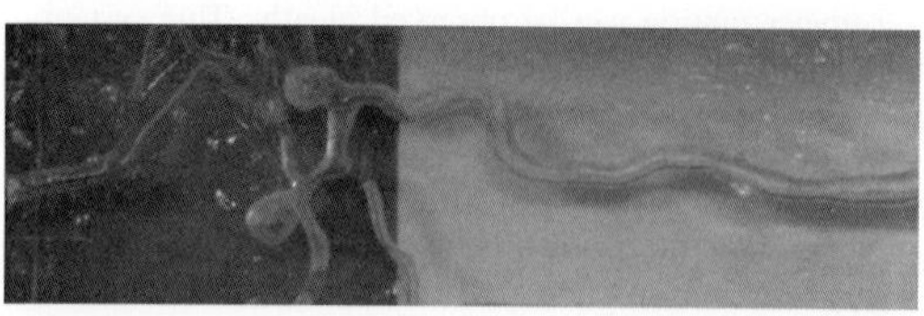

Figure 1. PVA-H biomodel for endovascular treatments.

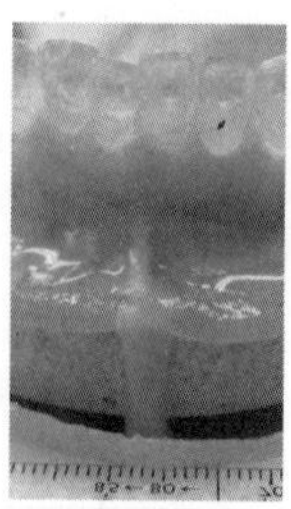

Figure 2. Biomodel of oral mucosa using PVA-H.

2. Materials and Methods

2.1. *Gelation of PVA and Lost-wax Technique*

PVA (JAPAN VAM & POVAL CO., LTD., Japan) was dissolved in a mixed solvent of dimethyl sulfoxide (DMSO) (Toray Fine Chemicals Co., Ltd., Japan) and distilled water (80/20, w/w) [9]. The PVA powder in the mixture solution was stirred for 2 hours at 100°C until dissolution. And then the solution was painted on the mold described in the following session (2.2) or cast in a transparent box (96 mm × 93 mm × 50 mm) with the mold described in the following sessions (2.3). The former model was used for the tracking of guide wire and the latter one was used for the tracking of catheter. These PVAs were maintained at -30°C for 24 hours to promote gelation of the PVA solution. After gelation, the mold materials were removed using water like a lost-wax technique.

2.2. *Tubular Model with a Realistic Geometry and Introduction of Guide Wire*

A tubular model with a realistic geometry made of PVA-H was constructed [11]. At first, an original geometry based on a silicone model (ELASTRAT Sali, Switzerland) was acquired using a rotational angiography. The silicone model of the cerebral vasculature was set in a conventional angiography with rotational data acquisition. Three-dimensional angiography was performed on a biplane C-arc unit (BV 3000, Philips Medical S ystems, the Netherlands). The intra-arterial bolus injection of contrast material was performed selectively for the artery of interest by hands. The rotational run was then transferred to the angiography workstation (INTEGRIS 3D-RA, Philips Medical Systems) and a 3-D reconstruction was performed. The 3-D geometry was transferred to a shape

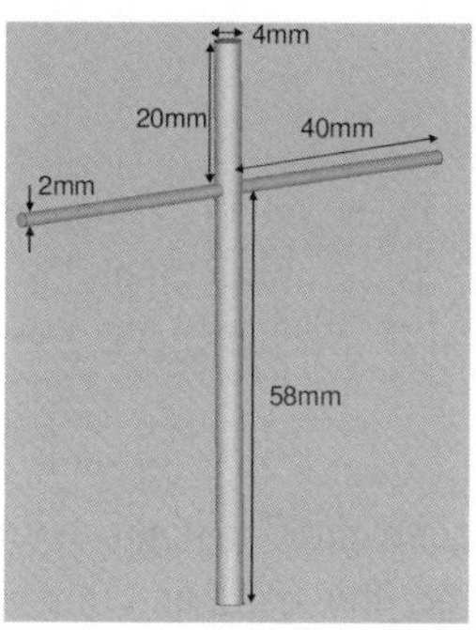

Figure 3. Cross model for tracking system.

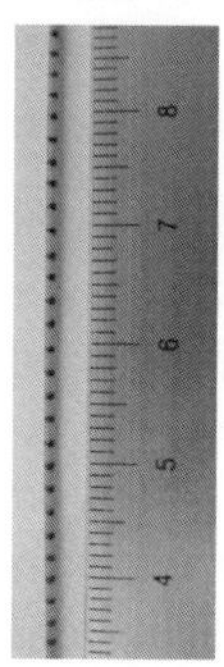

Figure 4. Catheter with zebra pattern for motion capture.

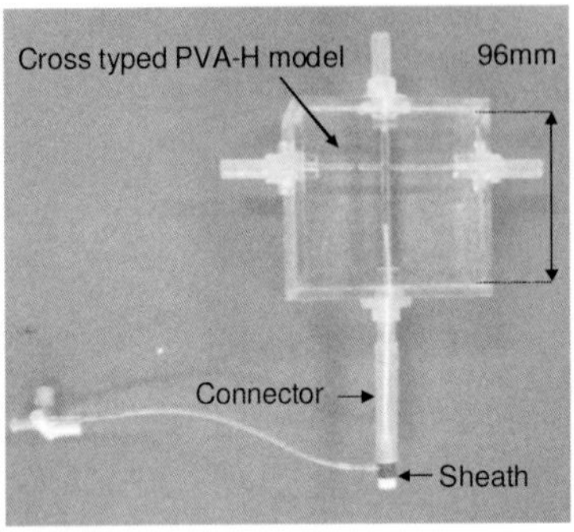

Figure 5. Circulation system for tracking motion of catheter.

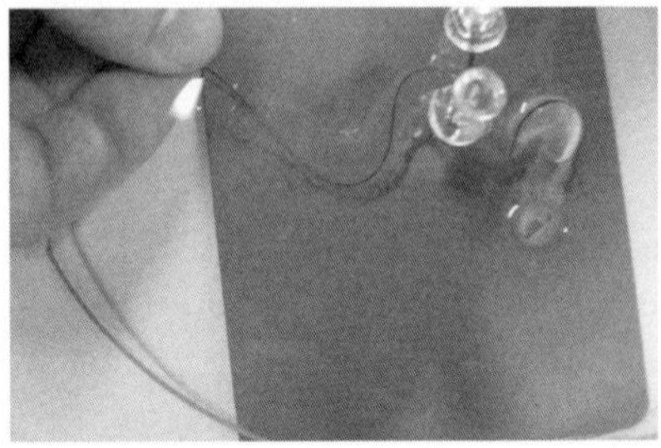

Figure 6. Guidewire in tubular PVA-H model with a realistic geometry.

made of gypsum, and PVA solution described in (2.1) painted on the model. After gelation, the gypsum materials were removed as a lost-wax technique.

2.3. *Cross Model for Tracking System and Motion Capture of Catheter*

A model with a cross branch was constructed using Magics software (Magics RP 8.5; Materialise, Leuven, Belgium) shown as Fig. 3. The diameter of main artery was 4 mm as a mean of intracranial artery [12,13]. A zebra pattern (2 mm distance and 2 mm thickness) was painted on a catheter (Marguerite, Micro Catheter; Cat.No.E102-130S1, GMA Co.,LTD., Japan) using an oily marker pen (MO-120-MC-BK, Zebra, Japan) shown as Fig. 4. The catheter was inserted into a circulation system made of acrylic resin with a sheath shown as Fig. 5 and was moved by hands slowly and the motion was recorded by a digital camera (Canon PowerShot G9, Canon, Japan).

3. Results

Figure 6 shows the photograph of the guidewire in the artery. The guide wire can be inserted into the artery smoothly. PVA-H is transparent and sufficient to observe the guidewire. The artery is soft and can be moved by the force from the guidewire.

Figure 7 shows sequential images of the catheter motion with time at the edge of PVA-H on the connector. Optical deformation of the marker on the catheter is observed. The connector is made of acrylic resin, and thus a refraction gap may cause the optical deformation at the edge.

Figure 7. Sequential images with time of catheter motion at the edge.

4. Discussions

Endovascular treatments using catheters have been applied for aneurysms or stenoses for the past several decades. As the motion of the catheter is captured using a medical imaging system with x-ray, this treatment is called as IGMIT (I mage Guided Minimally Invasive Treatment) [3]. The role of the catheter is to carry medicines and implants during operation. However, the use of a catheter requires highly skilled to operators and the medical imaging system can detect only the small markers on the catheter. Thus, medical operators must always image the motions of catheter based on their experience. For supporting the imaging, a navigator or a simulator has been developed for computer aided surgery [4,5]. However, that system only shows the results of the force interface between the operator and the system.

On the other hand, Takashima *et al.* have developed a computer-based simulator for catheter navigation for surgical planning based on force and balance in the catheter and the blood artery [6,7]. This simulator would be useful for the analysis of the structure of a catheter and may be useful in the design of a new catheter. The results indicate that the mechanical properties of the artery wall and the catheter affect the route of catheter and the ability to reach the desired target. To validate this indication, it may be necessary to develop a new *in vitro* model with geometry and mechanical properties similar those of a real artery. PVA-H may be useful realizing such an arterial model [8-10].

In this paper, we describe the development of an *in vitro* tracking system for catheter motion by applying the PVA-H model techniques and evaluate the system with observation of video recordings of catheter motion. PVA-H can possibly be used to reconstruct the shape of an artery as geometrical data of a patient. Also, the mechanical properties of the PVA-H such as Young's modulus or the surface of coefficient friction are suitable for modeling a realistic artery.

The transparency of PVA-H is sufficient to observe a catheter or a guidewire through the wall. Optical deformation of a catheter is observed at the edge of PVA-H on the connector. The difference of refraction of acrylic resin from that of PVA-H may cause this optical division at the edge. As a connector made of acrylic resin is necessary to fix a sheath to support the catheter, the edge will remain. As the refraction of PVA-H is also different from that of air, calibration should be performed for accurate measurement of the strain.

Because PVA-H contains water, its drying may limit the measurement time to around 30 minutes when catheter motion is observed for a long time. PVA-H containing DMSO maintains its humidity longer than that without DMSO. When a laser is used, care must be taken with regard to the surface temperature because the melting point of PVA-H is around 70°C [8].

5. Conclusion

An *in vitro* tracking system for catheter motion by applying a PVA-H model is developed and the transparency is enough to observe a catheter through the PVA-H wall.

Acknowledgments

This work is supported by Tohoku University Grobal COE Program "Global Nano-Biomedical Engineering Education and Research Network Centre", by Japan Society for the Promotion of Science "Core to Core Project 20001" and by GMA Co., LTD. .

References

1. A. Molyneux, R. Kerr, I. Stratton, P. Sandercock, M. Clarke, J. Shrimpton and R. Holman, *Lancet* **360**, 1267 (2002).
2. D. O. Wiebers, J. P. Whisnant, J. Huston, 3rd, I. Meissner, R. D. Brown, Jr., D. G. Piepgras, G. S. Forbes, K. Thielen, D. Nichols, W. M. O'Fallon, J. Peacock, L. Jaeger, N. F. Kassell, G. L. Kongable-Beckman and J. C. Torner, *Lancet* **362**, 103 (2003).
3. D. A. Rufenacht, M. Ohta, H. Yilmaz, C. Miranda, D. S. Ruiz, G. Abdo and P. Lylyk, *Swiss Archives of Neurology and Psychiatry* **155**, 348 (2004).
4. C. Tercero, Y. Okada, S. Ikeda, T. Fukuda, K. Sekiyama, M. Negoro and I. Takahashi, *Int J Med Robot* **3**, 349 (2007).

5. C. Tercero, S. Ikeda, T. Uchiyama, T. Fukuda, F. Arai, Y. Okada, Y. Ono, R. Hattori, T. Yamamoto, M. Negoro and I. Takahashi, *Int J Med Robot* **3**, 52 (2007).
6. K. Takashima, S. Ota, M. Ohta, K. Yoshinaka and T. Mukai, *Transactions of the Japan Society of Mechanical Engineers, Series C* **73**, 2988 (2007).
7. K. Takashima, S. Ota, M. Ohta, K. Yoshinaka and T. Mukai, *Journal of Japanese Society of Biorheology (B&R)* **22**, 1 (2008).
8. M. Ohta, A. Handa, H. Iwata, D. A. Rufenacht and S. Tsutsumi, *Technol Health Care* **12**, 225 (2004).
9. H. Kosukegawa, K. Mamada, K. Kuroki, L. Liu, K. Inoue, T. Hayase and M. Ohta, *Journal of Fluid Science and Technology* **3**, 544 (2008).
10. K. Mamada, H. Kosukegawa, K. Yamaguchi, N. Oikawa, Y. Katakura, Y. Shibaya, K. Kuroki and M. Ohta, *The 5th International Intracranial Stent Symposium,* 26 (2008).
11. S. G. Wetzel, M. Ohta, A. Handa, J. M. Auer, P. Lylyk, K. O. Lovblad, D. Babic and D. A. Rufenacht, *AJNR Am J Neuroradiol* **26**, 1425 (2005).
12. M. Ohta and D. A. Rüfenacht, *JSME, Fluid Engineering Conference 2006* **06-21**, 822 (2006).
13. M. Ohta, Y. Lachenal, G. Augsburger, G. Abdo, H. Yilmaz, N. Fujimura, D. Babic, P. Lylyk and D. Rüfenacht, *Journal of Biomechanics* **39S1**, S364 (2006).

ANALYSIS OF CHANGES IN THE GLIDING DIRECTION OF KINESIN-DRIVEN MICROTUBULES FOCUSING ON THEIR LENGTH AND KINESIN DENSITY

SHUKEI SUGITA*

Department of Biomedical Engineering, Graduate School of Biomedical Engineering, Tohoku University, 6-6-01 Aoba-yama, Aoba-ku, Sendai, Miyagi 980-8579, Japan

Kinesins, motor proteins moving along microtubules (MTs) in cells, can potentially be utilized as nano-scale transport systems with an inverted gliding assay, in which the MTs glide on a kinesin-coated surface. Although the key requirements include controls of the gliding direction of MTs, the details of motility properties of gliding MTs have not been elucidated. Here, the angular velocity of gliding MTs was quantitatively measured, particularly focusing on the effects of MT length and kinesin density. The gliding assay of MTs was performed on a substrate coated with kinesin densities of 7.5, 38, and 75 μg/ml that resulted in kinesin spacing of 7.8, 4.2, and 3.1 μm, respectively. The angular velocity for MTs shorter than kinesin spacing significantly decreased with increasing length, and that for MTs longer than kinesin spacing was not affected by their length. Moreover, the angular velocity for MTs longer than kinesin spacing was substantially higher at lower kinesin density. These results suggest that both the number of kinesins associated with MTs and the kinesin spacings may determine the gliding direction.

Keywords: Bioengineering; Motor Protein; Microtubule; Nano-scale Transport System; Gliding Direction.

1. Introduction

Kinesins are one of the nano-scale motor proteins that move unidirectionally along microtubules (MTs), and transport vesicles [1] and organelles [2] using energy derived from ATP hydrolysis in cells. Recently, many attempts utilizing kinesins have been made to develop nano-scale transport systems *in vitro* [3,4] because biological nanomachines have several advantages compared with artificial actuators. Firstly, the size of kinesin motors is nano-scale. Secondly, the energy conversion efficiency of kinesin motors is estimated to be as high as 50% [5]. Finally, MTs gliding on kinesins can be easily observed by constructing an *in vitro* system.

As the basic geometry of nano-scale transport systems, kinesin-MT systems have been employed in a so-called inverted gliding assay in which MTs are

* Shukei Sugita was a Tohoku University Global COE Research Assistant (2007-2008).

propelled by kinesins immobilized on the surface and act as carriers that transport attached cargos [6]. MTs used in the gliding assay system usually have a length ranging from less than one micrometer to more than tens of micrometers. It has been reported that a short MT captured by a single kinesin can rotate due to thermal diffusion depending on its size [7]. For a long MT, it is speculated that the leading tip can still fluctuate due to thermal diffusion, leaving the remaining part of the MT fixed on the kinesins due to crossbridges formed between the MT and the kinesins. Although this may potentially still lead to a change in the gliding direction of the MT, little is known about how the gliding direction of MTs can change with respect to their length and kinesin density.

Although knowledge of the fundamental motility properties of MTs is a prerequisite for the development of reliable and high efficient nano-scale transport systems, few studies on a characterization of such properties have been reported. Here, the author introduces the characterization of the motility properties of MTs gliding on kinesins by using an *in vitro* motility assay system based on a former study [8]. The characteristic parameter of angular velocity was determined, particularly focusing on the length of MTs and kinesin density, *i.e.*, the spacing between neighboring kinesins.

2. Materials and Methods

2.1. *Motility Assay*

The kinesin gene pGEX-DK411-BDTC was kindly provided by Prof. H. Higuchi (University of Tokyo, Japan). It includes the coding sequences for the 411 N-terminal residues of *Drosophila* kinesin heavy chain (DK-411) and biotin carboxyl carrier protein (BCCP) at the C-terminal. This construct was expressed in *E. Coli* and purified by glutathione-sepharose column.

Bovine tubulins (TL238, Cytoskeleton, Denver, CO, USA) containing 20% rhodamine-labeled tubulin (TL331M, Cytoskeleton, Denver, CO, USA) were polymerized into MTs by mixing with 1 mM GTP in BRB80 buffer (80 mM PIPES, 1 mM EGTA, 4 mM $MgSO_4$, pH 6.9 with KOH) and incubating at 37°C for 1 hour. This was followed by a dilution in a solution containing 100 μM paclitaxel.

Motility assay was performed as previously described [9] with some modifications. A flow cell was assembled by sandwiching 30-μm-thick double-sided tapes with two coverslips. The channel in the flow cell was then sequentially filled with 1 mg/ml biotinylated BSA solution, 1 mg/ml streptavidin solution, and 38 μg/ml kinesin solution. Two other different concentrations of kinesin solution, 75 μg/ml (high density) and 7.5 μg/ml (low density), were also used in order to assess the effect of the spacing between kinesins on the motility of MTs. The flow cell was then filled with 1 AMP-PNP solution containing MTs, and the solution was then replaced with 1 mM ATP solution containing oxygen scavenger additives (1.5% ß-mercaptoethanol, 1.5 mg/ml bovine serum albumin, 15 μM paclitaxel, 30 mM glucose, 120 μg/ml glucose oxidase, and 30 μg/ml catalase) and 0.1% methylcellulose. BRB80 was used as the buffer for all experiments. Time-lapse images of MTs were captured using an inverted fluorescent microscope equipped with a CCD camera and image acquisition software. Experiments were performed at 20°C.

2.2. *Characteristic Parameters*

Image analysis was performed with image analysis software (Image J, National Institutes of Health, Bethesda, MD, USA). Characteristic parameters such as angular velocity of MTs were determined with respect to the length of MTs (L) as follows. Firstly, the gliding velocity (V) was calculated from displacement of the leading tips of MTs every 5 s. The angular velocity (ω) was then determined as the rate of change in the orientation angle of the gliding velocity vector. The data were averaged over the period of observation and expressed as a mean. MTs shorter than 20 μm were analyzed. The spacing between neighboring kinesins was assessed by following the trajectories of the MTs as shown in Fig. 1. In the gliding assay, some MTs transiently showing a partial winding track (Fig. 1(b)) were observed. By superimposing the time-lapse images of those MTs, the points that the MTs always passed through were assumed to reflect the kinesin positions, and the distance between the points was measured (arrows in Fig. 1(c)).

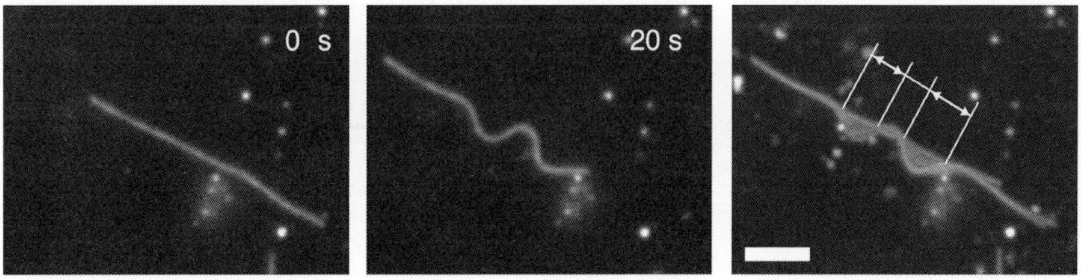

(a) (b) (c)

Figure 1. Typical images of gliding MTs for estimation of the spacing between kinesins. (a and b) Sequential images of a gliding MT. The MT traveling in a straight track (a) suddenly showed winding patterns (b). (c) The points MTs always passed through on superimposed time-lapse images for 0–20 s gave the position of kinesin, and the spacings between kinesins were measured as the distance between points (arrows). Bar = 5 μm

2.3. *Statistical Analysis*

Because of an unequal variance, mean angular velocities between kinesin densities were compared by the Steel-Dwass test. The association between MT length and mean angular velocity was assessed by the Pearson's correlation coefficient. Data were shown as mean ± standard deviation (SD).

3. Results and Discussion

Typical time-lapse images of the gliding MTs in 38 μg/ml of kinesin solution are shown in Figs. 2(a) and (b) for a long (8.0 μm) MT and a short (2.9 μm) MT, respectively. The images clearly show that the long MT glided over the surface along a rather straight path, whereas the short MT created a random path.

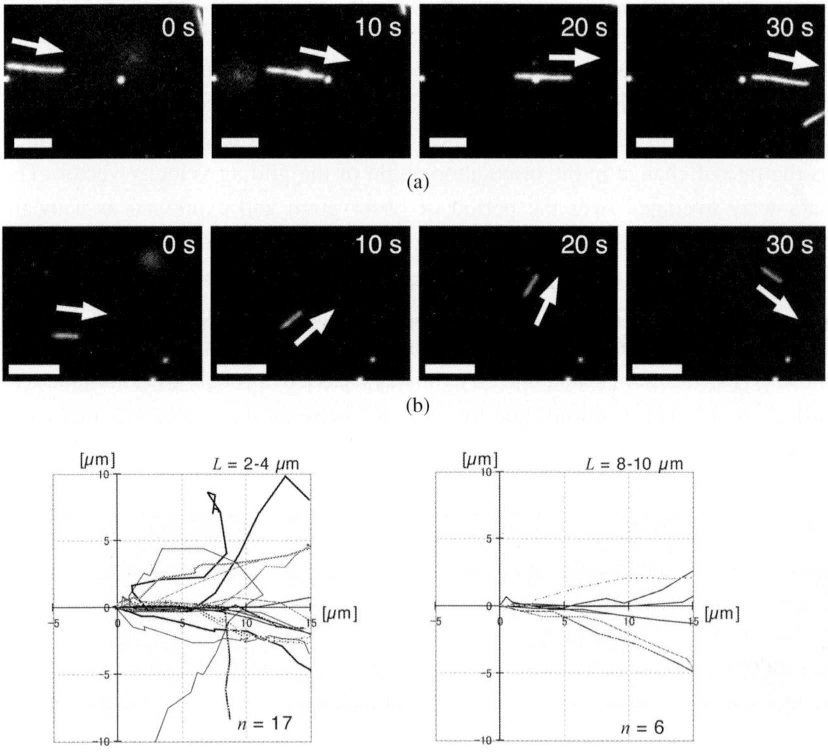

Figure 2. Time-lapse observation of gliding MTs. (a, b) Typical fluorescent images of (a) a long and (b) a short gliding MT. Bar = 5 μm. Arrows indicate the gliding directions of MTs. (c and d) Typical trajectories of gliding MTs for (c) L = 2–4 and (d) 8–10 μm. The gliding direction at the initial time is expressed from left to right.

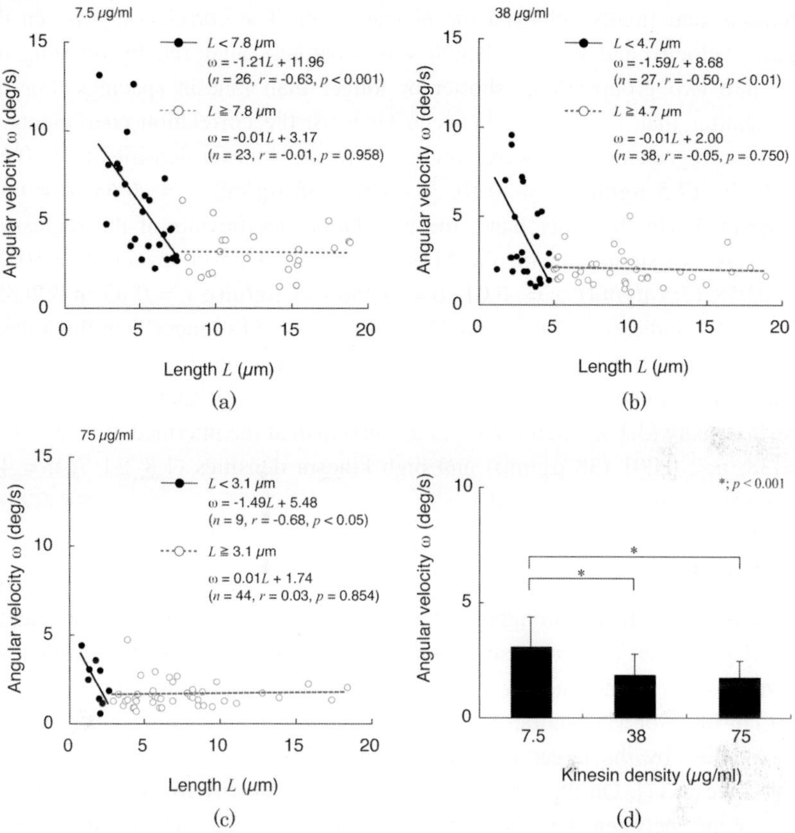

Figure 3. The angular velocity of gliding MTs. (a–c) The angular velocity plotted against MT length for kinesin densities of (a) 7.5, (b) 38, and (c) 75 μg/ml. The filled and open circles show the data for MTs shorter and longer than the estimated kinesin spacing, respectively. The solid and dotted lines were obtained by least squares regression. (d) The averaged angular velocity for the MTs longer than the estimated kinesin spacings.

Figures 2(c) and (d) show the trajectories of the gliding MTs in different ranges of length. The trajectories of MTs shorter than 4 μm showed that the movement is random over the period of time (Fig. 2(c)). In contrast, the trajectories of MTs longer than 8 μm showed straighter and smoother movement (Fig. 2(d)).

The measured kinesin spacings were 7.8 ± 2.7 (n = 29), 4.7 ± 1.3 (n = 43), and 3.1 ± 0.9 (n = 46) μm for 7.5, 38, and 75 μg/ml of kinesin densities, respectively. Figures 3(a) to (c) shows the length-dependency of the angular velocity at the three different kinesin densities. Under all the kinesin densities, the angular velocity substantially tended to decrease as the length of MTs increased, and finally achieved the plateau level. The correlation between the angular velocity and the MT length was separately analyzed by dividing the MTs into two groups: MTs shorter or longer than kinesin spacings. For the MTs shorter than the spacing between kinesins, the correlation coefficients of the linear regression lines were significant for all kinesin densities (r = -0.63, p < 0.001 (7.5 μg/ml); r = -0.50, p < 0.01 (38 μg/ml); r = -0.68, p < 0.05 (75 μg/ml)). On the other hand, the correlation coefficients of the regression lines were not significant for the MTs longer than kinesin spacing (r = -0.01, p = 0.958 (7.5 μg/ml); r = -0.01, p = 0.750 (38 μg/ml); r = 0.03, p = 0.854 (75 μg/ml)), indicating that the angular velocity of MTs longer than the kinesin spacing seemed to be independent of their length. The mean angular velocity of gliding MTs longer than the kinesin spacing was significantly higher at low kinesin density (3.1 ± 1.3, n = 23 (7.5 μg/ml)) than at the intermediate (1.9 ± 0.9, n = 38, p < 0.001 (38 μg/ml)) and high kinesin densities (1.8 ± 0.7, n = 44, p < 0.001 (7.5 μg/ml)) (Fig. 3(d)).

4. Discussion

For developing high throughput nano-scale transport systems, fundamental motility properties of carriers such as directionality should be well controlled. As shown in Fig. 3, the angular velocity significantly decreased with increasing length of MTs for MTs shorter than the estimated kinesin spacing. This result can be explained by the larger thermal fluctuation for longer MTs captured by a single kinesin [7]. On the other hand, the angular velocity for MTs longer than the spacing between kinesins was independent of the MT length. This is

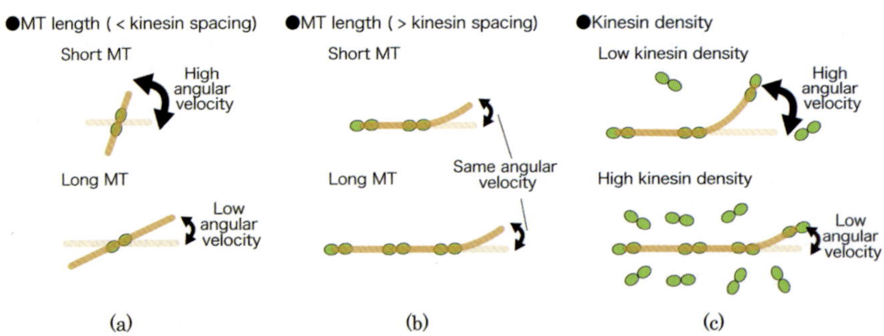

Figure 4. A possible underlying mechanism to explain the motility properties of MTs with respect to the their length and kinesin density. (a) Dependence of the MT length for MTs shorter than kinesin spacing and (b) independence of the MT length for MTs longer than kinesin spacing. (c) Dependence of kinesin density on the angular velocity of MTs. (d) Independence of the MT length and (e) dependence of kinesin density on the gliding velocity of MTs.

probably due to the fact that, within that range of the MT length, not a single kinesin but multiple kinesins could form crossbridges with the MTs and the crossbridges could work as fixed points against which reaction moments can be generated by the externally-driven thermal fluctuations. Although it is also possible that more crossbridges could decrease the deflection angle of the MT leading tip due to thermal fluctuations according to the strength of materials theory, the result shows that the effect was negligibly small. In addition, as shown in Fig. 3(d), the angular velocity significantly increased as the kinesin density decreased, in other words, as the spacing between kinesins increased. These results strongly indicate that the number of associated kinesins with MTs would be a critical factor in the gliding direction of MTs. It is thus possible that longer MTs associated with multiple kinesins at higher kinesin density may result in higher directionality of the gliding MTs.

From the results, the author proposes a concept to explain the possible mechanism for the angular velocity and the gliding velocity, as illustrated in Fig. 4. In Fig. 4(a), an MT shorter than the critical spacing between kinesins is hypothesized to be captured by a single kinesin and can rotate due to thermal diffusion depending on its length and shows high angular velocity. For an MT longer than the spacing captured by more than two kinesins, the angular velocity remains low and of constant value (Fig. 4(b)). In Fig. 4(c), the angular velocity is lower for higher kinesin density, probably because the increase of kinesin density results in an increase of the number of kinesins that MTs can be in contact with before showing large deflection.

In summary, the motility property of MTs gliding on a kinesin-coated surface was characterized with respect to their length and kinesin density. The angular velocity of the gliding MTs significantly decreased with increasing length for MTs shorter than kinesin spacing and achieved a plateau for MTs longer than kinesin spacing. Moreover, the angular velocity was substantially higher at lower kinesin density. These results suggest that both the number of kinesins associated with MTs and the kinesin spacing may contribute to the gliding direction. Towards development of high throughput nano-scale transport systems, the use of MTs longer than kinesin spacing would appear to be efficacious as they would have high directionality.

Acknowledgments

The authors acknowledge the support of Tohoku University Global COE Program "Global Nano-Biomedical Engineering Education and Research Network Center."

References

1. M. P. Sheetz, R. Vale, B. Schnapp, T. Schroer and T. Reese, Vesicle movements and microtubule-based motors. *J Cell Sci Suppl* **5**, 181–188 (1986).
2. T. A. Schroer, B. J. Schnapp, T. S. Reese and M. P. Sheetz, The role of kinesin and other soluble factors in organelle movement along microtubules. *J Cell Biol* **107**, 1785–1792 (1988).
3. C. Brunner, C. Wahnes and V. Vogel, Cargo pick-up from engineered loading stations by kinesin driven molecular shuttles. *Lab Chip* **10**, 1263–1271 (2007).
4. H. Hess, J. Clemmens, D. Qin, J. Howard and V. Vogel, Light-controlled molecular shuttles made from motor proteins carrying cargo on engineered surfaces. *Nano Lett* **1**, 235–239 (2001).
5. J. Howard, *Mechanics of Motor Proteins and the Cytoskeleton*, Sinauer Associates, Inc, Massachusetts, 254 (2001).
6. J. Howard, A. J. Hudspeth and R. D. Vale, Movement of microtubules by single kinesin molecules. *Nature* **342**, 154–158 (1989).
7. A. J. Hunt and J. Howard, Kinesin swivels to permit microtubule movement in any direction. *Proc Natl Acad Sci U S A* **90**, 11653–11657 (1993).
8. S. Sugita, N. Sakamoto, T. Ohashi and M. Sato, Characterization of motility properties of kinesin-driven microtubules towards nano-scale transporter: focusing on length of microtubules and kinesin density. *J Biomech Sci Engng.* (accepted).
9. J. Howard, A. J. Hudspeth and R. D. Vale, Movement of microtubules by single kinesin molecules. *Nature* **342**, 154–148 (1989).

REMODELING OF VASCULAR ENDOTHELIAL CELLS INDUCED BY STRETCHING FORCE TRANSMITTED THROUGH INTERCELLULAR JUNCTIONS

YOSUKE UEKI [1)*], NAOYA SAKAMOTO [1)], TOSHIRO OHASHI [1)],
MASAAKI SATO [1,2)†]

1) Department of Bioengineering and Robotics, Graduate School of Engineering,
2) Department of Biomedical Engineering, Graduate School of Biomedical Engineering,
Tohoku University, Aramaki-Aoba 6 6-01, Aobu, Sendai, Japan

It is well known that mechanical forces, vessels such as fluid shear and cyclic stretch, acting on the inner surface of blood induce the remodeling of endothelial cells (ECs). However, in contrast to global remodeling, little is known of how local mechanical forces are transmitted through cells and transduced into biochemical signals which induce cell remodeling leading to alteration in cell functions. In this study, we demonstrated that endothelial cell (EC) remodeling can be induced by local tension generated in a neighboring EC. In the technique employed, a glass microneedle was used to apply local stretch in an EC in a confluent monolayer and the tension was transmitted to a neighboring EC across intercellular junctions. Local stretch induced reorientation and EC elongation parallel to the direction of stretch associated with reorganization of stress fibers. These findings suggest that intercellular junctions cannot only transmit but also sense local forces, and are potentially involved in EC mechanotransduction pathways.

1. Introduction

Endothelial cells (ECs) the lining inner surfaces of blood vessels are exposed to many kinds of mechanical stimuli, including fluid shear stress due to blood flow, cyclic stretch due to the wall deformation, and hydrostatic pressure due to blood pressure. Since the relationship between endothelial cell (EC) morphology and mechanical conditions has been implicated in vascular pathology, *in vitro* model systems have been widely used to investigate the effects of mechanical stimuli on EC remodeling. Most previous studies have exclusively applied global mechanical stimuli, including fluid shear stress [1,2], cyclic stretch [3,4] and hydrostatic pressure [5], to a cultured EC monolayer. For example, under

---

* Yosuke Ueki is a Tohoku University Global COE Research Assistant.
† Masaaki Sato is a Tohoku University Global COE Member.

exposure to fluid shear stress, ECs exhibit elongation and alignment parallel to the direction of flow concomitant with reorganization of actin stress fibers (SFs) [2,4]. It is also well known that ECs exposed to cyclic stretch elongate and align perpendicular to the direction of stretch [3]. Thus, ECs respond very specifically to the type of mechanical stimuli, raising the question of how externally applied mechanical forces are transmitted in cells and where the transmitted forces are sensed and converted into biochemical signals. It has been suggested that mechanotransduction of ECs may occur at specific sites located in the cell membrane (*e.g.*, stretch activated channels (SACs)), points of EC attachment extracellular matrix (*e.g.* integrins), intercellular junctions (*e.g.*, platelet-endothelial cellular adhesion molecules-1, PECAM-1) and more likely, a combination of these sites. However, details of these mechanotransducers in cell signaling pathways remain unclear, because, in part, there is a possibility that these mechanotransducers can be simultaneously activated by globally applied mechanical stimuli.

Recently, increasing attention has been focused on the role of intercellular adhesion molecules with implications of EC mechanotransduction signalling pathways. Previous reports have revealed that ECs under sparse conditions do not show morphological changes in response to fluid shear stress [6]. Osawa *et al.* [7] have reported that PECAM-1 is rapidly tyrosine-phosphorylated when ECs are exposed to fluid shear stress or hyperosmotic shock. Tzima *et al.* [8] have reported that PECAM-1-mediated mechanotransduction is involved in the upstream region of the integrin signaling cascade, leading to cytoskeletal reorganization in flow-conditioned ECs. This leads to the idea that forces transmitted at adherens-junctions may stimulate PECAM-1 phosphorylation, leading to morphological changes of ECs. Thus, PECAM-1 may be responsible for morphological responses to mechanical stimuli transmitted through intercellular junctions serving as one of the principal mechanotransducers.

To test this hypothesis, in this study, local stretch was applied to an EC by using a microneedle to selectively stimulate intercellular junctions between the EC and a neighboring EC. After application of mechanical stimuli, morphological changes of the neighboring EC were fluorescently observed to test the hypothesis that local stimuli are transmitted via intercellular junctions and induce EC remodeling.

2. Materials and Methods

2.1. *Cell Culture and Transfection*

Human umbilical vein endothelial cells (HUVECs) were obtained from umbilical veins with trypsin treatment. Cells were cultured in Medium 199 containing 20% fetal bovine serum, 10 ng/ml basic fibroblast growth factor and penicilillin-streptomycin, and were used from the third to the sixth passage. For experiments, cells were plated on a glass base culture dish (ø = 35 mm) coated with 0.1% gelatin. A plasmid encoding enhanced green fluorescent protein (EGFP)-actin was transfected into HUVECs with a liposomal method.

2.2. *Local Stretch Experiment*

Figure 1 shows the experimental setup used to apply local stretch. A glass microneedle with a diameter of *ca.* 1 μm was made from a glass tube (ø = 1 mm) using a pipette puller and manipulated by a 3-D hydraulic micromanipulator. A pair of ECs (Cell 1 (C1) and Cell 2 (C2)) in the figure) expressing EGFP-actin were selected under a confocal laser scanning microscope (CLSM), and the nucleus of C2 was then moved horizontally by 10 μm by manipulating the glass microneedle to locally stretch C1. This procedure allows us to mimic cell deformation induced by externally applied local tension via intercellular junctions between C1 and C2. After application of local stretch, fluorescent images were obtained up to 60 min at intervals of 5 min with the CLSM to assess feartures of the C1 remodeling process. All procedures were performed in a CO_2 incubator mounted on the microscope stage to maintain the cell culture environment at 37°C and 5% CO_2.

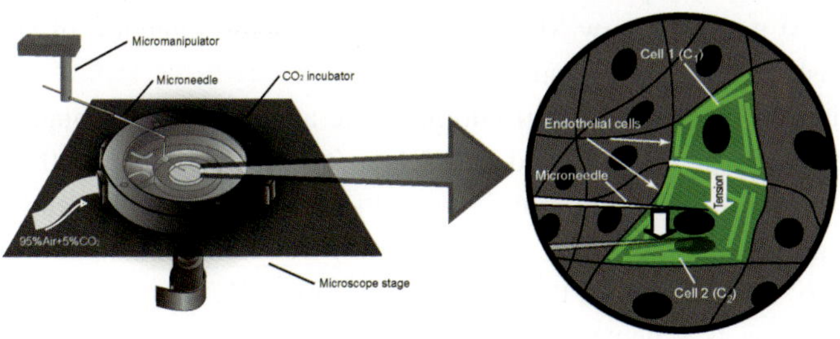

Figure 1. Schematic illustrations of experimental system for local stretch application. A microneedle was used to apply local stretch to C_2, possibly inducing remodeling of C_1 via intercellular junctions. The experiments were performed in a CO_2 incubator at 37°C and 5% CO_2. Reproduced with kind permission from Springer Science+Business Media [15]

2.3. *Image Analysis*

Image analysis was performed to evaluate morphological and cytoskeletal changes of C1 using the following parameters: angle of cell orientation, cell aspect ratio. An equivalent ellipse for the cell outline was automatically determined with a function of the software. An angle between the major axis of the ellipse and the direction of stretch was then defined as the angle of cell orientation. The ratio of length of the minor axis to the major axis was defined as the cell aspect ratio.

2.4. *Statistical Analysis*

Statistical comparisons of morphological parameters were performed using repeated measures analysis of variance (ANOVA) followed by Bonferroni's multiple comparison (0 min vs 30 min, 60 min). A p value less than 0.05 was considered to be significant. Data are expressed a mean ± SEM.

3. Results

3.1. *Morphological Changes of ECs after Application of Local Stretch*

Typical florescent images of EGFP-actin are shown in Fig. 2 at 0 min, 30 min and 60 min after application of local stretch. The observed cell C1 initially elongated perpendicular to the direction of stretch (Fig. 2A) exhibited contraction at 30 min (Fig. 2B) followed by elongation parallel to the direction of stretch at 60 min (Fig. 2C). Concurrent with these morphological changes, SFs were initially orientated perpendicular to the direction of stretch at 0 min (Fig. 2A, arrowheads), and then disappeared (Fig. 2B, arrowheads), while formation of new SFs parallel to the direction of stretch was observed (Fig. 2B, arrow) at 30 min. The new SF formation towards the direction of stretch was continuously observed at 60 min (Fig. 2C, arrow).

Figure 4 represents the distribution of the angle of cell orientation and the cell aspect ratio of the observed cell C1 at 0 min, 30 min and 60 min. From different types of initial cell morphology, the experimental data can be divided into two groups: Group A, ECs with an angle of cell orientation > 60° and an aspect ratio < 0.5; Group B, the remainder of Group A. Roughly, distribution of the two groups is schematically shown in the figure by ellipsoids. For both groups, the angle of cell orientation significantly decreased from 0 min (71.3 ± 4.4°, 53.3 ± 4.6° for Groups A and B, respectively) to 60 min (52.1 ± 6.8°, and 29.0 ± 3.7° for Groups A and B respectively, $p < 0.05$ vs 0 min for both groups). For Group A, the aspect ratio significantly increased from 0 min (0.42 ± 0.02) to

30 min (0.57 ± 0.07, $p < 0.05$ vs 0 min) and then decreased at 60 min (0.53 ± 0.08, NS vs 0 min). On the other hand, for Group B, the aspect ratio did not show a significant change from 0 min (0.57 ± 0.05), to 30 min (0.55 ± 0.07) and significantly decreased at 60 min (0.47 ± 0.06, $p < 0.05$ vs 0 min).

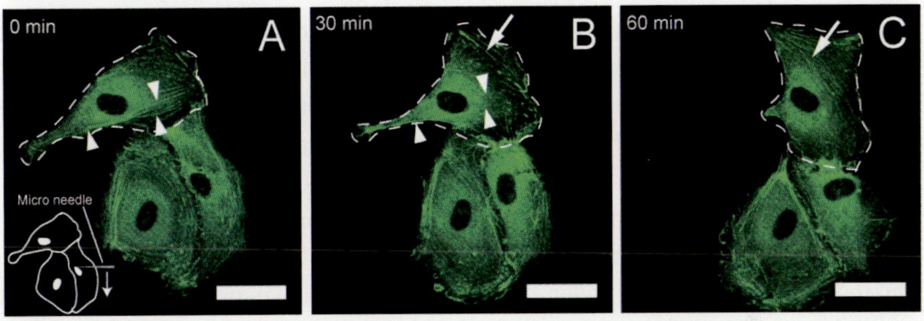

Figure 2. Typical examples of fluorescent images of EGFP-actin at (**A**) 0 min, (**B**) 30 min, and (**C**) 60 min after application of local stretch. Broken lines indicate the edge of C_1. The horizontally elongated C_1 at 0 min (**A**) showed contraction at 30 min (**B**) and re-elongated parallel to the direction of the stretch at 60 min (**C**). Preexisting SFs at 0 min (**A**) disappeared at 30 min (**B**) and formation of new SFs formation was observed at 60 min (**C**). Bar = 40 μm. Reproduced with kind permission from Springer Science+Business Media [15]

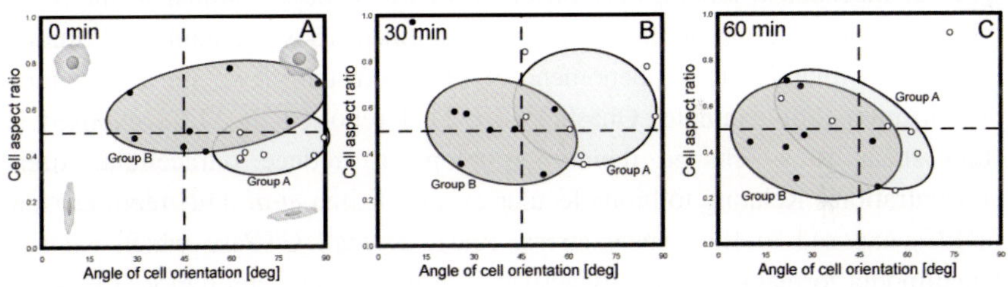

Figure 3. Changes in morphological parameters of ECs at (A) 0 min, (B) 30 min and (C) 60 min after application of local stretch. Group A represents cells with angle of cell orientation > 60° and aspect ratio < 0.5 and Group B, the rest of Group A. The two groups showed different time course of changes in the morphological parameters. Reproduced with kind permission from Springer Science+Business Media [15]

3.2. *Discussion*

In this study, local tension transmitted via intercellular junctions was applied to ECs to induce morphological changes in a neighboring EC. The result showed that ECs elongate and orient parallel to the direction of stretch concomitant with reorganization of stress fibers. Most previous studies relating to cell remodeling have utilized global mechanical stimuli including fluid shear stress [1,2] and cyclic stretch [3,4]. With these conventional techniques, it is difficult to identify both mechnotransmission pathways and mechanotransduction sites, probably because these events may occur by a combination of several candidate sensors. To overcome this difficulty, in the present study, a new experimental method was employed to locally apply mechanical stimuli to ECs to identify the role of intercellular junctions involved in EC remodeling.

Interestingly, ECs finally aligned with the direction of stretch at 60 min after application of mechanical stimuli. The mechanism in the present EC remodeling process is unclear. It is well known that ECs exposed to cyclic stretch show alignment perpendicular to the direction of stretch [3]. Sokabe *et al.* [9] have suggested that when cells are subjected to cyclic stretch, actin cytoskeletons are disassembled during the relaxation phase and disruption of actin fibers activates several downstream signals, leading to a cell shape change. On the other hand, ECs under continuous stretch have been shown to align parallel to the direction of stretch [10]. This is consistent with our results in which ECs aligned in the direction of local continuous stretch, strongly suggesting a critical role of relaxation phase in the EC alignment in response to the two different types of stretch. It is interesting to note that flow-imposed ECs are well known to show alignment in the direction of flow [1,2]. Although the types of mechanical loadings are different, this tendency is similar to the result of this study in terms of the directional alignment. Several numerical studies have well explained these phenomena of ECs from the viewpoint of structural optimization. For example, Ohashi *et al.* [11] reported that ECs exposed to fluid shear stress change their morphology to reduce intracellular stress concentrations. Relating to biomolecular events, Tzima *et al.* [12] reported that, in ECs exposed to fluid shear stress, Rac1, a small GTPase which controls lamellipodia formation, is locally activated in the downstream region. Likewise, these numerical and extra experimental approaches are worthwhile for elucidation of the underlying mechanism in the present EC remodeling process. Further investigations may, therefore, include more detailed quantification of the intracellular strain field and observations of local activity of Rho small GTPases, including Rac1, responsible for reorganization of actin-cytoskeletons.

The time course of change in EC morphology was investigated to characterize the EC remodeling process. The remodeling process was found to depend on the initial shapes of ECs and thus was divided into the two groups: Group A and B. For Group A, ECs initially elongated perpendicular to the direction of stretch exhibited spontaneous contraction prior to elongation/alignment in the direction of stretch. In contrast, for Group B, ECs initially aligned in the direction of stretch immediately exhibited elongation and alignment of SFs parallel to the direction of stretch. It has been reported that

ECs exposed to shear stress show the following time course of change: 1) contraction, 2) reorientation and 3) elongation [5]. Taking these results and our results into consideration, a specific time course of change depending on the type of externally applied mechanical stimuli may exist. It is beyond the scope of the present study whether the present remodeling process might be involved in the EC remodeling exposed to global mechanical stimuli including fluid shear stress. Noria *et al.* [14] suggested that flow-induced morphological changes of ECs were caused by reorganization of SFs. Moreover, Li *et al.* [14] reported that ECs transfected with dominant negative RhoA, responsible for SF formation, do not respond to fluid shear stress. Taken together, the various findings indicate that SF reorganization associated with cell morphological changes may play an important role in the EC remodeling process.

4. Conclusions

In this study, local stretch was applied to an EC to stimulate a neighboring EC via intercellular junctions using a micromanipulation technique. ECs reorientated and elongated parallel to the direction of stretch. This remodeling process varied from cell to cell, particularly depending on the initial cell shape. These results suggest that intercellular junctions can transmit mechanical forces between cells and transduce the forces into biochemical signals.

Acknowledgement

This work was supported in part by the Global COE Program "Global Nano-Biomedical Engineering Education and Research Network Centre" and Grants-in-Aid for Scientific Research from the Ministry of Education, Culture, Sports, Science and Technology (MEXT) in Japan (Nos. 15086203, 17200030, 17680036).

References

1. C. F. Dewey, S. R. Bussplari, M. A. Gimbrone and P. F. Davies, The dynamic response of vascular endothelial cells to fluid shear stress. *J Biomech Eng* **103**, 177–185 (1967).
2. C. G. Galbraith, R. Skalak and S. Chien, Shear stress induces spatial reorganization of the endothelial cell cytoskeleton. *Cell Motil Cytoskeleton* **40,** 317–330 (1998).
3. V. P. Shirinsky, A. S. Antonov, K. G. Birukov, A. V. Sobolevsky, Y. A. Romanov, N. V. Kabaeva, G. N. Antonova and V. N. Smirnov, Mechano-chemical control of human endothelium orientation and size. *J Cell Biol* **109,** 331–339 (1989).

4. J. H. Wang, P. Goldschmidt-Clermont, J. Wille and F. C. Yin, Specificity of endothelial cell reorientation in response to cyclic mechanical stretching. *J Biomech* **34,** 1563–1572 (2001).
5. Y. Sugaya, N. Sakamoto, T. Ohashi and M. Sato, Elongation and random orientation of bovine endothelial cells in response to hydrostatic pressure: comparison with response to shear stress. *JSME Int J Ser C* **46,** 1248–1255 (2003).
6. M. Masuda and K. Fujiwara, Morphological responses of single endothelial cells exposed to physiological levels of fluid shear stress. *Front Med Biol Eng* **5,** 79–87 (1993).
7. M. Osawa, M. Masuda, K. Kusano and K. Fujiwara, Evidence for a role of platelet endothelial cell adhesion molecule-1 in endothelial cell mechanosignal transduction: is it a mechanoresponsive molecule? *J Cell Biol* **158,** 773–785 (2002).
8. E. Tzima, M. Irani-Tehrani, W. B. Kiosses, E. Dejana, D. A. Schultz, B. Engelhardt, G. Cao, H. DeLisser and M. A. Schwartz, A mechanosensory complex that mediates the endothelial cell response to fluid shear stress. *Nature* **437,** 426–431 (2005).
9. M. Sokabe, K. Hayakawa and H. Tatsumi, Varieties of mechanotransduction: the cytoskeletal stress fibre as a force transmitter and a mechanosensor. *Proceedings of the Australian Physiological society,* Canberra, vol. 36, 95 (2005).
10. A. Sasamoto, M. Nagino, S. Kobayashi, K. Naruse, Y. Nimura and M. Sokabe, *Am J Physiol Cell Physiol* **288,** C1012–C1022 (2005).
11. T. Ohashi, H. Sugawara, T. Matsumoto and M. Sato, Surface topography measurement and intracellular stress analysis of cultured endothelial cells exposed to fluid shear stress. *JSME Int. J. Ser. C* **43,** 780–786 (2000).
12. E. Tzima, M. A. Del Pozo, W. B. Kiosses, S. A. Mohamed, S. Li, S. Chien and M. A. Schwartz, Activation of Rac1 by shear stress in endothelial cells mediates both cytoskeletal reorganization and effects on gene expression. *EMBO J* **21,** 6791–6800 (2002).
13. S. Noria, F. Xu, S. McCue, M. Jones, A. I. Gotlieb and B. L. Langille, Assembly and reorientation of stress fibers drives morphological changes to endothelial cells exposed to shear stress. *Am J Pathol* **164,** 1211–1223 (2004).
14. S. Li, B. P. Chen, N. Azuma, Y. L. Hu, S. Z. Wu, B. E. Sumpio, J. Y. Shyy and S. Chien, *J Clin Invest* **103,** 1141–1150 (1999).
15. Y. Ueki, N. Sakamoto, T. Ohashi and M. Sato, Morphological responses of vascular endothelial cells induced by local stretch transmitted through intercellular junctions. *Exp Mech.* (in press).

ESTIMATION OF UNSTEADY BLOOD FLOW RATE IN ULTRASONIC-MEASUREMENT-INTEGRATED SIMULATION

TAKAYUKI YAMAGATA *

Department of Bioengineering and Robotics, Tohoku University, 6-6-04 Aramaki Aza Aoba, Aoba-ku, Sendai, Miyagi 980-8579, Japan

TOSHIYUKI HAYASE †

Institute of Fluid Science, Tohoku University, 2-1-1 Katahira, Aoba-ku, Sendai, Miyagi 980-8577, Japan

To obtain accurate and detailed information on blood flow, Ultrasonic-Measurement-Integrated (UMI) simulation has been studied. Assuming that the flow rate of a blood flow is given, the UMI simulation reproduces the detailed velocity field of a complex blood flow by combining numerical simulation and measured Doppler velocity, which is a part of blood flow velocity. However, existing medical equipment has limitations for obtaining accurate flow rate of the complex blood flow. In this paper, a method of estimating unsteady flow rate using Doppler velocity is proposed. The applicability of the method was numerically investigated in UMI simulation for a flow in an aneurysmal descending aorta.

1. Introduction

Acquisition of multidirectional and time-dependent blood velocity data is essential to clarify the relation between blood flow and circulatory diseases, which are serious health problems in many developed countries. Aneurysm, a circulatory disease, is usually diagnosed by considering shapes and sizes of blood vessels detected by medical diagnostic imaging apparatuses [1]. Medical ultrasonography shows the cross section of a blood vessel and the Doppler velocity that is a part of blood flow velocity projected in the direction of an ultrasonic beam. However, more detailed information on blood flow is required to enable accurate diagnoses. The use of Ultrasonic-Measurement-Integrated (UMI) simulation has been proposed to reproduce the detailed velocity field of complex blood flow by combining measured Doppler velocity and numerical simulation [2,3]. A numerical experiment has shown that UMI simulation

* Takayuki Yamagata is a Tohoku University Global COE Research Assistant.

† Toshiyuki Hayase is a Tohoku University Global COE Member.

reproduces a standard solution of the blood flow even if the accurate velocity profile of the standard solution is unknown at the upstream boundary [2]. However, conventional UMI simulation requires the correct flow rate of the standard solution or real blood flow to be reproduced at each time step.

In this paper, a method of estimating unsteady blood flow rate in UMI simulation is proposed to overcome the difficulty of estimating the correct flow rate from the Doppler velocity [4]. The applicability of the proposed method was numerically investigated for a flow in an aneurysmal descending aorta.

2. Methods

Figure 1 shows a block diagram of UMI simulation with a diagnostic image of the descending aorta with an aneurysm treated in this study. The gray part indicates the blood vessel wall, where an aneurysm can be seen in the lower part of the image. The UMI simulation, which is a kind of the observer in control theory, reproduces real blood flow with the aid of the feedback signal which compensates the difference between the simulation and the real blood flow [2]. Governing equations of the UMI simulation are the Navier-Stokes equations and a continuity equation. It should be noted that the feedback signal f_v is added to the streamwise Navier-Stokes equation as well as to the pressure equation derived from the continuity equation at each monitoring point as an artificial force term. The feedback signal is defined as being proportional to the difference between the measured Doppler velocity V_s and the calculated velocity V_c as follows:

$$f_v = -K_v \rho (V_c - V_s) u_{in} \Delta S \tag{1}$$

where K_v is the feedback gain, ρ is the density, U_{in} is the average inlet velocity,

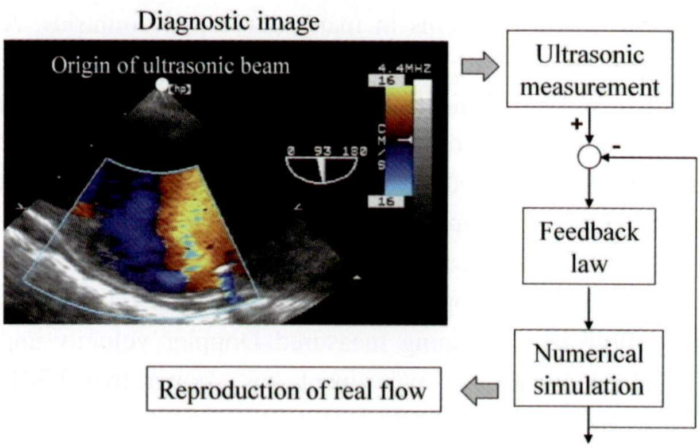

Figure 1. Block diagram of Ultrasonic-Measurement-Integrated simulation with a diagnostic image (Center frequency, 4.4 MHz; pulse repetition frequency, 4 kHz)

and ΔS is the interfacial area of the control volume. The details of the UMI simulation are described in Ref. [2].

The concept of estimating unsteady flow rate is given in the following. In the UMI simulation, reducing the error of Doppler velocity allows reproduction of the velocity field, but the velocity field can be properly reproduced only using the correct flow rate. Therefore, the UMI simulation probably shows the minimal error of the Doppler velocity at the correct flow rate. We determined the flow rate with the minimal error of the Doppler velocity with the golden section search, a technique for finding the minimum of a unimodal function by narrowing the search range.

Figure 2 shows a two-dimensional grid system with 67 × 42 grid points with grid spacing of 2.973×10^{-3} m. Monitoring points for the feedback are located in the domain containing the aneurysm. A numerical simulation with the upstream boundary condition of the parabolic velocity profile was used for the standard solution as a model of a real blood flow. We used a waveform of the flow rate from Ref. [5] and a mean flow rate of 3.85 l/min and a pulsatile cycle T = 1.15 s for the standard solution. UMI simulation was performed with an upstream boundary condition of a uniform parallel flow with unknown flow rate assuming that the real velocity profile is not available. The optimal feedback gain K_v = 0.5 was determined for the UMI simulation by test computation. In addition, ordinary simulation without feedback, whose feedback gain is K_v = 0.0, was conducted for comparison. In order to evaluate the error of the Doppler velocity, the error norm e_n is defined as follows:

$$e_n = \frac{1}{u_{ref}} \left(\frac{1}{N} \sum_i \left| V_{ci} - V_{si} \right| \right), \qquad (2)$$

where V_c and V_s are, respectively, the computed Doppler velocity and the standard one, N is the number of monitoring points, and u_{ref} is a reference

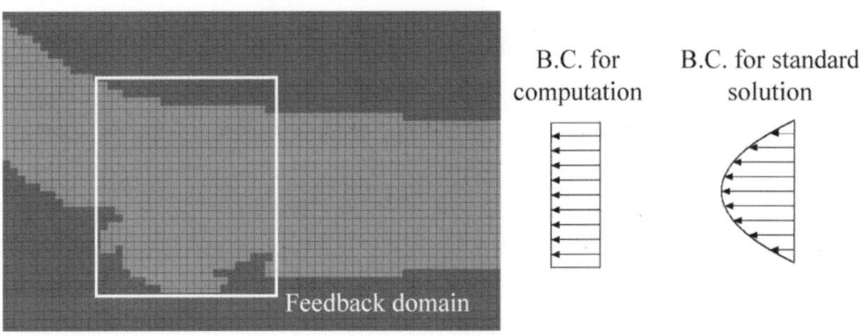

Figure 2. Computational grid system and feedback domain

velocity (u_{ref} = 0.64 m/s) corresponding to Reynolds number Re = 4500. The search range of the flow rate was defined from -9.2×10^{-4} to 9.2×10^{-4} m^3/s, the absolute value of which is double the maximum flow rate of the standard solution at systole. The golden section search was performed 20 times, the resolution being 6.1×10^{-8} m^3/s.

3. Results and Discussion

Figure 3 shows flow rate changes of the UMI simulation and the ordinary numerical simulation without feedback determined by golden section search with 20 iterations as well as those of the standard solution. Both the UMI and the ordinary simulations reproduced the flow rate change of the standard solution in systole (t = 0.05 - 0.4 s). It seems that a velocity field does not depend on the upstream velocity profile in systole. However, in diastole (t = 0.4 - 1.2 s), the ordinary simulation shows a different flow rate from the standard value, implying that a different flow structure explicitly appears due to the parabolic and the uniform upstream velocity profile. The UMI simulation correctly reproduces the flow rate all the time due to the effect of the feedback.

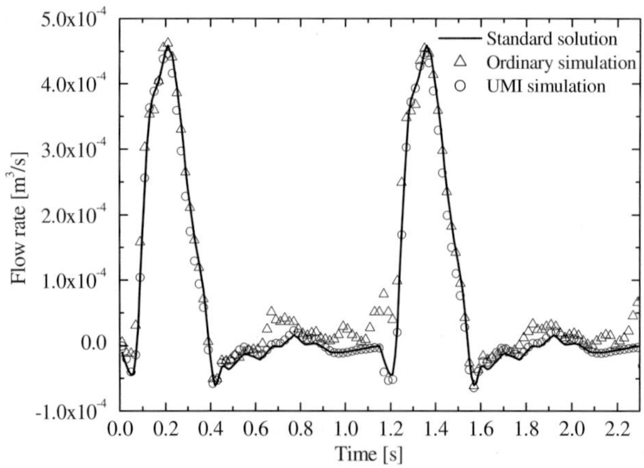

Figure 3. Time histories of estimated flow rate

Figure 4 shows the relationships between the flow rate and the error norm of the Doppler velocity defined as Eq. (2) at t = 0.21 s (a) and t = 1.20 s (b). The UMI simulation provides a minimal error norm of the Doppler velocity at the correct flow rate at each time step. In the ordinary simulation, the error norm shows a minimum value at the correct flow rate in Fig. 4(a), but the error norm remains larger than in the UMI simulation. In addition, the ordinary simulation does not necessarily estimate the correct flow rate at the minimal error norm (see Fig. 4(b)). The UMI simulation, which reduces the error norm of the Doppler velocity, can correctly estimate the flow rate of the standard solution.

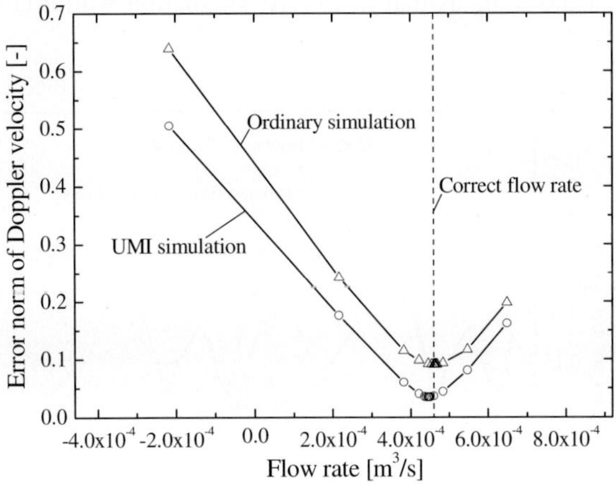

(a) $t = 0.21$ s

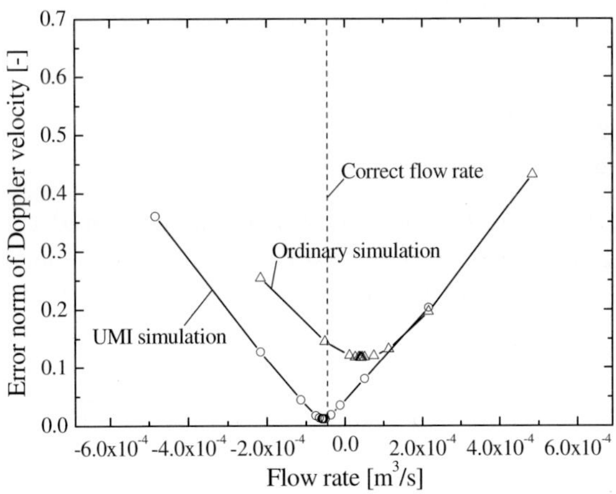

(b) $t = 1.20$ s

Figure 4. Relationship between the flow rate and the error norm of the Doppler velocity

Figure 5 compares time histories of the squared error norm of the velocity as against the standard solution between the ordinary and the MI simulations in two cases of the correct flow rate and the estimated one with 20 iterations. The ordinary simulation with the correct flow rate involves an error in time-average of 21% due to the difference of the inflow velocity profile. The error of the ordinary simulation with the estimated flow rate slightly decreases to 16%. The MI simulation with the estimated flow rate reduces the velocity error to 6%, which is almost the same as that of the MI simulation with a correct flow rate of 5%.

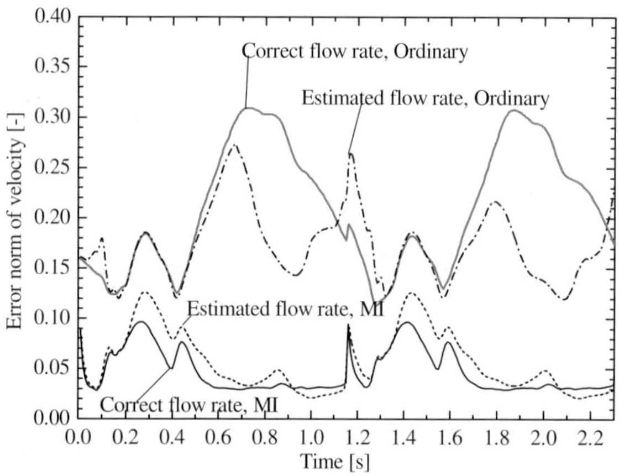

Figure 5. Time histories of the squared error norm of velocity

4. Conclusion

A method of estimating the blood flow rate was developed and validated by numerical experiment in order to reconstruct actual pulsatile blood flow from the Doppler velocity. UMI simulation effectively reproduced the flow rate of the standard solution with golden section search, reducing the error norm of the Doppler velocity in the monitoring domain. The UMI simulation reproduced the velocity of the standard solution with 6% error at the same level as in the case of the correct flow rate. Our future work includes the optimization of computational conditions and flow rate search conditions for real time operation of the UMI simulation.

Acknowledgments

The authors acknowledge the support from the Tohoku University Global COE Program "Global Nano-Biomedical Engineering Education and Research Network Centre."

References

1. H. Ujiie, Y. Tamano, K. Sasaki and T. Hori, Is the Aspect Ratio a Reliable Index for Predicting the Rupture of a Saccular Aneurysm? *Neurosurgery* **48**, 495–503 (2001).
2. K. Funamoto, T. Hayase, A. Shirai, Y. Saijo and T. Yambe, Fundamental Study of Ultrasonic-Measurement-Integrated Simulation of Real Blood Flow in the Aorta. *Ann Biomed Eng* **33**, 415–428 (2005).
3. K. Funamoto, T. Hayase, Y. Saijo and T. Yambe, Numerical Experiment for Ultrasonic-Measurement-Integrated Simulation of Three-Dimensional Unsteady Blood Flow. *Ann Biomed Eng* **36**, 1383–1397 (2008).
4. T. Yamagata and T. Hayase, Method of Determining Unsteady Blood Flow Rate Condition in Ultrasonic-Measurement-Integrated Simulation. *The 2nd Tohoku-NUS Joint Symposium on the Future Nano-Medicine and Bioengineering in the East Asian Region*, Singapore, 45–46 (2006).
5. M. S. Olfsen, C. S. Peskin, W. Y. Kim, E. M. Pedersen, A. Nadim and J. Larsen, Numerical Simulation and Experimental Validation of Blood Flow in Arteries with Structured-Tree Outflow Conditions. *Ann Biomed Eng* **28**, 1281–1299 (2000).

Section 2

Nano-Bioimaging

STRUCTURAL CHANGE OF THE HUMAN BRAIN WITH AGING IN HEALTHY SUBJECTS-A BRAIN MAGNETIC RESONANCE IMAGE (MRI) STUDY *

HIROSHI FUKUDA [1)†], YASUYUKI TAKI [2)], KENTARO INOUE [1)], SHIGEO KINOMURA [1)], KAZUNORI SATO [1)], RYOI GOTO [1)], KEN OKADA [1)], RYUTA KAWASHIMA [3)*]

1) Department of Nuclear Medicine and Radiology, 2) Division of Cognitive Neuroscience, 3) Department of Functional Brain Imaging, Institute of Development, Aging and Cancer, Tohoku University, 4-1 Seiryo-machi, Aoba-ku, Sendai, Miyagi 980-8575, Japan

We analyzed structural change of the human brain with aging using MRI of the 2,000 healthy Japanese subjects. Volumetric analysis of revealed that gray matter volume linearly decreased with age, while white matter volume remained unchanged during aging. While, longitudinal study in the same subjects with 8 year-interval revealed that there are sex difference in speed and pattern of gray matter loss, the loss being slower in women that that in men. Correlation analyses revealed that there were negative correlations between gray matter volume and cerebrovascular risk factors, such as hypertension, amount of alcohol intake and obesity. The anatomical network analysis using regional gray matter volume exhibited "small-world" attributes. Importantly, the results demonstrated that significant statistical differences exist in the small-world properties between different sex and among decades from 20 to 70.

1. Introduction

Recently, the importance of human neuroimaging database was recognized greatly. The most important purpose and needs for database are electronic data sharing in neuroscience community as well as the construction of more sophisticated and complete model of brain structure and function. This model of normal brain structure and function can be used as the references not only for neuroimaging study for humans but also for computer aided automated diagnosis of the brain diseases.

Most remarkable recently developed method for brain image analysis is a voxel based morphometry (VBM). This technique enables us to extract brain regions which show correlations between tissue volume and variables, such as

---

* Work partially supported by a grant in-aid from the Japanese Society for the Promotion of Sciences (No 20659183).

† Hiroshi Fukuda is a Tohoku University Global COE Member.

age, sex and subject's characteristics. We can analyze not only age-related normal changes but also diseased brain, such as dementia.

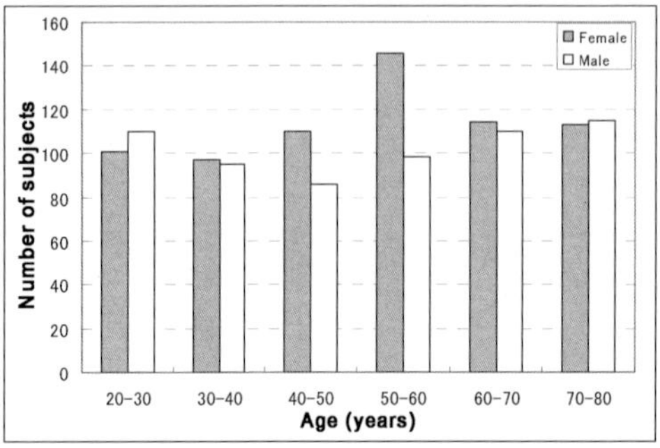

Figure 1. Number of subjects for each age group and sex in brain MRI database

We have collected 2,000 brain MRI of healthy Japanese and constructed an MRI database together with their characteristics such as age, sex, blood pressure, present and past disease history and cognitive functions [1] (Fig. 1). This is a largest database in Japan and one of the largest one in the world. The ages of the subjects ranged from 20th to 70th of age for men and women, and therefore the database is a good for brain aging analyses. Using this MRI database, we made the following analyses on age-related structural changes of the brain.

1.1. *Brain Volume Change with Aging*

Volume of the brain tissue segment, that is gray matter, white matter and cerebrospinal fluid (CFS) space were separately measured and their age-related changes were examined.

1.2. *Correlation between Brain Volume Change and Cerebrovascular Risk Factors*

Correlation between regional gray matter volume and characteristics of the subjects, such as age, sex and their medical, mental or social background was tested using VBM technique.

1.3. *Cross-generational Differences in Brain Shape Index*

Cephalic index, which is the ratio of length and breadth of the skull or scalp, is different among the races. Generally, the index in European people is smaller that that in Mongoroid including Japanese indicating that European skull is longer and narrower than that of Mongolian skull. Furthermore, even in a race, the index also changes depending on the era in which the people lived.

In the present study, we measured the ratio of length and width of the brain (brain shape index).

1.4. *Correlation between Fiber Connection and Cerebral Glucose Metabolism*

Activity in neural cells is coupled with glucose metabolism. Neural function of signaling are carried out by the interconnection of neurons via neuronal fibers. Diffusion-tensor imaging (DTI) allows *in vivo* visualization of white matter fiber tracts by measuring the anisotropy of water molecular diffusion. Several studies focusing on normal aging have reported significant reduction in fractional anisotropy (FA) in the corpus callosum. The purpose of this study was to examine whether degradation of microstructure of fiber tracts in the elderly was associated with change in the glucose metabolism in the cerebral cortex measured with ^{18}F-FDG-PET.

1.5. *Anatomical Network Analysis*

Recent research has revealed that human brain functional networks during behavior or even at resting state have the small-world property. This study was attempted to examine the network of anatomical connections in the human brain. We investigated large-scale anatomical connection patterns of human brain using regional gray matter volume from Japanese MR Image database.

2. Materials and Methods

2.1. *Image Processing and Statistical Analysis using VBM Technique*

All the brains with different size and shape were transferred into a standard stereotactic space (Talairach space) and their size and shape were transformed into a standard template brain using liner and non-linear parameters (anatomical standardization). Next step is brain tissue segmentation. A brain images were segmented into gray matter, white matter, cerebrospinal fluid space (CSF) and outer brain space depending on the differences of signal intensities on T1 weighted MRI of each tissue (Fig. 2). The volumes of each tissue segment were calculated by summing up the value of all the voxels which belong to each tissue segment.

The standardized and segmented gray matter images were smoothed by convoluting a 12-mm-FWHM isotropic Gaussian kernel. Then, the smoothed gray matter images were statistically analyzed by voxel based morphometry (VBM) technique using SPM2 package (Fig. 2). VBM was performed to investigate correlation between regional gray matter volume and attributes to the subject, such as age, cerebro-vascular risk factors, and scores of cognitive functions. This approach is not biased toward any one brain region and permits the identification of unsuspected potential brain structural abnormalities. Simple or multiple regression analysis was performed using SPM2. These attributes were used as dependent variables, and regional gray matter volume as an independent variable. We set the significance level at $P<0.05$ for multiple comparison. The method is a voxel based t-statistics extended to the three dimensional space based on the general linear model. Simple or multiple regression analysis and group comparison using t-test were performed.

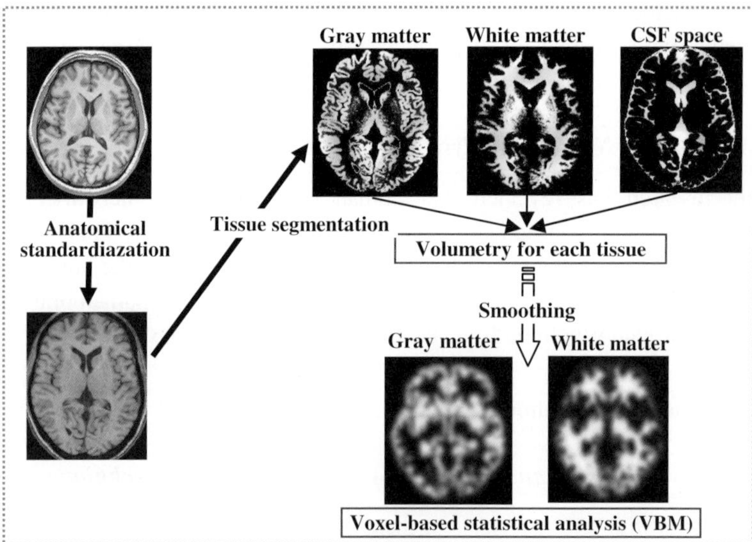

Figure 2. Anatomical standardization, tissue segmentation of the brain MRI and statistical analyses

2.2. *Measurement of Brain Shape Index on Brain MRI*

We randomly selected 270 male brains from the database with birth year from 1920 to 1980 (age range from 19 to 79). The antero-postal and left-to-right diameters of the brain were measured automatically and ratio of them (brain shape index) was calculated.

2.3. *Correlation between Fiber Connection and Cerebral Glucose Metabolism*

Fifteen healthy volunteers (male 8, female 7, age 73.0±2.2 yr) participated. A 10 min emission scan was performed with a PET scanner from 45 min after the injection of 217±32 MBq of ^{18}F-FDG. MRI measurement was performed using a 1.5T system. A volumetric T1-weighted image (T1WI) was acquired using a MPRAGE sequence. DTI was acquired using a single-shot diffusion-weighted echo planar imaging with six sets involving diffusion gradients placed along non-collinear directions (b = 1000 seconds/mm^2) and another set without diffusion weighting (b = 0).

2.4. *Anatomical Network Analysis Using Regional Gray Matter Volume*

To study on the properties of anatomical network among decades from 20-70 between different gender, we examined the correlation matrix using graph-theoretical analysis.

To analyze the anatomical network, we made an anatomical connection matrix by using the regional gray matter volume (RGMV) of 56 regions. We computed the Pearson correlation coefficient (PRC) between RGMV across subjects in each group to construct the interregional correlation matrix (N×N, where N is the number of gray matter regions, here N=56). Each connection matrix (12 in total) can be converted to a binarized and undirected graph G by considering a threshold T (range from 0 to 1). For increasing values of the threshold, more and more edges will be lost and the resulting graphs will become sparse, leading to a decrease of the mean degree. As the correlation threshold reaches the maximum Tmax, the mean degree of the resulting network will be less than the log of the number of nodes (here $K_{min}<\log(N)=4.025$) and the properties of small-world will become unestimable. Since the structure of the graph is generally biased by the number of existing edges, statistical measures should be calculated on graphs of equal degree K. Therefore, threshold T was chosen such for each connection matrix (12 in total) that all the produced graphs had a fixed mean degree ($T_{K=i}$, i=5, 6, 7; K=5, 6, 7> K_{min}). Thresholding each connection matrix with $T_{K=i}$, we obtained the anatomical connection matrix described as the binarized and undirected graph G.

We calculated the 3 important metrics of the graph G, namely, the mean degree K, the clustering coefficient C and the characteristic path length L [2]. Network topology was said to correspond to a ''small world'' if the network's clustering coefficient is much greater than that of equivalent random controls $C>C_{rand}$ or $\gamma= C/C_{rand}>1$, while their path lengths are comparable $L\approx L_{rand}$ or

$\lambda = L/L_{rand} \approx 1$. The small-worldness σ_{sw} is defined as $\sigma_{sw} = \gamma / \lambda > 1$. Comparisons are carried out against populations of n =1000 degree-matched random networks [2,3].

3. Results and Discussion

3.1.1. *Age related brain volume change*

In this study, we analyzed 860 men and 840 women. Their age and characteristics were described in Table 1.

Table 1. Description of the subjects

	Men (n=860)	Women (n=840)
Age	44.5±17.8	47.6±15.1
Body mass index (BMI)	23.4 ± 3.2	22.4 ± 3.1
Systolic blood pressure (mmHg)	129.5±16.5	125.7±18.2
Diastolic blood pressure (mmHg)	78.6±11.2	75.1 ± 11.9
Smoking habit (%)	62	17.1
Drinking habit (%)	91	65.4
Hypertension (%)	15.4	11.3
Diabetes mellitus (%)	4.3	2.1
Hypercholesterolemia (%)	7.2	12.1
Ischemic heart disease (%)	2.3	1.3

Figure 3 shows brain volume change with age in each tissue segment. Volume of each tissue segment was expressed as the ratio to whole brain volume. Gray matter ratio, which is the gray matter volume divided by whole brain volume, linearly decreased with age. White matter ratio slightly increased in younger age and remained unchanged during aging. However, there was no significant volume change with age. CSF volume ratio increased with age [4] (Fig. 3), in particular later age.

Gray matter ratio decreased linearly with age. Therefore, a brain with mean gray matter ratio can be a representative brain for the age group. We calculated regression line for the gray matter ratio change with age. We defined "a normally aged brain for his/her age", which is a brain with gray matter ratio on the regression line (Fig. 3). This is a first objective criterion for normally aged brain. We determined the normal range of the gray matter ratio as the value within ±2SD from the mean value on the regression line.

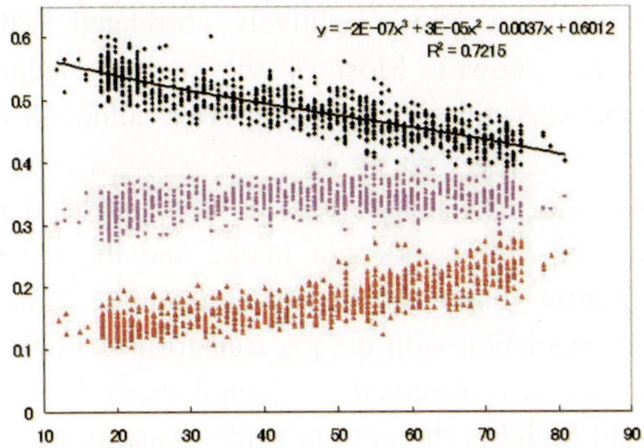

Figure 3. Age-related volume change of the human brain (gray matter: black, white matter: pink, CSF space: red)

3.1.2. *Longitudinal analysis of brain volume change during aging*

Brain volume change described above are obtained from cross sectional analysis and do not represent age related change in strict sense. We performed a longitudinal study in the same subjects between 8 years interval of MRI study. The results indicated that gray matter ratio for men decreased linearly with age as was in our previous cross sectional analysis. While those for women decreased slowly than that for men until the age of 50 and then went down as the similar slope for men (Fig. 4). White matter volume increased until age of 40 and went down thereafter in either men or women, although variation of the data was large (data are not shown).

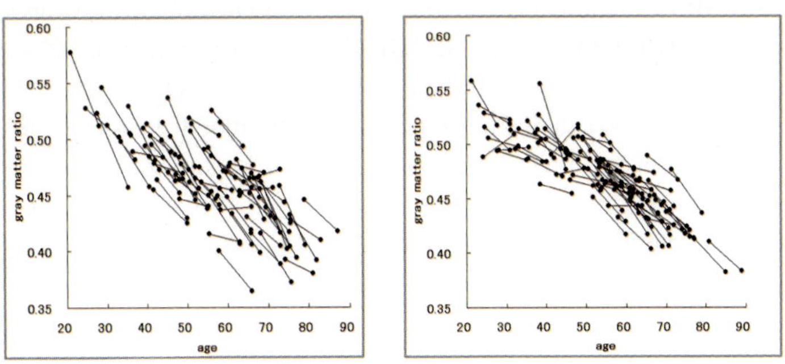

Figure 4. Longitudinal analysis of brain volume change with aging (right: women, left: men)

3.2. *Risk Factors for Brain Volume Decrease*

We analyzed correlation between regional gray matter volume and subject's characteristics using VBM technique. We found that total gray matter volume and regional gray matter volume negatively correlated with systolic blood pressure [2] (data not shown). Most of the regions mainly distributed in watershed regions between major cerebral arteries, although other region was included.

Figure 5 shows the gray matter regions that had a significant negative correlation between the lifetime alcohol intake and the regional gray matter volume. The gray matter volume of the bilateral middle frontal gyri showed a significant negative correlation with the log transformed lifetime alcohol intake [5]. These regions are also observed in alcohol dependent subjects, although hippocampus was not found in the present study. Amount of alcohol intake and severity of alcoholism may affect the gray matter loss in the hippocampus.

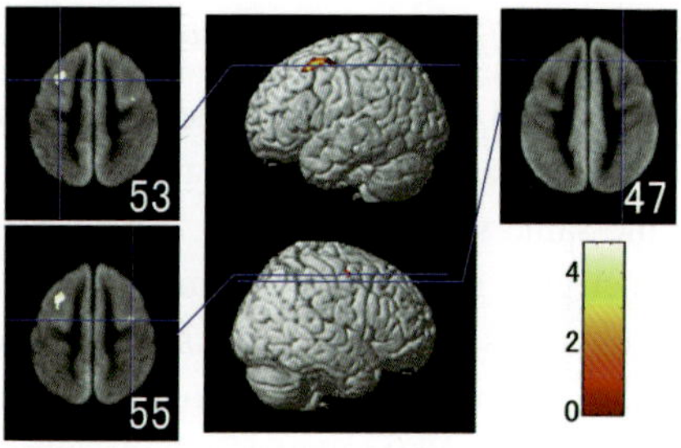

Figure 5. Correlation between gray matter volume and life time alcohol intake

We tested correlation between gray matter ratio and obesity. As an indicator of obesity, we used body mass index (BMI). Volumetric analysis revealed that there are significant negative correlation between BMI and the gray matter ratio only in men ($p < 0.001$, adjusting for age, systolic blood pressure, and lifetime alcohol intake). On the other hand, we could not find any correlation in women [4]. In men, the regional gray matter volume of the bilateral medial temporal lobe, bilateral anterior lobes of the cerebellum, bilateral fusiform gyrus, bilateral frontal lobes, bilateral precuneus, and midbrain showed significant negative correlations with BMI (Fig. 6) [6]. In addition, the regional gray matter volume of several regions shown as blue area in Fig. 6 showed significant positive correlations with BMI, after adjusting for age, lifetime alcohol intake, history of hypertension, and diabetes mellitus in men. However, biological significance of these results remained unknown.

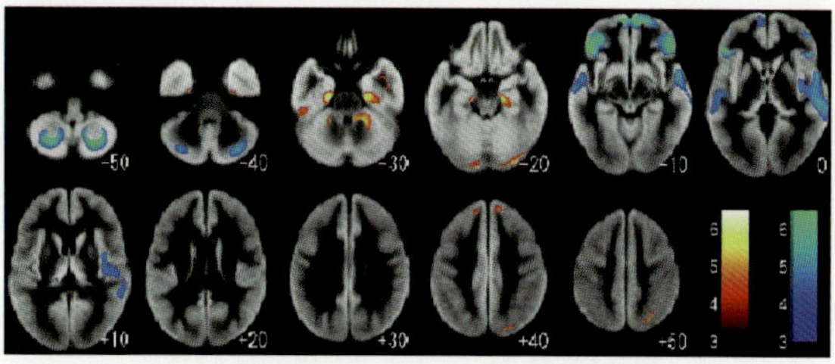

Figure 6. Brain regions that showed negative (red) and positive (blue) correlations between gray matter volume and body mass index (BMI) (Taki Y *et al.*, *Obesity* **16**, 2008)

3.3. *Cross Generational Differences in Brain Shape*

Figure 7 shows mean values of brain shape index against birth year of the subjects. The value of brain shape index was small in older generation, increased in younger generation and reached plateau in around 1960-1970. The similar trend was observed for the change of cephalic index, which is a ratio of length and breadth of the head. Kochi [7] reported the cephalic index in Japanese gradually increased during recent 100 years. However the index remained constant during 1960 to 1980 and she concluded that the increase ceased around year of 1960. This suggests that the shape of head synchronizes with the shape of brain. Biological meaning of the temporal change of brain shape in the secular time scale is not unclear. Further investigation is required.

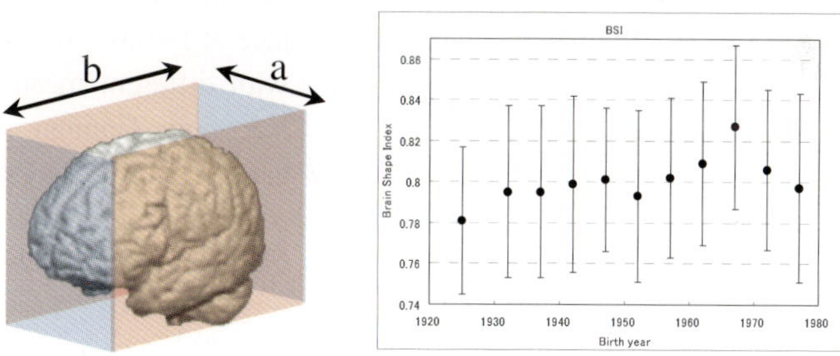

Figure 7. The secular changes of brain shape index (a/b) in Japanese male

3.4. *Correlation between Fiber Connection and Cerebral Glucose Metabolism*

There was no statistically significant correlation of FA with age, FA with GM concentration, or GM concentration with age. On the other hand, we found statistically significant positive correlation of ^{18}F-FDG accumulation in the lateral frontal cortex bilaterally with FA of the genu of the corpus callosum (Fig. 8). There was no age effect in FDG accumulation [8]. The results suggest that neuronal activity in the frontal cortices may decrease with the disruption of the microstructures of the CC without corresponding gray matter atrophy.

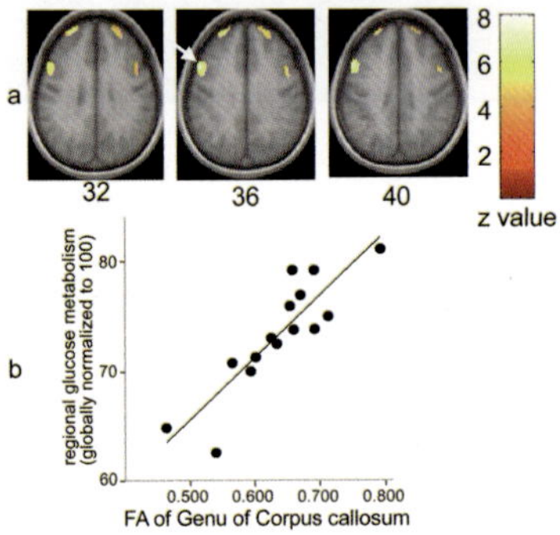

Figure 8. Areas that showed a significant positive correlation of glucose metabolism with the FA of the genu of the corpus callosum (a). A relationship between glucose metabolism and the FA on the right middle frontal gyrus (arrow) (b) (Inoue K *et al.*, *Human Brain Mapping* **29**, 2008)

3.5. Gender and Among Decade Differences in Anatomical Network Pattern

There exist significant statistical difference in all the small-wolrd properties by the group of the fixed mean degree (K=5, 6, 7). Gamma and Lamda show the significance at P<0.05. Sigma shows the significance at P<0.01. In addition, we analyzed the small-world properties by gender with different mean degree K (K=5, 6, 7). With the fixed mean degree K=5, only Gamma shows the significance at P<0.01. Gamma and Sigma show the significance at P<0.05 both with K=6 and K=7. Moreover, Lamda from K=6 has higher F-score than that from K=7. In conclusion, the choice for K=6 was identical to this study.

Table 2. Small-Worldness in each gender

	Female		Male			
	Mean	S. D.	Mean	S. D.	F	Sig.
C	.5136	.0365	.6278	.0818	9.741	.011 *
L	2.6818	.2119	3.0549	.4526	3.345	.097
Gamma	2.4137	.2300	3,7912	1.0722	9.468	.012 *
Lamda	1.0885	.0880	1.2540	.1882	3.809	.080
Sigma	2.2353	.3256	3.0444	.7386	6.030	.034 *

▲ K=6;

* Correlation is significant at the 0.05 level (2-tailed).

With the fixed mean degree K=6, the results by the group of gender show that significant difference exist in C, Gamma, Sigma between female and male at level of $P<0.05$, with the probability of 0.011, 0.012, 0.034 respectively (Table 2).

4. Conclusion

In this brain image database project, we aimed to understand normal (healthy) brain aging through the analysis of structural change of the human brain MRI, and to clarify the risk factors which accelerate brain aging. Finally, we intended to give the results back to clinical medicine and apply it to the prevention of brain aging.

Acknowledgments

This study was supported by the 2007 Tohoku University Global Centre of Excellence (GCOE) Program (Ministry of Education, Culture, Sports, Science and Technology; MEXT), entitled “Global Nano-Biomedical Engineering Education and Research Network Centre”.

References

1. K. Sato, Y. Taki, H. Fukuda and R. Kawashima, Neuroanatomical database of normal Japanese brains. *Neural Networks* **16**, 1301–1310 (2003).
2. Y. He, Z. J. Chen and A. C. Evans, *Cerebral Cortex*. **17**, 2407 (2007).
3. O. Sporns, C. J. Honey and R. Kötter, *PLoS ONE*. **2**, e1049 doi:10.1371/journal.pone.0001049 (2007).
4. Y. Taki, R. Goto, A. Evans *et al.*, Voxel-based morphometry of human brain with age and cerebrovascular risk factors. *Neurobiol Aging* **25**, 455–463 (2004).
5. Y. Taki, S. Kinomura, K. Sato *et al.*, Both global gray matter volume and regional gray matter volume negatively correlate with lifetime alcohol intake in non-alcohol-dependent Japanese men: A volumetric analysis and a voxel-based morphometry. *Alcohol Clin Exp Res* **30**, 1045–1050 (2006).
6. Y. Taki, S. Kinomura, K. Sato *et al.*, Relationship between body mass index and gray matter volume in 1,428 healthy individuals. *Obesity* **16**, 119–124 (2008).
7. M. Kouchi, Brachycephalization in Japan has ceased. *American Journal of Physical Anthropology* **112**, 339–347 (2000).

8. K. Inoue, H. Ito, S. Uchida *et al.*, Decrease in glucose metabolism in frontal cortex associated with deterioration of microstructure of corpus callosum measured by diffusion tensor imaging in healthy elderly. *Human Brain Mapping*. **29**, 375–384 (2008).

ELASTICITY-BASED TISSUE CHARACTERIZATION OF ARTERIAL WALL

HIDEYUKI HASEGAWA [1,2], KENTARO TSUZUKI [2], MASATAKA ICHIKI [3], FUMIAKI TEZUKA [4], HIROSHI KANAI [2,1]*

1) Graduate School of Biomedical Engineering, Tohoku University, 6-6-05 Aramaki-aza-Aoba, Aoba-ku, Sendai 980-8579, Japan

2) Graduate School of Engineering, Tohoku University, 6-6-05 Aramaki-aza-Aoba, Aoba-ku, Sendai 980-8579, Japan

3) Sendai Hospital of East Railway Company, 1-1-5 Itsutsubashi, Aoba-ku, Sendai 980-8508, Japan

4) Sendai Medical Center, 2-8-8 Miyagino, Miyagino-ku, Sendai 983-8520, Japan

Pathological changes in arterial walls significantly influence their mechanical properties. We have developed a correlation-based method, the *phased-tracking method*, for measurement of the regional elasticity of the arterial wall. Using this method, elasticity distributions of lipids, blood clots, fibrous tissue, and calcified tissue were measured by *in vitro* experiments of excised arteries (mean ± SD: lipid, 89 ± 47 kPa; blood clot, 131 ± 56 kPa; fibrous tissue, 1022 ± 1040 kPa; calcified tissue, 2267 ± 1228 kPa). It was found that arterial tissues can be classified into soft tissues (lipids and blood clots) and hard tissues (fibrous tissue and calcified tissue) on the basis of their elasticity. However, there are large overlaps between elasticity distributions of lipids and blood clots and those of fibrous tissue and calcified tissue. Thus, it was difficult to differentiate lipids from blood clots and fibrous tissue from calcified tissue when a threshold for a single elasticity value was set. Therefore, we developed and optimized a tissue classification method using the elasticity distribution in each small region. In this method, the elasticity distribution of each small region of interest (ROI) (not a single pixel) in an elasticity image is used to classify lipids, blood clots, fibrous tissue, and calcified tissue by calculating the likelihood function for each tissue. The ratio of correctly classified pixels to the total number of classified pixels was 29.8% when the size of a small region was 75 μm × 300 μm (a single pixel), but became 54.2% when the size of a small region was 1,500 μm × 1,500 μm (100 pixels).

1. Introduction

Noninvasive measurement of elasticity of the arterial wall is useful for diagnosis of atherosclerosis because there are significant differences between the elasticity of a normal arterial wall and that of an arterial wall affected by atherosclerosis

* Hiroshi Kanai is a Tohoku University Global COE Member.

[1]. In particular, mechanical properties of atherosclerotic plaque are important because plaque rupture may cause acute myocardial infarction and cerebral infarction [2-4]. Magnetic resonance imaging (MRI) and intravascular ultrasound (IVUS) are promising technologies for directly imaging plaque morphology [5,6]. On the other hand, the dynamic change of artery diameter due to the pulsation of the heart can be measured noninvasively by conventional methods with ultrasound [7-10]. Some parameters related to artery-wall elasticity can be obtained by the measured change in diameter of the artery [11-13]. However, in the derivation of these parameters, the artery is assumed to be a cylindrical shell with an uniform wall thickness and, thus, the elasticity of atherosclerotic plaque cannot be evaluated.

For measurement of the mechanical properties of the arterial wall, including the case with atherosclerotic plaque, we previously developed a method, namely, the *phased-tracking method*, for measuring small vibrations in the heart wall or arterial wall with transcutaneous ultrasound [14]. For a number of years, we have been measuring the displacement and small change in thickness of the arterial wall caused by the heartbeat using this method for elasticity imaging [15-17]. Elasticity images of the human carotid artery have been obtained by the measured distribution of changes in thickness, and the potential for transcutaneous tissue characterization has been shown by classifying the elasticity images using elasticity reference data obtained by *in vitro* experiments [16,18,19].

We have already measured the elasticity distributions for lipids, blood clots, fibrous tissue (mixture of the smooth muscle and collagen), and calcified tissue. In these previous studies, it was found that arterial tissues can be classified into soft tissues (lipids, blood clots) and hard tissues (fibrous tissue, calcified tissue) on the basis of their elasticity. However, it was difficult to differentiate lipids from blood clots and fibrous tissue from calcified tissue. We thus proposed a tissue classification method using the elasticity distribution in a small region [20]. In this method, the elasticity distribution of each small ROI (not a single pixel) in an elasticity image was used to classify lipids, blood clots, fibrous tissue, and calcified tissue. The precision of tissue classification was improved using the elasticity distribution in each small region.

However, the accuracy of this method in relation to the size of an ROI has not yet been thoroughly investigated. In the present study, to determine the optimum size of an ROI, the accuracy of tissue classification (including calcified tissue) was quantitatively investigated in relation to the size of the ROI. In addition, in the proposed classification method, an ROI is classified into one of the four tissue components, *i.e.* lipids, blood clots, fibrous tissue, and calcified tissue, even when the maximum likelihood is low. In the present study, such a region is defined as an unclassified region by setting a threshold for the likelihood. From these investigations, tissue classification was much improved in comparison with that in the previous study [21,22].

2. Materials and Methods

2.1. *Experimental Setup and Specimens*

Change in the pressure inside an excised artery was realized by circulating a fluid using a flow pump. The fluid inside the artery and that circulating in the flow pump were separated by a rubber membrane to prevent the flow pump from being contaminated, and only the change in internal pressure propagated to the inside of the artery. The change in internal pressure was measured by a pressure transducer (Model 110-4, Camino, San Diego, CA, USA).

In the acquisition of ultrasonic echoes, excised arteries (eight iliac and ten femoral) were measured with a conventional 7.5 MHz linear-type ultrasonic probe (SSH-140A, Toshiba, Japan). The quadrature demodulated signals of RF echoes were acquired at 10 MHz at a frame rate of 200 Hz. The elasticity of the arterial wall was defined as the tissue strain calibrated by the average stress of the entire wall thickness, namely, the circumferential elastic modulus E_θ^h [17]. The strain distribution was obtained by applying the *phased-tracking method* to the measured demodulated signals [17]. During the ultrasonic measurement, a needle was attached to the external surface of the artery for identification of the measured section so that a pathological image of the same section could be obtained after the ultrasonic measurement.

2.2. *Tissue Classification Using the Likelihood Function*

Each pixel in an elasticity image was classified into one of 5 categories, namely, lipids, blood clots, fibrous tissue, calcified tissue, and unknown using the likelihood function $\{L_i\}$ (i = 1: lipid; 2: blood clot; 3: fibrous tissue; 4: calcified tissue) of the elasticity distribution in the small region around the pixel. To obtain the likelihood function $\{L_i\}$, the elasticity distribution of the i-th tissue was translated into the normal distribution to describe the probability distribution by the mean and the standard deviation as described below [20].

From *in vitro* experiments, the elasticity distribution of each tissue i was obtained as illustrated in Fig. 1(a). The elasticity distribution of the i-th tissue consists of J_i data points with the respective elastic moduli. Using all data of J_i

points (J_1: 228, J_2: 179, J_3: 19,121, J_4: 1,101) with the respective elastic moduli, the ascending sequence is constructed for tissue i as shown in Fig. 1(b). In this sequence, the j-th datum ($j = 1, 2, \ldots, J_i$) has the corresponding elastic modulus E_j ($E_j \leq E_{j+1}$), where j was termed the elasticity number. The probability distribution of each tissue was obtained by allocating all the data of J_i points of each tissue i to boxes of the normal distribution. The box numbers, $\{B_i\}$, of the normal distribution are determined so that the number of data in the box at each end is only one. As shown in Fig. 1(c), the number of data, $D_{i,h}$ ($h = 1, 2, \ldots, B_i$), included in box B_i is determined so as to follow the profile of the normal distribution. Thus, the (J_i / 2)-th datum is included in the box with the highest probability. By allocating all the data of J_i points of each tissue to boxes of the corresponding normal distribution, the mean elasticity $\bar{E}_{i,h}$ of the data included in each box was obtained.

ROIs were assigned to an elasticity image which was obtained by ultrasonic measurement. The likelihood function $L_i(m,n)$ is defined as the joint probability that all the elasticity values in ROI $R_{m,n}$ (center of ROI: n-th sampled point along the m-th beam) simultaneously belong in the i-th category as follows:

$$L_i(m,n) = \left(\prod_{(k,l)\in R_{m,n}} p_i(E_{k,l}) \right)^{1/N_0} \qquad (i = 1, 2, 3, 4), \tag{2.1}$$

where $p_i(E_{k,l})$ is the probability density which shows the probability that elasticity value $E_{k,l}$ in the k-th row and l-th column in an ROI belongs to the i-th tissue category, and N_0 denotes the number of pixels in an ROI $R_{m,n}$. The multiplier $1/N_0$ shows the geometric mean for compensation of the effect of the size of an ROI. The pixel at the center of an ROI is classified into the class which has the maximum likelihood.

In this classification, there may be a region which has an extremely small value for the maximum likelihood. Such regions are classified as unclassified regions by setting threshold T_o to the maximum likelihood. Thus, the category $C(R_{m,n})$, to which an ROI $R_{m,n}$ belongs, is expressed as follows:

$$C(R_{m,n}) = \begin{cases} \arg \max\limits_{1\leq i\leq 4} L_i(m,n) \\ \quad (\text{if} \max\limits_{1\leq i\leq 4} L_i(m,n) \geq T_o). \\ \text{unknown} \qquad\qquad (\text{otherwise}). \end{cases} \tag{2.2}$$

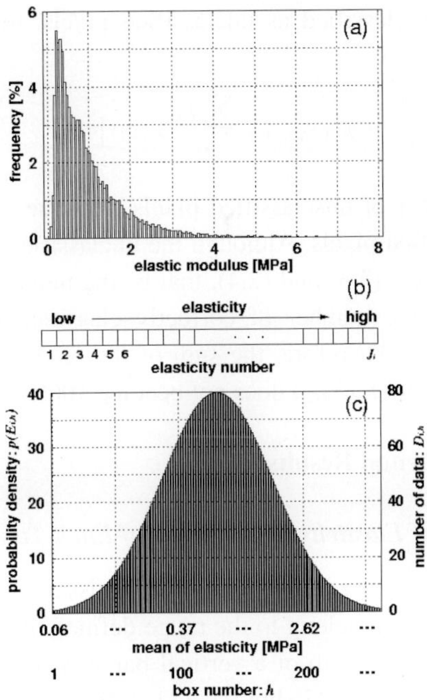

Figure 1. (a) Original elasticity distribution of the tissue. (b) Ascending sequence of the elastic modulus in an elasticity distribution. (c) Normal distribution whose number of boxes depends on the number of data points of (a).

By comparing the pathology-based classification images with the tissue classification images obtained by the proposed method, the recognition rate $R_r(S_{\mathrm{ROI}})$ for all tissues in the arterial wall was defined as the ratio of the number of correctly classified pixels to the number N of all pixels in the image as follows:

$$R_r(S_{\mathrm{ROI}}) = \frac{\sum_i N_i}{N} \times 100\,[\%], \tag{2.3}$$

where N_i is the number of correctly classified pixels of tissue i and S_{ROI} is the size of an ROI. Recognition rate $R_r(S_{\mathrm{ROI}})$ was used to determine the optimum size of an ROI.

For an ROI with a very low likelihood for all classes (i = 1, 2, 3, 4), the pixel which is located at the center of the ROI should be defined as an unclassified pixel by thresholding. For determination of the optimum threshold for the likelihood function, the false recognition rate $F_r(S_{\mathrm{ROI}})$ for all tissues in the arterial wall was defined as the ratio of the number of misclassified pixels, except for the pixels classified as unclassified pixels, to the number N of all pixels as follows:

$$F_r(S_{\text{ROI}}) = \frac{\sum_i F_i}{N} \times 100\,[\%], \tag{2.4}$$

where F_i is the number of misclassified pixels of tissue i, except for the pixels classified as unclassified pixels. Although the unclassified pixels are included in the denominator of Eqs. (2.3) and (2.4), that is, the number of all pixels N, they are not included in the number of correctly classified pixels N_i nor that of misclassified pixels F_i. Therefore, the sum of the recognition rate $R_r(S_{\text{ROI}})$ and the false recognition rate $F_r(S_{\text{ROI}})$ does not become 100%.

3. *In Vitro* Experimental Results

3.1. *Measurement of Elasticity Distribution of Each Tissue*

Figure 2 shows the elasticity distribution of each tissue, that is, the frequency of the elasticity values which belong to the range defined by the position and width of each vertical bar. The width of a vertical bar was set at 50 kPa. Means and standard deviations are 89 ± 47 (lipids), 131 ± 56 (blood clots), 1,022 ± 1,040 (fibrous tissue), and 2,267 ± 1,228 kPa (calcified tissue). Although similarities were found in the elasticity distributions of lipids and blood clots and in those of fibrous and calcified tissues, differences in the elasticity distributions of these tissues were also found.

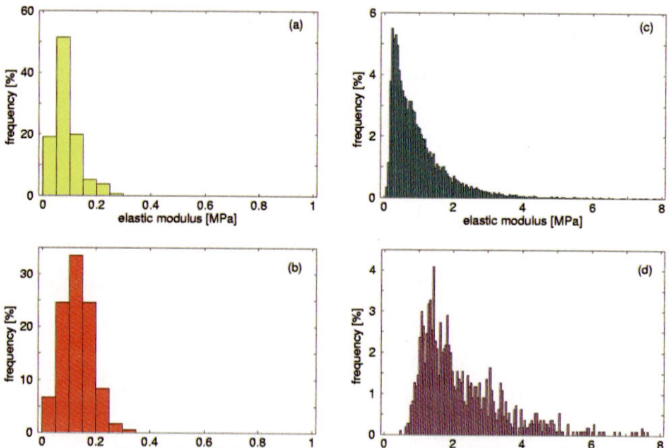

Figure 2. Elasticity distribution of each tissue. (a) Lipids (N = 288), (b) blood clots (N = 178), (c) fibrous tissue (N = 19,120), and (d) calcified tissue (N = 1,101).

3.2. *Results of Classification*

Figures 3(c) and 3(d) show an example of tissue classification results obtained by the proposed method for an iliac artery. The regions classified as lipids, blood clots, fibrous tissue, and calcified tissue were stained yellow, red, blue, and purple, respectively. Figure 3(c) graphically shows the tissue classification image obtained with an ROI size of 1 × 1 pixel. Although arterial tissues were roughly classified as soft tissues (lipids and blood clots) and hard tissues (fibrous tissue and calcified tissue), the classified tissue distributions are scattered, and the misclassified regions are prominent. Alternatively, Fig. 3(d) shows the result of classification with an ROI size of 1,500 μm (= 20 pixels) in the radial direction and 1,500 μm (= 5 pixels) in the longitudinal direction. Moreover, the region with low likelihood for all tissue components is colored gray. The threshold T_o for the maximum of the likelihood functions $\{L_i\}$ was set at 0.21. As shown in Fig. 3(d), the region with the maximum likelihood which is higher than threshold T_o is accurately classified as the corresponding tissue identified by referring to the pathological image.

Figure 4 shows the relationship between the size $S_{\rm ROI}$ of an ROI and the recognition rate $R_r(S_{\rm ROI})$. The ROI size $S_{\rm ROI}$ was changed with its square shape being maintained. The horizontal axis shows the width $W = (S_{\rm ROI})^{-1/2}$ of an ROI in the longitudinal direction. An ROI consists of a single pixel when width W in Fig. 4 is 0.3 mm. Only in this specific case is an ROI not square (75 μm × 300 μm).

Figure 4(b) shows the relationship between width $W = (S_{\rm ROI})^{-1/2}$ of an ROI in the longitudinal direction and the recognition rate $R_r(S_{\rm ROI})$ in arteries which are composed of a single type of tissue, such as fibrous tissue. In such case, the recognition rate $R_r(S_{\rm ROI})$ is monotonically improved by increasing the size of an ROI because an elasticity image is uniformly classified as the corresponding tissue using a large ROI, which results from the worsening spatial resolution in tissue classification. Figure 4(a) shows the relationship between width W of an ROI and the recognition rate $R_r(S_{\rm ROI})$ in arteries composed of different types of tissues. For this case, tissue classification using a certain number of pixels in an ROI is superior to that using a single pixel. However, the improvement of tissue classification by the enlargement of an ROI is limited because the classification using a large ROI provides a uniform tissue classification image, whereas the arterial wall is composed of different kinds of tissues. Therefore, there should be an optimum size of an ROI. As shown in Fig. 4(a), the recognition rates became maximum in most arteries when the size of an ROI was 1,500 μm × 1,500 μm.

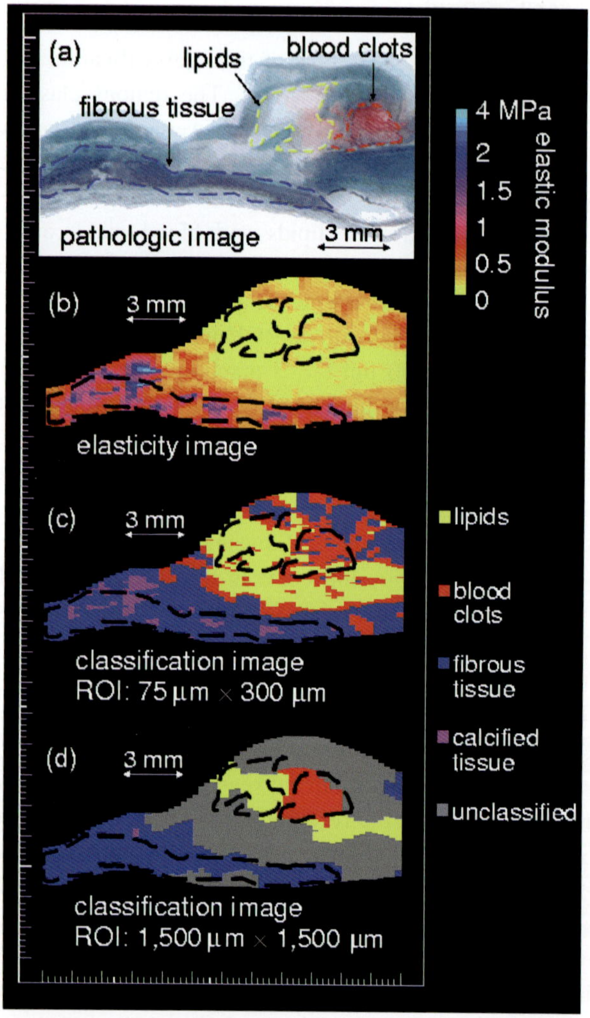

Figure 3. Example of tissue classification. (a) Pathological image of an arterial wall subjected to elastica-Masson staining. (b) Elasticity image. (c) Tissue classification image (ROI size: 1 × 1 pixel). (d) Tissue classification image (ROI size: 5 × 20 pixels).

In this study, the optimum threshold T_o was determined by considering the ratio (CMR(T_o)) of the number of correctly classified pixels (the numerator of Eq. (2.3)) to the number of misclassified pixels (the numerator of Eq. (2.4)) for all arteries composed of different types of tissues, which was evaluated as follows:

$$\mathrm{CMR}(T_o) = \frac{\sum_i N_{i,\mathrm{all}}}{\sum_i F_{i,\mathrm{all}}}, \tag{3.1}$$

where $N_{i,\mathrm{all}}$ and $F_{i,\mathrm{all}}$ are the sum of correctly classified pixels of tissue i and that of misclassified pixels of tissue i, respectively, for all arteries composed of different types of tissues. Figure 5 shows the relationship between threshold T_o for the likelihood function and the average CMR(T_o) of all arteries composed of different types of tissues. As shown in Fig. 5, the CMR(T_o) reached the maximum when the threshold T_o was 0.21.

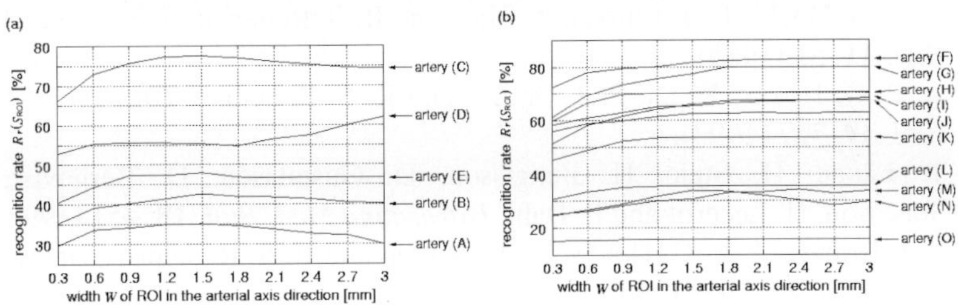

Figure 4. Relationship between width W of an ROI in the longitudinal direction and the recognition rate $R_r(S_{\mathrm{ROI}})$. (a) Arteries composed of several types of tissues. (b) Arteries composed of a single tissue. Each line shows the recognition rate $R_r(S_{\mathrm{ROI}})$ of the corresponding artery.

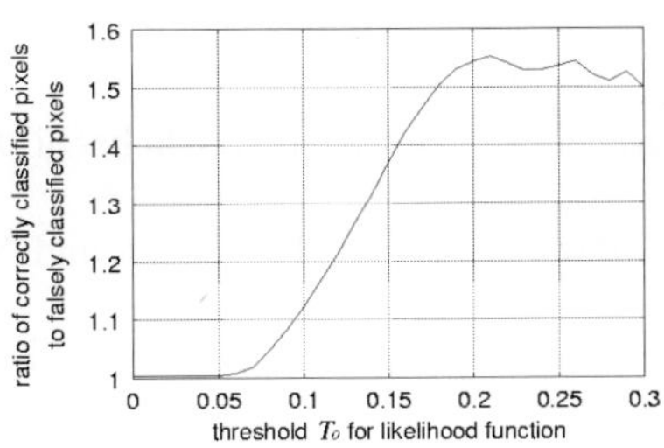

Figure 5. Relationship between threshold T_o for the likelihood function and the ratio of the number of correctly classified pixels to the number of misclassified pixels in all arteries composed of several types of tissues.

4. Conclusions

In this study, tissue classification based on the likelihood function with the configured appropriate ROI size (not a single pixel) and a lower limit of likelihood were investigated. Using the elasticity distribution of an ROI, the differentiation of lipids from blood clots and that of fibrous tissue from calcified tissue were improved.

References

1. R. T. Lee, A. J. Grodzinsky, E. H. Frank, R. D. Kamm and F. J. Schoen, *Circulation* **83**, 1764 (1991).
2. E. Falk, P. K. Shah and V. Fuster, *Circulation* **92**, 657 (1995).
3. M. J. Davies, *Circulation* **94**, 2013 (1996).
4. J. Golledge, R. M. Greenhalgh and A. H. Davies, *Stroke* **31**, 774 (2000).
5. M. V. McConnell, M. Aikawa, S. E. Maier, P. Ganz, P. Libby and R. T. Lee, *Arterioscler. Thromb. Vasc. Biol.* **19**, 1956 (1999).
6. B. N. Potkin, A. L. Bartorelli, J. M. Gessert, R. F. Neville, Y. Almagor, W. C. Roberts and M. B. Leon, *Circulation* **81**, 1575 (1990).
7. A. P. G. Hoeks, C. J. Ruissen, P. Hick and R. S. Reneman, *Ultrasound Med. Biol.* **11**, 51 (1985).
8. A. P. G. Hoeks, X. Di, P. J. Brands and R. S. Reneman, *Ultrasound Med. Biol.* **19**, 727 (1993).
9. T. Länne, H. Stale, H. Bengtsson, D. Gustafsson, D. Bergqvist, B. Sonesson, H. Lecerof and P. Dahl, *Ultrasound Med. Biol.* **18**, 451 (1992).
10. P. J. Brands, A. P. G. Hoeks, M. C. Rutten and R. S. Reneman, *Ultrasound Med. Biol.* **22**, 895 (1996).
11. D. H. Bergel, *J. Physiol.* **156**, 445 (1961).
12. L. H. Peterson, R. E. Jensen and J. Parnel, *Circ. Res.* **8**, 622 (1960).
13. K. Hayashi, H. Handa, S. Nagasawa, A. Okamura and K. Moritake, *J. Biomech.* **13**, 175 (1980).
14. H. Kanai, M. Sato, Y. Koiwa and N. Chubachi, *IEEE Trans. Ultrason. Ferroelect. Freq. Contr.* **43**, 791 (1996).
15. H. Hasegawa, H. Kanai, Y. Koiwa and N. Chubachi, *Electron. Lett.* **33**, 340 (1997).
16. H. Kanai, H. Hasegawa, M. Ichiki, F. Tezuka and Y. Koiwa, *Circulation* **107**, 3018 (2003).
17. H. Hasegawa and H. Kanai, *IEEE Trans. Ultrason. Ferroelect. Freq. Contr.* **53**, 2050 (2006).
18. H. Hasegawa, H. Kanai, N. Hoshimiya and Y. Koiwa, *J. Med. Ultrason.* **31**, 81 (2004).
19. J. Inagaki, H. Hasegawa, H. Kanai, M. Ichiki and F. Tezuka, *Jpn. J. Appl. Phys.* **44**, 4593 (2005).
20. J. Inagaki, H. Hasegawa, H. Kanai, M. Ichiki and F. Tezuka, *Jpn. J. Appl. Phys.* **45**, 4732 (2006).
21. K. Tsuzuki, H. Hasegawa, H. Kanai, M. Ichiki and F. Tezuka, *Ultrasound Med. Biol.* **34**, 573 (2008).
22. K. Tsuzuki, H. Hasegawa, H. Kanai, M. Ichiki and F. Tezuka, *Jpn. J. Appl. Phys.* **47**, 4180 (2008).

RADIOLOGIC ANATOMY OF THE RIGHT ADRENAL VEIN: PRELIMINARY EXPERIENCE WITH MDCT

SHOKI TAKAHASHI *, KEI TAKASE, TOMONORI MATSUURA

Department of Diagnostic Radiology, Tohoku University School of Medicine, 1-1 Seiryo-machi, Aoba-ku, Sendai, Miyagi, 980-8574, Japan

The purpose of our study was to determine how frequently the right adrenal vein could be identified on MDCT and the spectrum of anatomic variations seen in the right adrenal vein. Contrast-enhanced MDCT obtained in 104 patients using an 8-MDCT scanner was reviewed. The following points regarding the right adrenal vein were evaluated: degree of visualization; relationship to accessory hepatic or other veins; anatomy, including location of the orifice in relation to the surrounding structures. The right adrenal vein was detected in 79 (76%) of 104 patients. The right adrenal vein formed a common trunk with the accessory hepatic vein in six (8%) of the 79 patients. The orifice was craniocaudally located between the level of vertebrae T11 and L1. MDCT enabled the identification of the right adrenal vein and delineation of its anatomy, including its position and relationship to surrounding structures.

1. Introduction

Primary aldosteronism is the most common form of secondary hypertension [1]. Unilateral aldosterone-producing adenoma and bilateral idiopathic hyperaldosteronism are the two most common subtypes of primary aldosteronism. Distinguishing between the two is critical for treatment planning because the former is treated with adrenalectomy and the latter is treated medically [2]. For that purpose, adrenal venous sampling is often undertaken [3].

Catheterization of the right adrenal vein for selective adrenal venous sampling generally remains difficult, whereas catheterization of the left adrenal vein is a simple procedure [3]. The difficulty associated with catheterization of the right adrenal vein appears to result from its small size and variable anatomy: It is a vein that usually drains directly into the inferior vena cava (IVC) at a variable angle [4,5].

Multi-detector CT (MDCT) could possibly guide adrenal venous sampling if it were capable of delineating the anatomy of the right adrenal vein. A detailed analysis regarding visualizing the anatomy of the right adrenal vein with MDCT

* Shoki Takahashi is a Tohoku University Global COE Member.

has never been made to our knowledge. We undertook this study to determine how frequently the right adrenal vein can be identified on MDCT and what spectrum of anatomic variations is seen.

2. Materials and Methods

2.1. *Patients*

We performed a retrospective analysis of CT images from 104 consecutive patients (69 men, 35 women; mean age, 67 years) who underwent contrast-enhanced MDCT. Informed consent for contrast-enhanced CT had been obtained from all patients before their CT examination.

2.2. *CT Examinations*

The scanner was an Aquilon 8-MDCT scanner (Toshiba). Scans were obtained with the following parameters: 0.5 second per rotation, 1-mm collimation. Patients were asked to hold their breath for approximately 40 seconds during scanning immediately after the inhalation of oxygen.

Before scanning was started, 100 mL of contrast material containing 300 mg I/mL ([iopamidol] Iopamiron, Schering) was injected into an antecubital vein at a rate of 3.5 mL/s.

Transverse sections were reconstructed with a 1-mm section thickness at 0.5-mm intervals. We used the image data of a 15-cm field of view for retrospective evaluation of the right adrenal vein in this study.

2.3. *Definition of the Right Adrenal Vein on CT Images*

The extraglandular part of the right adrenal vein was identified according to the following criteria: an enhanced tubular or linear structure that arose from the right adrenal gland and eventually entered the IVC either directly or indirectly.

2.4. *Points of Evaluation*

We evaluated the following points regarding the right adrenal vein: the degree of visualization. For describing the direction of the right adrenal vein, we used a 3D coordinate system and aligned the rectangular coordinate axes with the body axes (*i.e.*, anteroposterior = x-axis, transverse = y-axis, and vertical = z-axis).

Degree of visualization of the right adrenal vein - The rate of visualization and number of the right adrenal veins were checked. The degree of visualization was arbitrarily graded on a 5-point scale as: excellent, good, fair, poor, and none. We defined the former three groups (excellent, good, and fair) as those with an unequivocally identified right adrenal vein in which the following anatomic analyses were made; the latter two groups were excluded from the following analyses.

Relationship of the right adrenal vein to accessory hepatic or other veins - We examined whether a common trunk for the right adrenal vein and the accessory hepatic or other veins was formed before entering the IVC.

Location of the right adrenal vein orifice in relation to surrounding structures - The craniocaudal level of the right adrenal vein orifice was specified relative to vertebral bodies and disks. The position of the orifice was also evaluated along the circumference of the IVC as an angle - θ - in the xy plane (Fig. 1).

Length and diameter of the right adrenal vein - We measured the length of the right adrenal vein between its exit from the right adrenal gland and its entry into the IVC, and its diameter at the junction with the IVC.

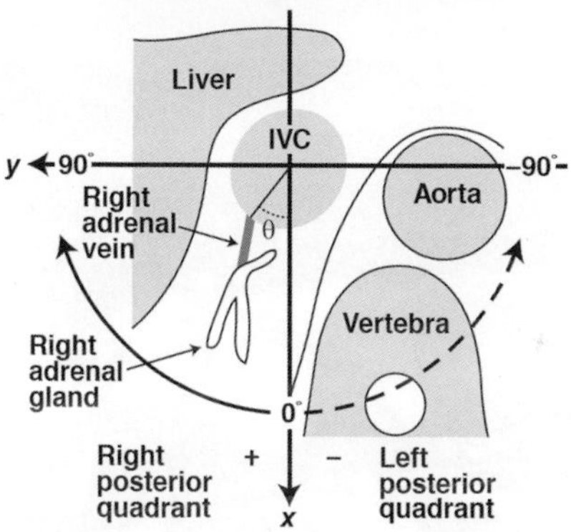

Figure 1. Position of right adrenal vein orifice and direction of right adrenal vein. Transverse plane seen inferiorly shows position of right adrenal vein orifice, which was estimated along circumference of inferior vena cava (IVC) as an angle - θ - between radius through orifice and x-axis in xy plane. Intersection of x- and y-axes is specified as center of IVC. Angle θ was measured positively in clockwise direction from x-axis, with a negative angle being in counterclockwise direction. When θ falls between 0° and 90°, right adrenal vein is described as entering IVC in right posterior quadrant; when θ falls between -90° and 0°, right adrenal vein is described as entering in left posterior quadrant.

3. Results

Degree of Visualization of the Right Adrenal Vein The degree of visualization was excellent in 36 (35%; Fig. 2A) patients, good in 30 (29%; Fig. 2B), fair in 13 (13%; Fig. 2C), poor in seven (7%), and none in 18 (17%) of 104 patients. Therefore, the right adrenal vein was detected in 79 (76%) of 104 patients according to our identification criteria (excellent to fair).

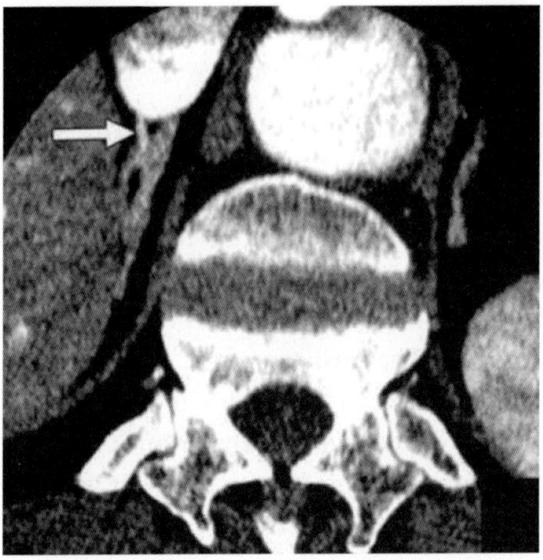

A

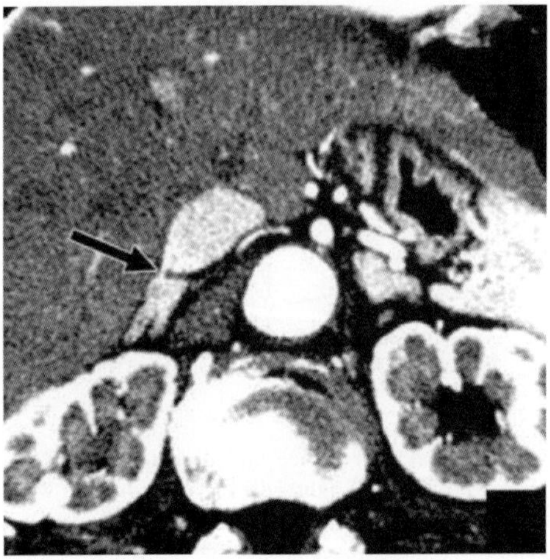

B

Figure 2. Examples of right adrenal vein show variable degrees of visualization. (A) 54-year-old man with aortic dissection and right adrenal mass. Paraaxial multiplanar reformatted image shows excellent visualization of right adrenal vein (arrow) running through intervening adipose tissue to join right posterior quadrant of inferior vena cava (IVC). Length of right adrenal vein is 2.9 mm. (B) 87-year-old woman with thoracic and abdominal aortic aneurysm. Paraaxial multiplanar reformatted image shows right adrenal vein (arrow) and right adrenal gland located close to IVC. Although right adrenal vein is a small structure, measuring only 2.9 mm in length, good visualization of it is attained because of dense enhancement. (C) 54-year-old man with aortic dissection. Paraaxial multiplanar reformatted image shows right adrenal gland close to IVC. Linear structure connecting adrenal gland with IVC displays only slightly denser enhancement than right adrenal gland and is regarded as right adrenal vein (arrow). Visualization is graded fair. Length of right adrenal vein is 3.5 mm.

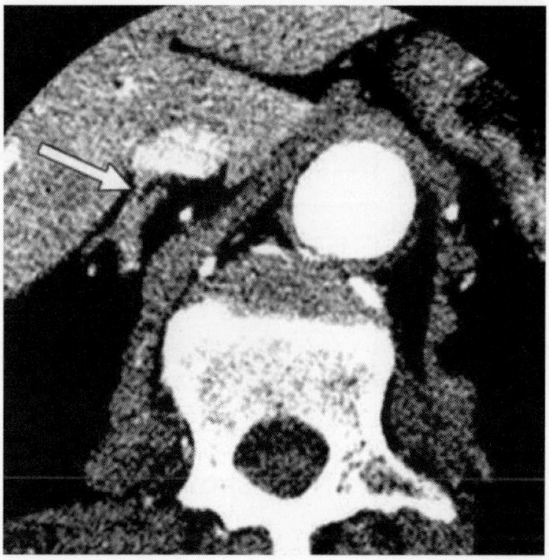

C

Figure 2. (*Continued*)

Relationship of the Right Adrenal Vein to Accessory Hepatic or Other Veins Among the 79 patients who had an identifiable right adrenal vein, the right adrenal vein and accessory hepatic vein formed a common trunk before entering the IVC in six (8%) patients, the right adrenal vein entered the IVC directly but almost shared a common orifice with an accessory hepatic vein in seven (9%), and it entered the IVC independently of other veins in the remaining 66 (84%) patients. In the six patients with a common trunk, the mean length of common trunk was 4.7 +/- 2.0 mm.

Location of the Right Adrenal Vein Orifice in Relation to Surrounding Structures The orifice was craniocaudally located between the level of the T11 and L1 vertebrae (Fig. 3A). In 50 (69%) of the 73 patients, the right adrenal vein joined the IVC at a level ranging from the middle third of T12 to the superior third of L1.

Angle θ, representing the position of the orifice along the circumference of the IVC, is shown in Fig. 3B. Angle θ ranged from -7° to 71° (mean, 39° ± 16°). The right adrenal vein joined the IVC in the right posterior quadrant in 71 (97%) of the 73 patients, most frequently in a range of angulation between 20° and 60°.

Length and Diameter of the Right Adrenal Vein The mean length was 3.8 ± 1.7 mm, and the mean diameter at the junction with the IVC was 1.7 ± 0.6 mm.

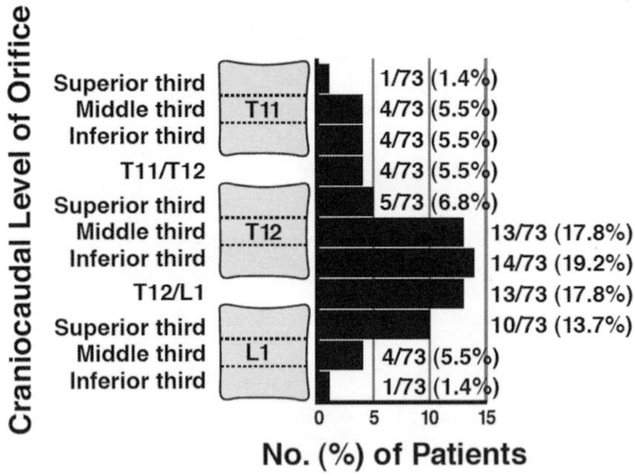

A

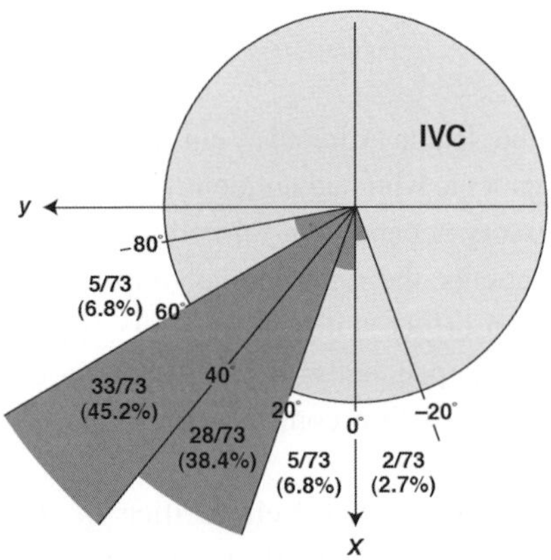

B

Figure 3. Location of right adrenal vein orifice and direction of right adrenal vein from inferior vena cava (IVC). Data are numbers (%) of right adrenal veins. (A) Craniocaudal level of orifice of right adrenal vein in relation to vertebral bodies and disks. (B) Position of orifice of right adrenal vein along circumference of IVC evaluated as angle θ (dark gray).

4. Discussion

MDCT enabled the identification of the right adrenal vein in most patients. The enhanced structure of the right adrenal vein was easily detected when it was surrounded by abundant adipose tissue, although its detection was difficult in patients having poor adipose tissue and only modest contrast enhancement of the right adrenal vein. The results concerning the length, diameter, and craniocaudal level of the orifice were all similar to those of the earlier studies using autopsy or venography [4,6].

The position of the orifice along the circumference of the IVC was most commonly in the right posterior quadrant of the IVC in our study, which is in accord with studies in the literature. A common trunk of the right adrenal vein with an accessory hepatic vein was found in 8% of the patients in our series. Recognition of such variations should be important preoperative information because the tip of the catheter might better be advanced selectively into the right adrenal vein.

Catheterization of the right adrenal vein remains difficult. Its success rate generally remains at approximately 60-70% [2,4]. The difficulty may arise because of the small size of the right adrenal vein, its anatomic variations in the junction with the IVC, or confusion with accessory hepatic veins that may form a common stem with the right adrenal vein [4,5].

Our study suggests the potential for MDCT to provide detailed information on the right adrenal vein anatomy similar to other small arteries [5]. Such information would be useful in planning adrenal venous sampling. For patients with suspected primary aldosteronism, MDCT with 1-mm collimation not only for imaging of the adrenal gland but also for mapping the right adrenal vein before venous sampling, which allows us to catheterize easily and efficiently.

In conclusion, MDCT enabled the identification of the right adrenal vein and delineation of its anatomy, including its position and relationship to the surrounding structures such as the IVC, in a high percentage of patients. This preoperative information may be helpful in the catheterization of the right adrenal vein for adrenal venous sampling.

References

1. R. D. Gordon, M. Stowasser, T. J. Tunny, S. A. Klemm and J. C. Rutherford, High incidence of primary aldosteronism in 199 patients referred with hypertension. *Clin Exp Pharmacol Physiol* **21**, 315–318 (1994).
2. W. F. Young, Jr., Primary aldosteronism: management issues. *Ann N Y Acad Sci* **970**, 61–76 (2002).
3. S. B. Magill, H. Raff, J. L. Shaker *et al.*, Comparison of adrenal vein sampling and computed tomography in the differentiation of primary aldosteronism. *J Clin Endocrinol Metab* **86**, 1066–1071 (2001).
4. J. K. Davidson, P. Morley, G. D. Hurley and N. G. Holford, Adrenal venography and ultrasound in the investigation of the adrenal gland: an analysis of 58 cases. *Br J Radiol* **48**, 435–450 (1975).

5. N. Daunt, Adrenal vein sampling: how to make it quick, easy, and successful. *RadioGraphics* **25** (Suppl 1), S143–S158 (2005).
6. C. G. Mikaelsson, Venous communications of the adrenal glands: anatomic and circulatory studies. *Acta Radiol Diagn (Stockh)* **10**, 369–393 (1970).

POTENTIALS OF NANO-BIO-IMAGING WITH POSITRON EMISSION TOMOGRAPHY AND RADIOPHARMACEUTICALS *

MANABU TASHIRO †

Division of Cyclotron Nuclear Medicine, Cyclotron and Radioisotope Center, Tohoku University, Sendai, Miyagi 980-8578, Japan

KAZUHIKO YANAI

Department of Pharmacology, Graduate School of Medicine, Tohoku University, Sendai, Miyagi 980-8575, Japan

"Nano-bio-imaging" encompasses various kinds of imaging techniques supported by "nanotechnology". Nuclear medicine techniques are very useful for nano-bio-imaging, where a very small amount of radiopharmaceuticals is injected into the body of the subjects to investigate "endo-phenotypic alterations". The initial system was called scintigraphy, and later progressive innovation fused this method with computed tomography, producing "single photon emission computed tomography (SPECT)", and "positron emission tomography (PET)". One of the most important merits of these techniques is their ability to visualize pharmacodynamic/kinetic information in the living tissue with only a tiny amount of radiopharmaceuticals. In this article, our research activities are reviewed in terms of 1) an application for evaluating intensities and elucidating the mechanisms of psychiatric disorders and drug side effects, as well as 2) an application to health sciences, using PET.

1. Introduction

1.1. *Nano-bio-imaging*

"Nano-bio-imaging" encompasses various kinds of imaging techniques supported by "nanotechnology". It enables further progress of research in nanomedicine and nanotechnology. In the nuclear medicine technique, one of the most useful techniques for nano-bio-imaging, we administer a very small amount of radiopharmaceuticals to investigate "endo-phenotypic alterations"

---

* Work partially supported by Grants-in-Aid for Scientific Research from the Japan Society of Promotion of Science (JSPS) and the Ministry of Education, Culture, Sports, Science and Technology in Japan, as well as by a grant from the Japan Society of Technology (JST) on research and education in "molecular imaging".

† Manabu Tashiro is a Tohoku University Global COE Member.

taking place in a living body from the outside. The foundation of this technique dates back to the early 20th century when Dr. George von Hevesy, a Nobel laureate in chemistry in 1943, originally developed it. This technique continued to make progress and from the latter half of the 20th century, it was more popularly referred to as nuclear medicine and was applied to human studies. The initial system was called scintigraphy, and subsequent progressive innovation continued. The technique was later fused with the technology of computed tomography, and has been established as "single photon emission computed tomography (SPECT)" and "positron emission tomography (PET)" in the late 20th century, both of which have been most suitable tools for "nano-bio-imaging". One of the most important advantages of these techniques have been the fact that they can visualize pharmacodynamic/pharmacokinetic information in the tissue of living subjects by injecting only a tiny amount of radiopharmaceuticals (1 nano-mole or so). Here, the term "tracers" mean radiopharmaceuticals to be injected for the purpose of obtaining signals regarding phenomena in the living body, and sometimes can also be termed "a probe", meaning that we can probe the presence of extremely small amounts of biological substances. Thus, in nuclear medicine, radio-labeling with high specific radioactivity is very important for visualizing the presence of small amounts of bio-active substances at a "nano-" to "pico-" mole level. Therefore, this technique is the right tool optimized for nano-bio-imaging.

1.2. *Information Available from the Living Human Brain*

Human mental function is based on the actions of our brain network. On one hand, an ultimate and fundamental aim of this study topic is the "elucidation of psychiatric functions of the human brain". In the modern society, the incidence of neuropsychiatric disorders such as depression, anxiety disorder and cognitive disorders (dementia) is constantly increasing. A large amount of "stress" in our daily lives has been associated with the increased rate of various psychiatric disorders. When we think about a structured strategy for overcoming such problems, we find that imaging technology can provide basic data for developing a practical strategy.

As for the available methodology for nano-bio-imaging, we can use PET to measure cerebral energy (glucose) metabolism by injecting a small amount of radio-labeled glucose ([^{18}F]fluorodeoxyglucose: FDG) (Fig. 1). Due to activated regional brain metabolism, the demand for glucose and oxygen increases, inducing dilation of brain capillaries, which is observed as an increase in regional cerebral perfusion. Cerebral perfusion can be measured using radiolabeled water ([^{15}O]H_2O: Fig. 1). More recently, a method using magnetic resonance imaging (MRI) has also been applied for the measurement of brain regional perfusion (functional MRI: fMRI). In addition, another technique using near-infrared light (NIRS) has been introduced for measuring brain regional perfusion. PET is still commonly used, although mainly for measuring regional brain glucose consumption and for evaluating neuro-transmission function (Fig. 2). In the human brain, neurotransmitters can manifest their effects even in very small amounts. It is not easy to visualize the actions of neurotransmitters in the living human brain externally without using a highly sensitive technique such as PET. In addition, with PET, it is possible to quantify interactions between neurotransmitters and neuroreceptors, as well as tissue metabolism in the living brain using the time course data of radioactivity in the blood (plasma time activity curve: pTAC) and brain tissue (tissue time activity curve: tTAC).

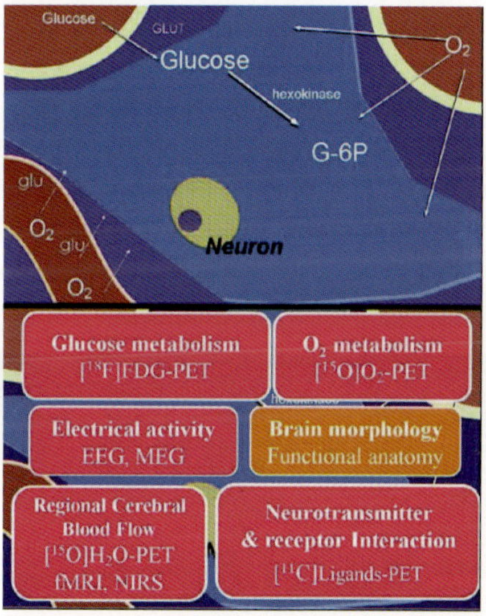

Figure 1. Information available from the living human brain. The most important energy resource of the human brain is glucose. Oxygen is necessary for glucose metabolism. These substances are supplied by the blood stream. Brain regions with increased activity are accompanied by increased regional cerebral blood flow. Information regarding glucose and oxygen metabolism can be obtained using PET. Presently, regional cerebral blood flow can be measured using various methods. The interactions of neurotransmitters and receptors can be measured mainly using PET.

2. Imaging of a Neural Transmission for Elucidation of Psychiatric Disorders and Drug Side Effects

2.1. *Functional Imaging of the Histaminergic Nervous System*

Stress in our daily lives has been associated with an increase in the number of various psychiatric disorders such as depression, schizophrenia, and cognitive disorders. To date, we have conducted many studies to elucidate the pathophysiological mechanism of the above-mentioned disorders [1-4], placing emphasis on alteration in neural transmission of the histaminergic nervous systems [1]. For this purpose, [^{11}C]doxepin, a tracer of choice for imaging histamine H1 receptors (H1Rs), has been a suitable tool, and increasing evidence has accumulated regarding the role of the histaminergic neuronal system in the pathophysiology of these disorders. Histamine H1R binding was measured using PET and [^{11}C]doxepin in 10 normal male subjects and 10 patients with schizophrenia [2] and major depression [3], as well as in 10 normal female subjects and 12 female patients with anorexia nervosa [4]. Interestingly, in these studies a significant reduction in H1R binding was observed in the patients with schizophrenia and major depression while a significant increase was observed in the patients with anorexia nervosa [4] (Fig. 2). Further investigation will elucidate the mechanism of these neuropsychiatric disorders in terms of stress responses in the brain histaminergic neuronal system [4]. Thus, PET is very useful for elucidation of the pathophysiological mechanisms underlying various disorders.

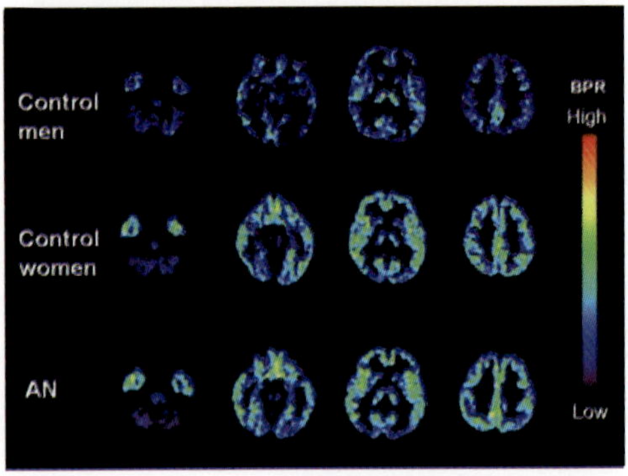

Figure 2. Brain distribution of [^{11}C]doxepin radioactivity in control subjects and anorexia nervosa (AN) patients. Modified from Ref. [4].

In addition, PET and [^{11}C]doxepin are also useful for the evaluation of drug-induced side effects and their mechanism. One of the most frequently used therapeutic drugs for allergies such as seasonal pollinosis (or "hay fever") is a H1R antagonist (antihistamine). There are many available antihistamines but some of them have sedative side effects. Therefore, it is important to develop an objective and reliable method for measuring the strength of such sedative side effects [1,5]. To date, we have studied the mechanism of functional suppression in signal transmission through H1Rs in the brain. Usually, antihistamines are used for suppressing the actions of mast cells in the peripheral blood and for controlling allergic reactions. However, some of these drugs may enter the brain and suppress the signal transmission of intra-cerebral H1Rs. As a result, it becomes difficult to maintain arousal (sedative effects), and sometimes these drugs might cause people to make mistakes during work or while driving, resulting in decreased work efficiency or traffic accidents. Considering such a background, objective measurement of the sedative effects of these drugs becomes very important.

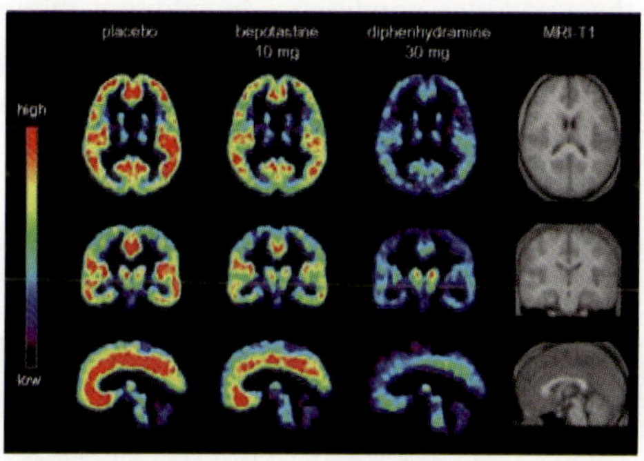

Figure 3. Binding potential ratio (BPR) images of [^{11}C]doxepin in the human brain. BPR of [^{11}C]doxepin was measured in healthy subjects using PET after oral administrations of placebo (left), bepotastine (10 mg) or diphenhydramine (30 mg) for each treatment condition were compared. Modified from Ref. [8] courtesy of Blackwell Publishing Company.

We have succeeded in quantifying the strength of the sedative effects of antihistamines in terms of H1R occupancy (rate) in the brain using PET, and we have measured this clinically. Previously, investigators performed macroscopic behavioral techniques such as measurement of psychomotor performance including psychomotor speed and accuracy as well as the measurement of subjective sleepiness using many volunteer subjects to evaluate drug sedative effects [6-9]. Recently, we have conducted a clinical test to evaluate the sedative profiles of bepotastine besilate, a new antihistamine developed in Japan [8]. The basic pharmaceutical classification of this antihistamine has been regarded as “a mildly sedative antihistamine”. We succeeded in obtaining supporting data regarding this classification using PET [8] (Fig. 3).

From the viewpoint of cognitive neuroscience, the category of cognitive function (psychomotor performance) measured in antihistamine studies is mainly “vigilance and attention”. The sedative side effects of antihistamines are recognized as potentially dangerous for daily tasks, such as car driving; however, the mechanisms underlying these effects have not yet been well elucidated We, therefore, attempted to elucidate the brain mechanism of impaired performance using a car-driving simulator and $[^{15}O]H_2O$ PET [9]. We examined regional cerebral blood flow (rCBF) responses during a simulated car-driving task following oral administration of d-chlorpheniramine using $[^{15}O]H_2O$ PET. Results of the performance evaluation revealed that one part of the driving performance (lane deviations) significantly increased under d-chlorpheniramine conditions compared with the placebo conditions, while subjective sleepiness was not significantly different between the two drug conditions. In addition, brain imaging analysis suggested that d-chlorpheniramine tended to suppress the regional brain activities associated with visuo-spatial cognition and visuo-motor coordination [9].

Thus, the nuclear medicine technique is very useful for the objective evaluation of intensities and mechanisms of the brain effects of various drugs. These data can be used to develop new drugs with reduced side effects.

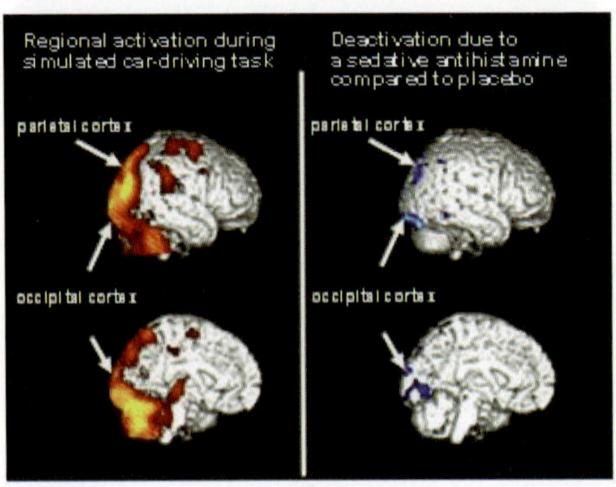

Figure 4. Results of PET analysis demonstrating regional activation during active driving compared with resting (yellow & red), and regions with diminished activation under the influence of sedative antihistamine (blue). Modified from Ref. [9] courtesy of John Wiley & Sons, Ltd.

2.2. *Imaging Study for Improvement of Quality of Life (QOL) in Patients*

In addition, functional imaging techniques such as PET have also been used for studies of neurological and psychiatric disorders in patients with physical diseases, such as malignant tumors. Relatively mild psychiatric abnormalities could also be detected using PET. Since cancer and cancer treatments have various effects on the central nervous system, the diagnosis of psychiatric symptoms in cancer patients is in part problematic. Nano-bio-imaging could be used as a supplementary diagnostic tool. Previously, we proposed the use of FDG-PET in the neuropsychiatric evaluation of cancer patients, and have since then been performing a series of studies to examine whether the images of a cancer patient's brain are in fact normal or not.

It is now widely accepted that psychological factors are equally important as external factors in disease progression. Thus, psychological evaluation and patient care are very important, not only for improving quality of life (QOL), but also for prolonging survival. If we suppose a certain psychological disturbance truly exists to the extent of affecting the systemic functions of patients, it is also possible for their brain activity to have significant alterations. Thus, it seems reasonable to think that cancer patients may manifest a corresponding abnormality in their functional brain images. Our preliminary works demonstrated regional hypometabolic findings, mainly in the prefrontal cortex, limbic structures and striatum of Japanese cancer patients [10]. Later, both Japanese and German patients manifested abnormal regional metabolism compared with benign disease patients [11]. Common findings in these studies were hypometabolism in the prefrontal cortex, anterior and posterior cingulate gyri, insular cortex and striatum. These regions showed similar changes to those demonstrated in previous neuroimaging studies of patients with major depression, demonstrating commonly repeated findings of hypometabolism in the prefrontal cortex, anterior cingulate gyrus and basal ganglia. Typical findings of hypometabolism are observed in the anterior cingulate gyrus, prefrontal cortex, lateral prefrontal cortex, basal ganglia, and the temporoparietal regions. Thus, considering the similarity to known lesions in major depression, it is possible to postulate that the regional metabolic reduction in cancer patients starts at very mild stages of depression. Even relatively mild disorders, such as adjustment disorder, would be accompanied by hypometabolism.

Our additional cross-sectional study in Japanese patients showed that the depth of hypometabolic findings tended to fluctuate in some regions and constantly decreased in other regions [12]. The hypometabolic levels of the orbitofrontal, basolateral prefrontal, and ventral anterior cingulated cortices and insula tended to be constantly low in all phases. Those of other regions, such as the prefrontal, anterior cingulated, and posterior cingulated cortices, as well as that of the subcortical nuclei tended to fluctuate. The fluctuations gave the impression that these findings were state-dependent, and more likely caused by non-organic factors. Finally, we demonstrated that regional hypometabolism in the prefrontal cortex negatively correlated with the subjective measure of depression (Zung's Self-Rating Depression Scale: SDS) in cancer patients. Although patients with negligible SDS scores manifested negligible hypometabolism in FDG brain images, those with mild or moderate SDS scores manifested significant hypometabolism in wide areas in the frontal, temporo-parietal, and cingulated cortices among others.

In particular, metabolic brain activity negatively correlated with SDS score [12]. Thus, PET is expected to be a useful tool for further evaluation of the psychological and behavioral aspects of cancer patients. Replications from other Japanese groups based on refined protocols are presently available. PET has shown its usefulness in the observation of effective therapy in consideration of a psychosomatic, and especially a psycho-immune, correlation [13]. The purpose of this study was to examine the relationship between psychological factors, regional brain activity, and natural killer cell activity (NKA). Eight patients with malignant diseases were studied using FDG PET under resting conditions. NKA and the degree of depression were measured using SDS, and anxiety was measured using Taylor's manifest anxiety scale (MAS). The linear correlation of NKA and psychological measures to regional brain metabolism in cancer patients was examined. A positive linear correlation between NKA and regional metabolic rate ratios was identified in the visual cortex, anterior cingulate gyrus and the sensorimotor area. A negative correlation was identified in the prefrontal, orbitofrontal and anterior temporal cortices. The NKA and MAS scores positively correlated with each other ($p<0.001$). The results might serve as supporting data for a hypothesis that psycho-immune interaction is also mediated by the cerebral cortex and limbic system.

3. Application of Nano-bio-imaging Technology to Health Science

3.1. *Definition of Health*

According to the definition given by the World Health Organization (WHO), "health" is "a complete state of physical, mental and social well-being, and not merely the absence of disease or infirmity." Needless to say, QOL is very important in all procedures of medical treatment and care. The definition suggests that attention should be paid to QOL in daily activities. A highly sensitive PET measurement system has enabled examinations at minimal radiological doses, below the annual environmental exposure (2.4 mSv/year). This has allowed PET to be applied to elucidating the mechanisms in our bodily functions that achieve and maintain a healthy state. This could be viewed as a new application of PET in the field of preventive medicine and health promotion. Our studies on brain activities of patients who suffer from malignant diseases have yielded interesting results. The brain metabolic pattern of cancer patients show abnormalities that correlate to the intensity of depressive mood. Also, the psychological state correlates to immune functions, possibly via regional brain functions [13]. PET appears to be useful not only for the diagnosis of cancer, but also for evaluation of the psychological status of cancer patients. Furthermore, we have conducted studies regarding the effects of exercise such as running and bicycle riding to observe regional activities not only in the brain [14-16], but also in the skeletal [17,18] and cardiac muscles [19]. PET is useful for evaluating the effects of various therapeutic interventions including exercise for metabolic disorders, such as diabetes mellitus, hypertension, cerebro- and cardiovascular diseases, in terms of interactions between the mind, brain and body. Recently, alternative medicine has become popular among patients, and the scientific evaluation of these therapeutic techniques would be of clinical importance. PET can be very useful for such evaluations based on scientific evidence.

In the future, we plan to investigate the relationship between psycho-behavioral factors and achieving a healthy state. Our ultimate goal is to establish a model of whole-body organ interactions.

3.2. *Exercise and Brain Imaging Related to "Physical and Mental Health"*

In the scope of applying PET to health promotion science, we have performed imaging studies during various forms of exercise, such as running and bicycle riding, scanning not only in the brain [14-16], but also in the skeletal [17,18] and cardiac muscles [19]. By conducting whole-body scanning, we can obtain a whole-body map of energy metabolism in the living human body [17]. To date, there are few study groups using PET for health promotion science, and this topic will have great relevance in the future. Here, it would be useful for readers to know about the brief history of brain imaging in sports science [15].

Regional cerebral metabolic changes induced by exercise was examined in animals employing an autoradiography technique using [^{14}C]deoxyglucose

([^{14}C]2-DG) as a tracer [20, 21]. These studies provided the first functional indices of brain activity with respect to exercise. [^{14}C]2-DG has been a useful tracer for exercise studies because it does not require the simultaneous scanning of subjects during an exercise [22]. Using this technique, Sharp and coworkers [23] demonstrated a regional increase in glucose uptake in the cerebellar vermis of swimming rats. A similar study on free-running rats showed no activation in the cerebellar vermis but in the entire cerebellum [24]. Human studies were conducted later. The first study on human brain activity during exercise, as far as we know, was conducted by Herholz and coworkers in the late 1980s, using the ^{133}Xe clearance method for studying regional changes in brain activity [25]. They demonstrated the largest increase in blood flow induced by an ergometer bicycle task in the frontal region, although spatial resolution was limited (a few centimeters). Later, Fink and coworkers demonstrated regional activation during and immediately after an ergometer task by PET using [^{15}O]H_2O. They succeeded in demonstrating activation in the motor cortex associated with leg and arm motion [26]. Because muscle fibers and motor neurons jointly form "motor units", [^{18}F]FDG uptake in muscles may correlate with activity in their corresponding cerebral regions. Mishina and coworkers applied [^{18}F]FDG-PET to the neuropathological evaluation of patients with olivo-pontine-cerebellar atrophy manifesting as gait disturbances, who exhibited decreased responses to a walking task in the cerebellar vermis compared with normal subjects [27].

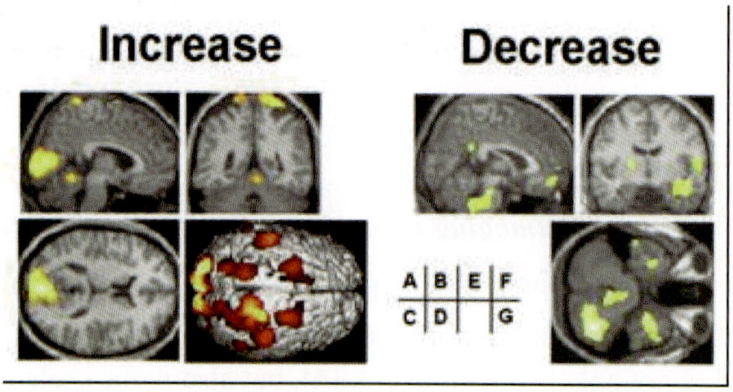

Figure 5. Relative glucose metabolic differences between Controls (resting) and Running subjects demonstrated in terms of adjusted regional glucose metabolic rate ratios. Brain regions of statistically significant differences are shown in this figure. In all areas, $p < 0.001$, compared Controls (resting) and Running subjects. Symbols: ** = Z score > 4.0, * = Z > 3.0. Modified from Ref. [11] courtesy of Minerva Medica.

Then, we first applied [^{18}F]FDG-PET to human subjects during a running task in the upright posture, and demonstrated augmented energy consumption in the parieto-occipital region during the task compared with the motor area [14]. This was probably due to the higher energy consumption necessary for integrating multimodal sensory information. Our results also showed that frontal activity was reduced during running compared with resting. In addition, our study showed a trend of relative decrease in whole brain mean activity during exercise compared with the resting condition. In parallel, our group also examined whole-body energy (glucose) redistribution and the effect of exercise. We showed no significant changes in relative glucose metabolism between exercise and resting conditions [28]. Based on our work, Kemppainen and coworkers examined absolute glucose consumption in the human brain of healthy volunteers. They confirmed, for the first time using the [^{18}F]FDG-PET technique, that the glucose consumption level decreases during strenuous exercise, especially in the cingulate gyrus [29].

We demonstrated the relative increase in glucose uptake in the temporo-parietal association cortex, occipital cortex, premotor cortex, primary sensorimotor cortex and the cerebellar vermis [14]. Relative reduction in glucose uptake was detected in the prefrontal cortex, temporal cortex, cerebellar hemisphere, brain stem, and striatum. Mean values of global brain glucose uptake was relatively lower in runners than in resting controls [14]. Kemppainen and coworkers later demonstrated significant reductions of regional glucose metabolic rate in all cortical regions in relation to exercise intensity [29]. They also pointed out that exercise could be associated with adaptive metabolic changes in the frontal cortex [29]. Thus, global and regional brain metabolic declines were observed using [^{18}F]FDG PET, particularly in the limbic and frontal regions [14,29]. It is easy to explain the metabolic increase in the regions directly associated with the execution of an exercise task, while it is not so easy to explain the mechanisms of relative decreases in the regions not involved in an exercise task. Previous imaging studies in anxiety disorders demonstrated increased glucose metabolism in these regions [30,31]. We speculated that the metabolic reduction in the frontal and limbic regions was associated with emotional changes in runners, including the phenomenon called runner's high [32].

Dietrich and Sparling reported that endurance exercise impaired prefrontal-dependent cognitive ability in healthy volunteers [33]. Dietrich later proposed a new theory (transient hypofrontality theory: THT) to explain the metabolic reduction in the prefrontal region [34], where the prefrontal activity is suppressed indirectly due to the limitation in energy supply to the brain [34].

Interestingly, this theory also explains a neural mechanism regarding the mental health benefits of exercise [33,34]. Here, it is of interest to also point out that Kemppainen and coworkers have suggested that substrates other than glucose, most likely lactate, are used by the brain as an energy source in order to compensate the increased energy demand to maintain neuronal activity during high-intensity exercise, since lactate availability during exercise tended to correlate negatively with the brain glucose uptake measured using [^{18}F]FDG PET in their study [29].

Acknowledgments

This work was supported by the Global COE program. The authors thank the support of all the staff of the Cyclotron and Radioisotope Center, Tohoku University, for their support during the study.

References

1. K. Yanai and M. Tashiro, *Pharmacol Ther.* **113**, 1–15 (2007).
2. K. Iwabuchi, C. Ito, M. Tashiro, M. Kato *et al.*, *Eur Neuropsychopharmacol.* **15**, 185–191 (2005).
3. M. Kano, S. Fukudo, A. Tashiro, A. Utsumi *et al.*, *Eur J Neurosci.* **20**, 803–810 (2004).
4. M. Yoshizawa, M. Tashiro, S. Fukudo, K. Yanai *et al.*, *Biol Psychiatry.* (2008).
5. M. Tashiro, H. Mochizuki, K. Iwabuchi, Y. Sakurada *et al.*, *Life Sci.* **72**, 409–414 (2002).
6. M. Tashiro, Y. Sakurada, K. Iwabuchi, H. Mochizuki *et al.*, *J Clin Pharmacol.* **44**, 890–900 (2004).
7. M. Tashiro, H. Mochizuki, Y. Sakurada, K. Ishii *et al.*, *Br J Clin Pharmacol.* **61**, 16–26 (2006).
8. M. Tashiro, X. Duan, M. Kato, M. Miyake *et al.*, *Br J Clin Pharmacol.* **65**, 811–821 (2008).
9. M. Tashiro, Y. Sakurada, H. Mochizuki, E. Horikawa *et al.*, *Hum Psychopharmacol.* **23**, 139–150 (2008).
10. M. Tashiro, K. Kubota, M. Itoh, T. Yoshioka *et al.*, *Psychooncology.* **8**, 283–286 (1999).
11. M. Tashiro, F. D. Juengling, M. J. Reinhardt, I. Brink *et al.*, *Psychooncology.* **9**, 157–163 (2000).
12. M. Tashiro, F. D. Juengling, M. J. Reinhardt, M. Mix *et al.*, *Med Sci Monit.* **7**, 687–695 (2001).
13. M. Tashiro, M. Itoh, K. Kubota, H. Kumano *et al.*, *Psychooncology.* **10**, 541–546 (2001).

14. M. Tashiro, M. Itoh, T. Fujimoto, T. Fujiwara *et al.*, *J Sports Med Phys Fitness*. **41**, 11–17 (2001).
15. M. Tashiro, M. Itoh, T. Fujimoto, M. M. Masud *et al.*, *Methods*. **45**, 300–306 (2008).
16. H. Rikimaru, M. Itoh, M. Tashiro and M. Watanabe, *Brain mapping of muscle-brain interaction during chewing by SPM analysis: an FDG-PET study*, in *Brain Imaging using PET*, edited by M. Senda, Y. Kimura and P. Herscovitch, Academic Press, San Diego, CA, USA, 269–275 (2002).
17. T. Fujimoto, M. Itoh, H. Kumano, M. Tashiro *et al.*, *Lancet*. **348**, 266 (1996).
18. M. Tashiro, T. Fujimoto, M. Itoh, K. Kubota *et al.*, *J Nucl Med*. **40**, 70–76 (1999).
19. M. Iemitsu, M. Itoh, T. Fujimoto, M. Tashiro *et al.*, *Advances in Exercise and Sports Physiology*. **7**, 53–58 (2001).
20. M. Reivich, L. Sokoloff, H. Shapiro, M. des Rosiers *et al.*, *Trans Am Neurol Assoc*. **99**, 238–240 (1974).
21. R. J. Schwartzman, J. Greenberg, M. Revich, K. J. Klose *et al.*, *Exp Neurol*. **72**, 153–163 (1981).
22. B. M. Gallagher, J. S. Fowler, N. I. Gutterson, R. R. MacGregor *et al.*, *J Nucl Med*. **19**, 1154–1161 (1978).
23. F. R. Sharp, *Brain Res*. **110**, 127–139 (1976).
24. J. Vissing, M. Andersen and N. H. Diemer, *J Cereb Blood Flow Metab*. **16**, 729–736 (1996).
25. K. Herholz, W. Buskies, M. Rist, G. Pawlik *et al.*, *J Neurol*. **234**, 9–13 (1987).
26. G. R. Fink, L. Adams, J. D. Watson, J. A. Innes *et al.*, *J Physiol*. **489**, 663–675 (1995).
27. M. Mishina, M. Senda, K. Ishii, M. Ohyama *et al.*, *Acta Neurol Scand*. **100**, 369–376 (1999).
28. M. Iemitsu, M. Itoh, T. Fujimoto, M. Tashiro *et al.*, *Med Sci Sports Exerc*. **32**, 2067–2070 (2000).
29. J. Kemppainen, S. Aalto, T. Fujimoto, K. K. Kalliokoski *et al.*, *J Physiol*. **568**, 323–332 (2005).
30. L. R. Baxter, *J Clin Psychiatry*. **51** (Suppl), 22–25 (1990).
31. D. Perani, C. Colombo, S. Bressi, A. Bonfanti *et al.*, *Br J Psychiatry*. **166**, 244–250 (1995).
32. H. Boecker, T. Sprenger, M. E. Spilker, G. Henriksen *et al.*, *Cereb Cortex*. (2008).
33. A. Dietrich and P. B. Sparling, *Brain Cogn*. **55**, 516–524 (2004).
34. A. Dietrich, *Psychiatry Res*. **145**, 79–83 (2006).

HIGH INTENSITY FOCUSED ULTRASOUND TREATMENT ENHANCED WITH NANO- TO MICRO-PARTICLES [*]

SHIN-ICHIRO UMEMURA [*,†]

Graduate School of Biomedical Engineering, Tohoku University
Aoba 6-6-05, Aramaki, Aoba-ku, Sendai, Miyagi 980-8579, Japan

SHIN YOSHIZAWA

Department of Information and Intelligent Systems, Tohoku University
Aoba 6-6-05, Aramaki, Aoba-ku, Senda,i Miyagi 980-8579, Japan

KAZUAKI SASAKI [*]

Department of Veterinary Medicine, Tokyo University of Agriculture and Technology
Saiwai-cho 3-5-8, Fuchu, Tokyo 185-8601, Japan

KEN-ICHI KAWABATA

Central Research Laboratory, Hitachi Ltd.
Higashi-Koigakubo 1-280, Kokubunji, Tokyo 185-8601, Japan

The bioeffect of ultrasound can be enhanced by orders of magnitude with microbubbles. Enhancement of ultrasonic tissue absorption with a microbubble agent is demonstrated both in theory and in a vivo experiment. The ultrasonic power absorbed by a microbubble in its continuous wave response is estimated through numerically solving a version of the Rayleigh-Plesset equation. At an ultrasonic frequency of 3 MHz, a resonant microbubble, approximately 1.1 μm in radius, showed an absorption cross section of about 0.005 mm^2 in its low power response. This estimation predicts that the tissue ultrasonic absorption will be doubled when such microbubbles are delivered to the tissue at a concentration of about 8 bubbles /mm^3 in tissue. An exteriorized murine kidney was exposed to focused ultrasound at 3.2 MHz in degassed saline and the tissue temperature change was measured. With an intravenous bolus administration of 0.2 ml/kg Optison™, the ultrasonically induced temperature elevation was multiplied by up to five times. The enhancement in temperature elevation gradually decreased as the microbubble agent was eliminated from the body. This effect will be useful in therapeutic application of ultrasound if a microbubble agent is delivered selectively to the target tissue.

[*] This work is partly supported by New Energy and Industrial Technology Development Organization, Japan.

[†] Shin-ichiro Umemura is a Tohoku University Global COE Member.

1. Introduction

Ultrasound has moderate tissue attenuation and absorption coefficients at an appropriate wavelength for penetrating intervening tissues to reach, vibrate, and heat non-superficial tissue objects while maintaining the ability to focus energy into small volumes. This is a unique advantage when compared to electromagnetic modalities such as laser beams in the application to non-invasive treatment of non-superficial tumors. Bioeffects of ultrasound, which can be potentially be used for such a kind of treatment, are shown in Fig. 1.

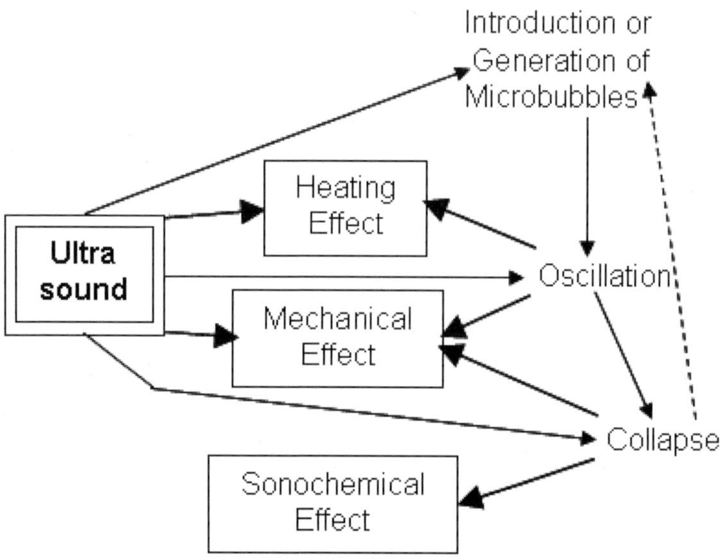

Figure 1. Bioeffects of ultrasound and their enhancement with microbubbles.

In the existence of microbubbles, the ultrasonic vibration amplitude is enhanced in their vicinity by orders of magnitude, and both mechanical and heating effects can be thereby accelerated. Furthermore, sonochemical effects can be induced when microbubbles collapse.

Recently, the heating effect was started being clinically used as the thermal coagulation treatment with high intensity focused ultrasound (HIFU). In this paper, the acceleration of heating [1] is chosen as an example of the ultrasonic bioeffects enhanced with microbubbles.

2. Mechanism of Enhancement of Ultrasonic Heating with Microbubbles

Microbubbles, subjected to ultrasonic pressure in the same frequency range with their resonant frequency, convert the acoustic energy to heat significantly through their volume oscillation. This type of energy conversion has two mechanisms: (1) viscous heating and (2) a pressure-volume hysteresis loop. In the mechanism (1), heat is generated by the shear motion of the liquid surrounding a volumetrically oscillating microbubble.

The mechanism (2) originates from the delay of the temperature change of the gas inside a microbubble: the temperature increase during its compression and the decrease during its expansion with some delay. If the change is either completely isothermal or adiabatic, there is no energy conversion due to this mechanism. The heat transfer taking place inside the bubble during the temperature change results in a phase discrepancy between its pressure change and its volume change from the phase in the quasi-static case, 180 degrees. Accordingly, a part of volumetric oscillation energy of the bubble is converted to heat in the heat cycle, which is opposite to that of a heat engine converting heat to mechanical energy such as the well-known Carnot cycle.

The temperature changes due to compression and expansion of a microbubble increase as the specific heat of gas inside the bubble increases. The amount of acoustic energy converted to heat in a heat cycle also increases as the specific heat increases. A heat cycle of a microbubble containing air (with a specific heat ratio of 1.40) is shown in Fig. 2 compared with that of Ar with a large specific heat ratio of 1.67 and that of SF_6 with a small specific heat ratio of 1.09. The area inside the pressure-volume hysteresis loop equals to the heat converted from acoustic energy in a heat cycle, oppositely to that of a heat engine. The loop area of Ar is approximately twice that of air. In contrast, that of SF_6 is negligibly small in comparison with air. The heat cycle of SF_6 is very close to isothermal change.

All the stabilized microbubbles of a new generation ultrasound contrast agent, developed for diagnostic purpose, contain a polyatomic gas such as SF_6 and perfluorocarbon. Optison^{TM}, used in the experiment, is such a new generation contrast agent and contains perfluorocarbon. The specific heat ratio of a polyatomic gas is as small as 1, because its molecule has an internal degree of freedom other than that of transitional motion. Therefore, the heat cycle of a microbubble of a new generation contrast agent, a microbubble containing a polyatomic gas, is very close to isothermal., and its heat generation of the mechanism (2) is negligible in comparison with that of the mechanism (1).

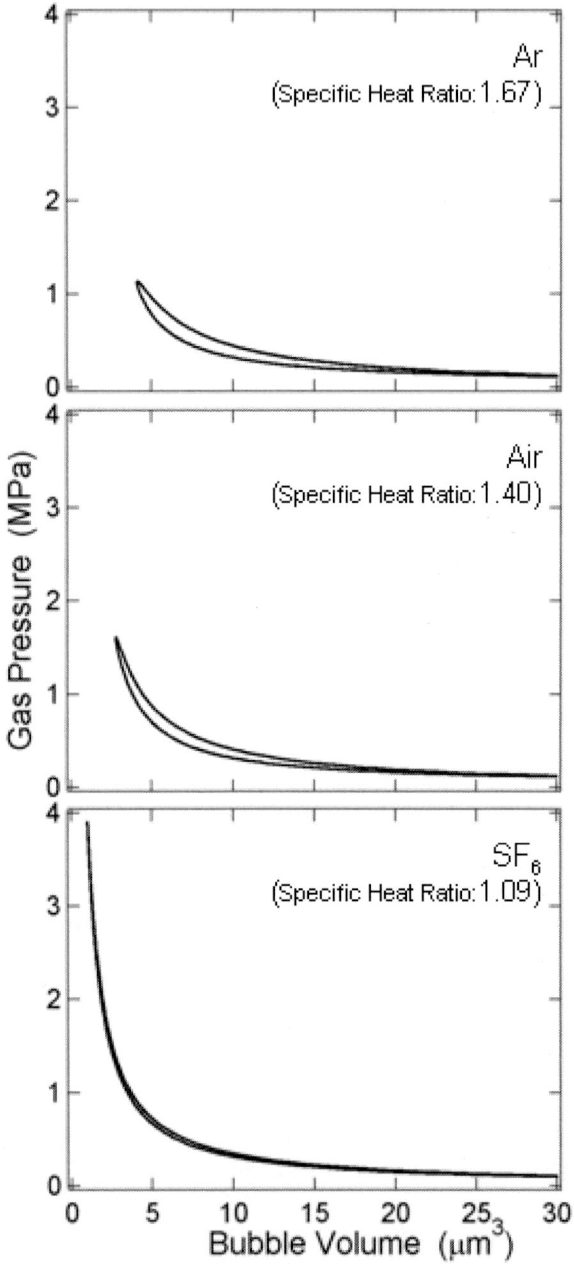

Figure 2. Thermal cycle of resonant microbubbles containing different gas species subjected to ultrasonic pressure at 2 MHz and 0.5 MPa (peak).

The acoustic power converted to heat by a microbubble in isothermal heat cycles is numerically calculated and the power per bubble volume is plotted against the bubble radius in Fig. 3. The ultrasonic intensity is geometrically varied from 1 mW/cm^2 to 1 W/cm^2.

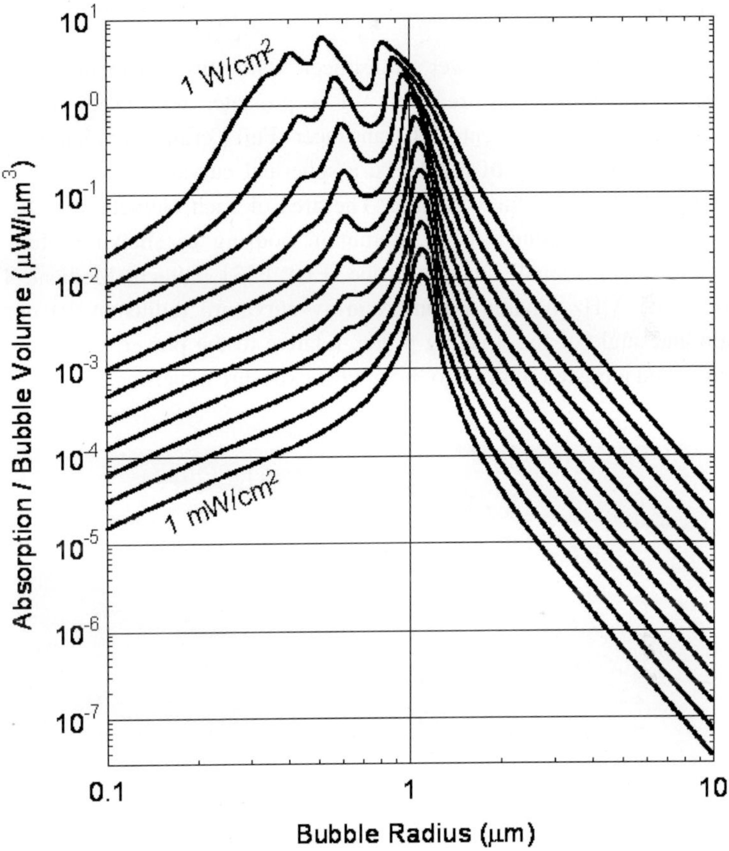

Figure 3. Ultrasonic abosorption by microbubble at 3 MHz, theoretically predicted by numerically solving the Rayleigh-Plesset equation. Ultrasonic intensity is varied as a parameter from 1 mW/cm2 to 1 W/cm2 in a geometric series.

A microbubble has a single resonant radius and shows an approximately linear response at an ultrasonic intensity of 1 mW/cm^2. In contrast, it starts showing its nonlinear response and having plural of resonant radii such as a half and a third of the fundamental, as the ultrasonic intensity increases as high as 1 W/cm^2. A stabilized microbubble agent normally has a size distribution in the order of twice. Owing to this nonlinear response, most of the microbubbles varying in size can contribute to then ultrasonic energy conversion to heat. It is estimated that the ultrasonic absorption by tissue is doubled in the existence of approximately 10 microbubbles in 1 mm^3 of tissue.

3. Methods

A prototype split-focus power transducer with two elements, originally constructed for transrectal treatment of a prostate, was used in exposure experiments. The dual element PZT transducer (Fuji Ceramics, Shizuoka, Japan) had a resonant frequency of 3.2 MHz, a spherical curvature radius of 35 mm, and an aperture of 40 mm x 20 mm. The area of each element was 40 mm x 10 mm. It was contained in an aluminum housing as shown in Fig. 4, in combination with a small imaging probe (EUP-F331, Hitachi Medical, Tokyo, Japan) at 6.5 MHz having a convex array curvature radius of 10 mm. The position and angle of the imaging probe, relative to the power transducer, were calibrated and adjusted in prior to the *in vivo* exposure experiment.

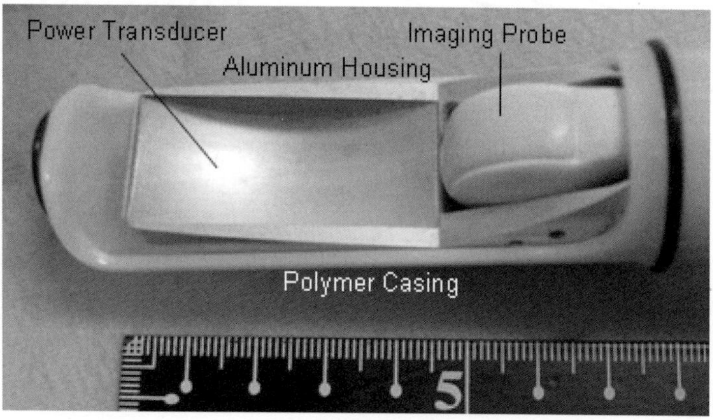

Figure 4. HIFU transducer at 3.2 MHz used in *in vivo* experiment.

When the two elements of the power transducer were driven at opposite phases, the focused beam is split into two. This dual split-focus beam was used instead of the conventional single-spot focal beam for suppressing the spatial peak intensity to minimize the possibility of excessively destroying the microbubble agent and inducing irreversible changes in the tissue. The schlieren images of the focal field from the prototype transducer are shown in Fig. 5. The side image and the front images of both single-spot and split-focus fields are shown. In the split-focus mode, the two focal beams were separated about a millimeter from each other.

Figure 5. Focal ultrasonic field of the HIFU transducer at 3.2 MHz.

The total acoustic power output from the transducer was calibrated against the drive voltage by measuring the radiation force on a hollow aluminum plate in degassed water. The acoustic intensity distribution pattern on the focal plane was measured in degassed water with a 0.5-mm-diameter needle-type hydrophone (Imotec, Coesfeld, Germany) at a low drive voltage. From these results, the spatial peak acoustic intensity was obtained as a function of drive voltage assuming that the intensity distribution pattern does not significantly change as the drive voltage increases. This peak acoustic intensity in water was used to describe the exposure condition in the *in vivo* experiments, ignoring the ultrasonic attenuation in traveling a short distance in tissue.

Optison™, a suspension of microspheres of human serum albumin with perflutren, C_3F_8, (Amersham Health, Princeton, NJ) was used as the microbubble agent in the experiment. The concentration, size distribution, and total volume of microbubbles in the suspension were measured using a Coulter Multisizer III instrument (Beckman Coulter, Fullerton, CA), fitted with an orifice tube of 30 μm aperture after diluted with phosphate buffer saline, pH 7.4, by 200 times.

After surgical anesthesia with sodium pentobarbital was given to a female Sprague-Dawley rat (approximately 250 g), the left kidney was mobilized and exteriorized through an incision. A polyethylene tube cannula, 0.5 and 0.8 mm in inner and outer diameter, respectively, was inserted into a jugular vein for systemic administration of the microbubble agent. The rat was held vertically in degassed saline at 33°C as shown in Fig. 6.

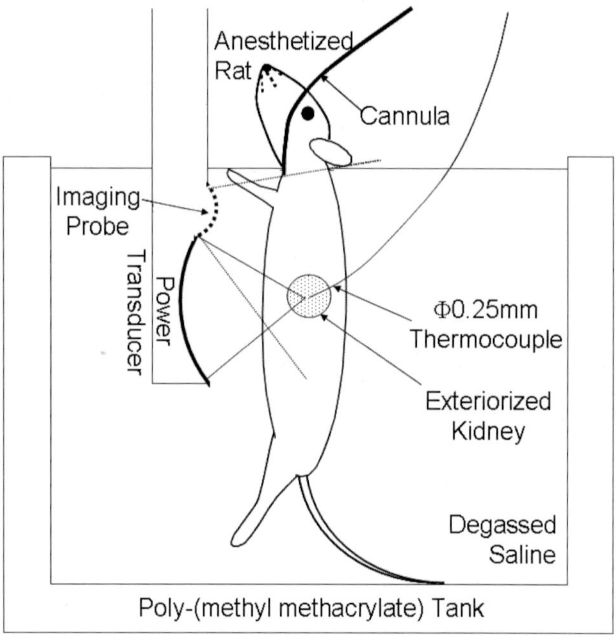

Figure 6. Experimental setup of HIFU exposure.

A very-thin sheathed thermocouple (Sukegawa Electric, Ibaraki, Japan) was chosen to minimize the potential artifacts in tissue temperature measurement. The 0.25-mm diameter thermocouple consisted of a pair of 0.05-mm diameter chromel and alumel wires running through a 0.025-mm thick stainless-steal sheath filled with packed, ultra fine magnesium oxide powder. Its longitudinal heat conductance is estimated to be 0.36 μWm/°C, which is equivalent to that of a 0.9-mm diameter cylindrical rod of tissue. It was inserted into the renal cortex tissue. The position of the rat in saline, relative to the transducer, was adjusted so as to locate the thermocouple right in the middle of the focal heating zone, first by using the B-mode images taken with the 6.5-MHz probe, and then by using the thermocouple output at low ultrasonic exposure intensity.

Three animals were used in each experiment to check the repeatability. The experimental animals were treated following the guideline proposed by the Science Council of Japan.

4. Results

In prior to the series of HIFU exposure experiments, the size distribution of the microbubble agent, Optison, which was used in the experiments, was measured and plotted in Fig. 7. Regarding the number of microbubbles, there are two peaks in radius around 0.6 μm and 1.6 μm, but there is only one gradual peak in radius around 3.2 μm regarding their volumes. The total bubble volume content of Optison was calculated to be 2.2%.

In Fig. 8, the temperature change in the kidney due to three-time ultrasonic exposure at an ultrasonic intesnsity of 290 W/cm^2 for 10 s is plotted. Optison with a dose of 0.2 ml/kg was injected at time 0 in the horizontal axis. In Fig. 9, the temperature curves right before and after the injection are compared by expanding the horizontal time axis. Optison multiplied the tissue temperature rise due to HIFU by 4-5 times.

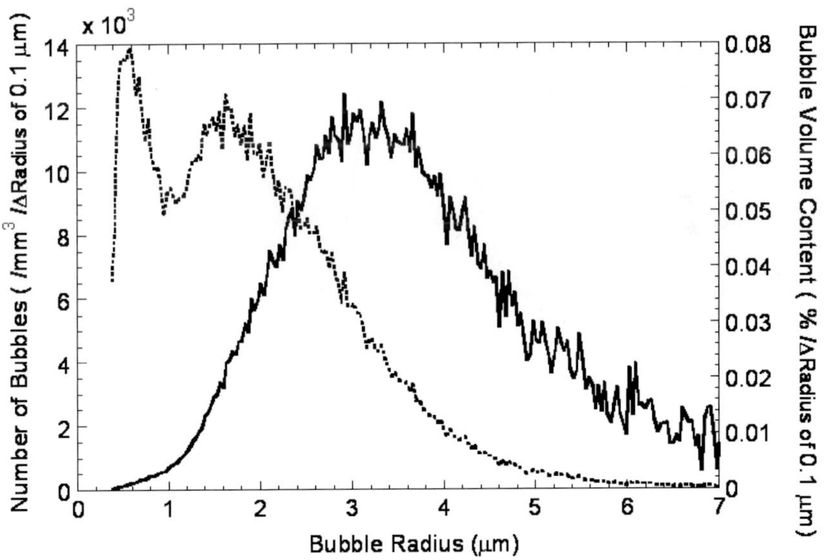

Figure 7. Size distribution of Optison (dotted line) and distribution in volume content (solid line). It was measured using a Coulter Multisizer III instrument (Beckman Coulter, Fullerton, CA), fitted with an orifice tube of 30 mm aperture after diluted with phosphate buffer saline, pH 7.4, by 200 times.

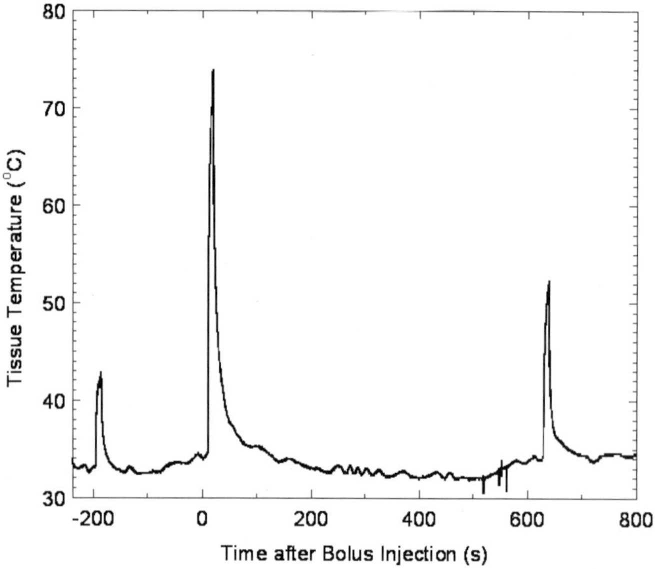

Figure 8. Murine kidney tissue temperature with HIFU exposure at 290 W/cm^2 before and after intravenous injection of Optison (0.2 ml/kg) at time 0.

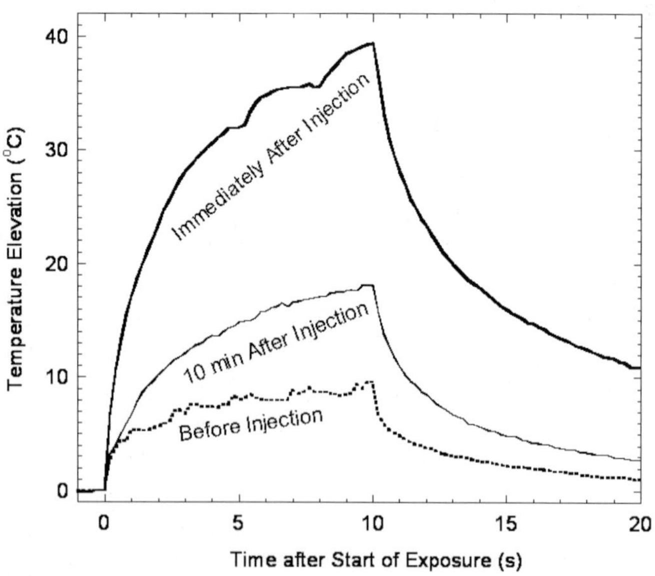

Figure 9. Tissue temperature elevation with HIFU with/without Optison.

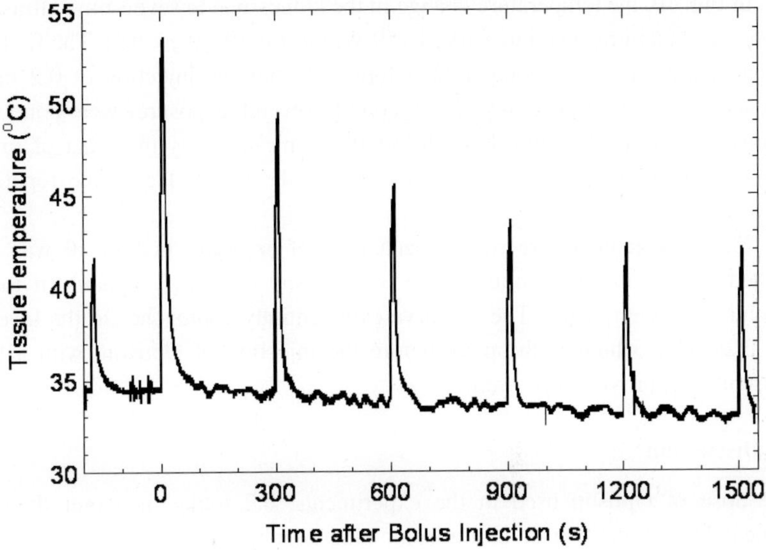

Figure 10. Murine kidney tissue temperature with HIFU exposure at 150 W/cm^2 before and after intravenous injection of Optison (0.2 ml/kg) at time 0.

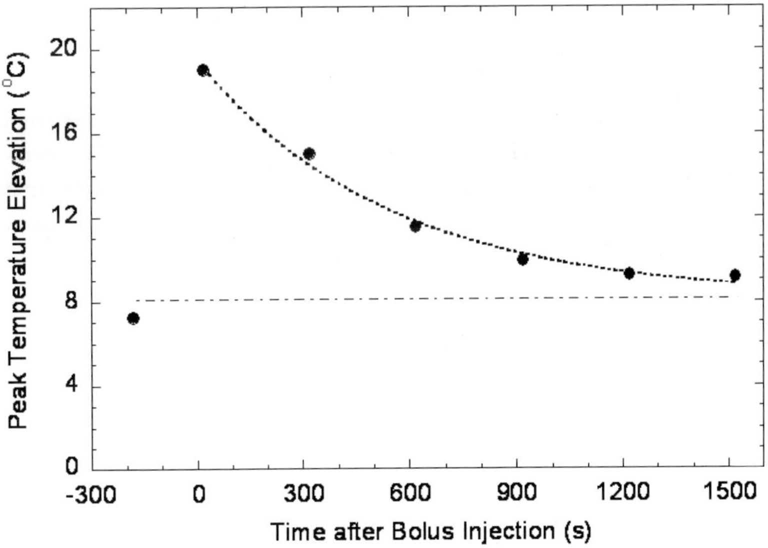

Figure 11. Peak tissue temperature elevation during each ultrasonic exposure plotted against time after bolus injection of 0.2 ml/kg Optison. A least-squares fit monoexponential curve (dotted line) and its final value (dotted broken line) are also shown.

In Fig. 10, the temperature change in the kidney due to seven-time ultrasonic exposure at an ultrasonic intensity of 150 W/cm^2 for 10 s is plotted. The first and second exposures were done right before and after the injection of 0.2 ml/kg Optison at time 0, respectively. The second to seventh exposures were done with a constant period of 5 min. Even lower ultrasonic intensity than that of Fig. 8 was chosen to prevent irreversible change of the tissue due to the repetitive exposure.

The peak temperature rise at each time of exposure in Fig. 10 was also plotted in Fig. 11. The dotted curve is the result of a least squares fit to the second to seventh peaks. The fit curve exponentially approaches to the level of the first peak, which is the peak before the injection of Optison, with a time constant of approximately 9 min.

5. Discussion

The dose of Optison used in the experiments, 0.2 ml/kg, is larger than the clinical dose normally used in ultrasonic diagnosis by an order of magnitude, but still within the clinically approved safe dose, with which the biological safety was confirmed. The observed 4-5 times enhancement of ultrasonic tissue heating with a safe dose of a microbubble agent demonstrated the high potential of this type of approach.

The gas contained in a stabilized microbubble agent such as Optison is excluded from the body by the gas exchange in circulating through lungs. The exclusion time constant is estimated to be approximately 5 min for a human form gas-chromatography experiments. Considering the difference between the species, it is reasonable to interpret that the time constant of approximately 9 min, seen in Fig. 11, corresponds to that of excluding Optison from the body. This result strongly supports the hypothesis that the observed enhancement of ultrasonic tissue heating is proportional to the concentration of Optison in blood.

The magnitude of the observed enhancement of ultrasonic tissue heating with microbubbles agreed well with the theoretical prediction, in which the heat converted by a microbubble was numerically calculated.

In the theoretical prediction, the nonlinear effect was seen even at an ultrasonic intensity of 1 W/cm^2, which is much lower than the intensity normally used as HIFU. Because of this nonlinear effect, many peaks of ultrasonic absorption other than the fundamental resonance peak are also seen as a function of bubble radius in Fig. 3. Microbubbles in a stabilized microbubble agent normally have a relatively broad distribution in their radius as shown in Fig. 7. If each bubble has had to be excited at its resonant frequency, the exposure should have had to be done at many different ultrasonic frequencies. However, the real microbubbles have a broad frequency response due to their nonlinear character as seen in Fig. 3. This should be the reason why the sufficiently large enhancement of ultrasonic heating was observed even at a single ultrasonic exposure frequency in the experiment.

Discussions on a point of view a little different from above, to interpret the results, are also written below.

A microbubble in liquid automatically collapses in a short period of time because its high inner pressure to compete against the surface tension results in fast gas diffusion outward. Therefore, surfactants reducing the surface tension or a shell resisting to gas diffusion are used in a stabilized microbubble agent. In Option, used in the experiments, an albumin shell is used to stabilize a perfluorocarbon microbubble. This albumin shell cannot stand the high acoustic pressure of HIFU, and it may have been destroyed when the microbubble started converting acoustic energy into heat. Assuming this situation, the theoretical results shown in Fig. 3 were calculated for a microbubble without a shell.

After the shell was destroyed, the shell fragments may start working as cavitation nucli. The cavitating bubbles may contain not only perfluorocarbon, carried over from Optison, but also air, originally contained in blood. For air as the inner gas, the mechanism (2) has to be taken into account, and the theoretical prediction may produce a little different results.

In the series of events on this hypothesis, the shell fragments rather than the perfluorocarbon gas enhance the ultrasonic heating. Therefore, the time constant of approximately 9 min, seen in Fig. 11, should correspond to time to exclude the albumin shell fragments rather than the perfluorocarbon gas from the body.

This hypothesis can be tested by using Optison after destroying its shells and sucking perfluorocarbon away under low pressure.

There may be another hypothesis that the perfluorocarbon gas itself is working as cavitation nucli. The cavitated microbubbles may contain air also in this hypothesis. Perfluorocarbon gas can work as cavitation nucli because it is extremely hydrophobic. The mechanism (2) should also be taken into account in the theoretical prediction on this hypothesis.

6. Conclusion

It was demonstrated that a microbubble agent, in its clinically approved dose range, multiplies the heating effect of ultrasound in both theory and experiment.

However, if it is distributed in the whole body at a similar tissue concentration, the ultrasonic attenuation in the intervening tissue will be multiplied at the same time and ultrasound will be mostly attenuated before it reaches to the target tissue in the focal zone. Therefore, a microbubble agent should be delivered selectively to the target tissue, to fully utilize the observed effect. This type of approach [2] should be pursued in the future work.

References

1. S. Umemura, K. Kawabata and K. Sasaki, *IEEE Trans. Ultrasonics, Ferroelectrics, and Frequency Control* **52**, 1690–1698 (2005).
2. K. Kawabata, N. Sugita, H. Yoshikawa, T. Azuma and S. Umemura, *Jpn. J. Appl. Phys.* **44**, 4448–4852 (2005).

MOLECULAR IMAGING AND ITS APPLICATION TO DRUG DEVELOPMENT AT TOHOKU UNIVERSITY

KAZUHIKO YANAI*, NOBUYUKI OKAMURA

Department of Pharmacology, Tohoku University School of Medicine, Sendai 980-8575, Japan

REN IWATA, MANABU TASHIRO*

Cyclotron and Radioisotope Center, Tohoku University, Sendai 980-8578, Japan

Positron emission tomography (PET) comprises the administration of carbon-11- or fluorine-18-labeled tracers to human subjects in order to describe the concentration-time profile in body tissues targeted for treatment. As PET involves the administration of only microgram amounts of unlabeled drug, the potential risk to human subjects is very limited. Consequently, required preclinical safety testing is reduced as compared to conventional human studies. PET molecular imaging studies are gaining increasing importance in clinical drug research, as they have the potential to shorten time-lines and cut costs along the critical path of drug development. We have developed several PET tracers for molecular imaging such as amyloid imaging and central nervous system drug research.

1. Introduction

Our research group carried out basic and clinical research on molecular PET imaging in humans. We measured the brain activities using [^{15}O]H_2O-PET or functional MRI in order to clarify the brain mechanism for emotion and cognition. We also have developed PET molecular imaging techniques in order to examine the neurochemistry in humans. For instance, we have applied [^{11}C]doxepin, a potent H1 antagonist, to humans in order to reveal the changes in the histaminergic neurotransmission in both normal volunteers and psychiatric patients. The molecular PET imaging is expected very useful for drug development and to shorten the total development time as shown in Fig. 1. For instance, we developed several PET ligands for imaging acetylcholine esterase and amyloid Aβ protein in order to examine the therapeutic efficacy in Alzheimer disease. [^{11}C]Doxepin-PET study is also useful to evaluate the sedative properties

* Kazuhiko Yanai and Manabu Tashiro are Tohoku University Global COE Members.

of 2nd generation antihistamines, which is often used in allergic disease such as atopic dermatitis, seasonal rhinitis, and urticaria.

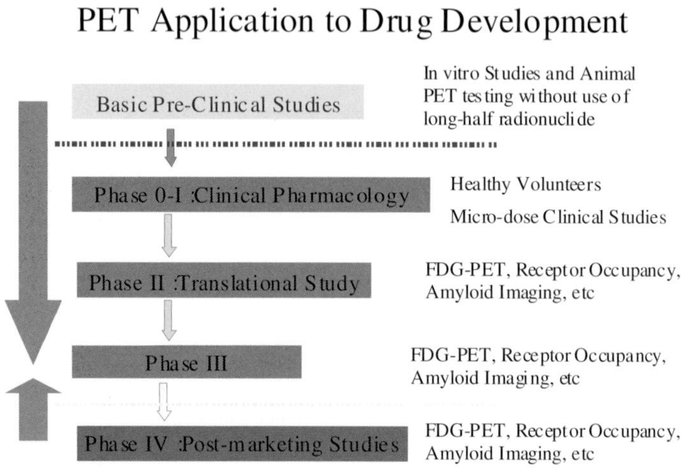

Figure 1. Molecular PET Imaging in drug development.

2. Methodological Development

Certain prerequisites for suitable compounds have to be met. First of all, the ligand should possess high affinity and selectivity for the molecular targets. It should be able to cross blood–brain barrier for high potency in the CNS, and it must at its final step of synthesis contain an atom or functional group suitable for radiolabeling such as $-N^{11}CH3$, $-S^{11}CH3$ or $-O^{11}CH3$, $-^{18}F$. Because of low radiation doses to humans, carbon-11 (half life, 20 min) labeled radiopharmaceuticals are preferable to fluorine-18 (half life, 110 min) labeled ones in early phase PET studies. However, fluorine-18 labeled compounds are much better than those of carbon-11 for their commercial development. As shown in Fig. 2, we have labeled several molecular probes with carbon-11 for clinical application using the automated synthesis system.

The recent development of automated synthesis system is the miniaturization of the total synthesis system because all system should be put in the small shield box. As the synthesis scales are very small with the total amounts of labeled compounds of less than 10 microgram, a very fine synthesis system, a micro-reactor synthesis will be developed for the future. Using such a miniaturized micro-reactor for synthesizing many different tracers, order-made PET diagnosis will be realized in the future as shown in Fig. 3.

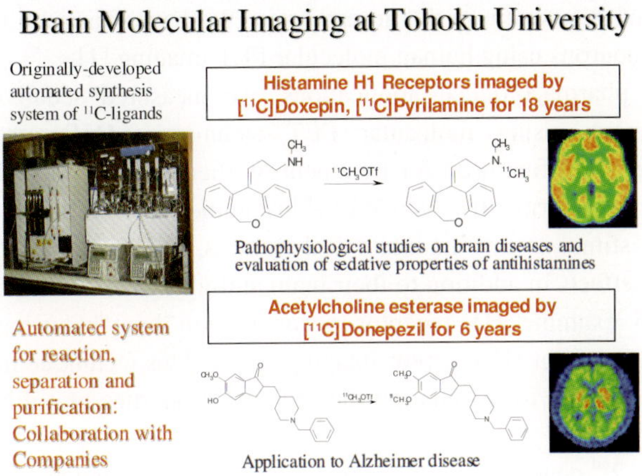

Figure 2. Carbon-11 labeled probes for molecular imaging.

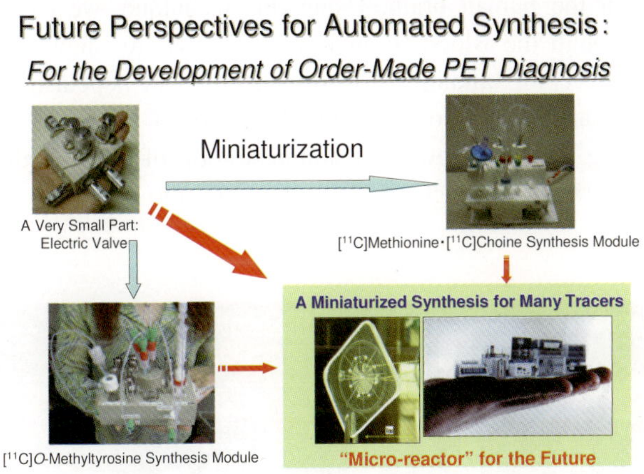

Figure 3. A "Micro-reactor" system for synthesizing many positron-labeled tracers in the future.

3. Molecular Imaging of Histaminergic Transmission in the Human Brain List

Histamine neurons are exclusively located in the posterior hypothalamus, and project their fibers to almost all regions of the human brain. The histaminergic neuron system modulates wakefulness, the sleep–wake cycle, appetite control, learning, memory and emotion. Although a significant amount of research has been done to clarify the functions in animals, a few studies have been reported on the roles of this system in the human brain. We have examined the functions of histamine neurons using human molecular PET imaging [1].

For the pharmacological applications, we measured sedative properties of antihistamines using molecular PET techniques. H1R antagonists, or antihistamines, are often used for treatment of allergic diseases such as allergic rhinitis and atopic dermatitis. As neuronal histamine is one of the most important systems that stimulate and maintain wakefulness, antihistamines often induce sedative side effects in addition to their main anti-allergic effects. We developed a method to examine H1 receptor occupancy in humans using PET with $[^{11}C]$doxepin, a potent H1 receptor imaging tracer. This technique has proved to be a reliable method for evaluating the sedative properties of antihistamines in humans.

In Fig. 4, 2nd generation antihistamine, bepotastine 10 mg and 1st generation antihistamine, diphenhydramine 30 mg were orally administered, and the time of Tmax (90 min in this case), $[^{11}C]$doxepin was injected to measure the remaining H1 receptor in the human brain. Using this technique, we can estimate the penetration through the Blood-Brain Barrier (BBB). As shown in Fig. 4, the remaining H1 receptor binding after administration of bepotastine 10 mg was essentially the same as that of placebo. On the other hand, the remaining H1 receptor binding was very few after administration of diphenhydramine 30mg [2]. The penetration through BBB represents the sedative properties of antihistamines.

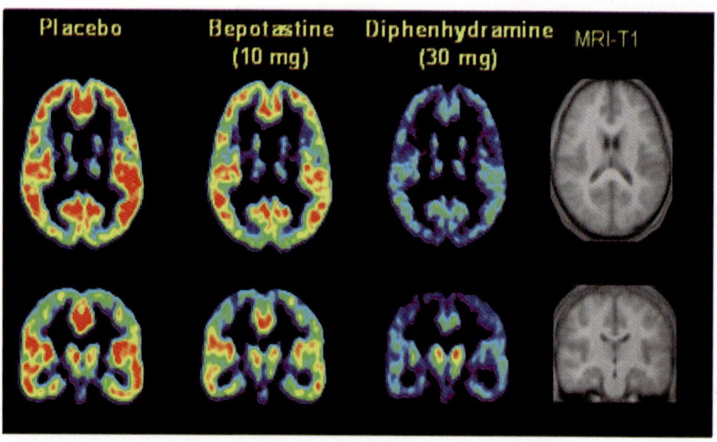

Figure 4. Histamine H1 receptor binding after administration of non-sedative (bepotastine 10 mg) and sedative (diphenhydramine 30 mg) antihistamines.

Figure 5 summarizes the histamine H1 receptor occupancy after administration of many antihistamines sold in Japanese Market. Antihistamines vary in their histamine H1 receptor occupancy, and can be classified into three categories: Non-sedative, less-sedative and sedative ones. Molecular PET imaging is the most sensitive method to measure the sedative properties of antihistamines.

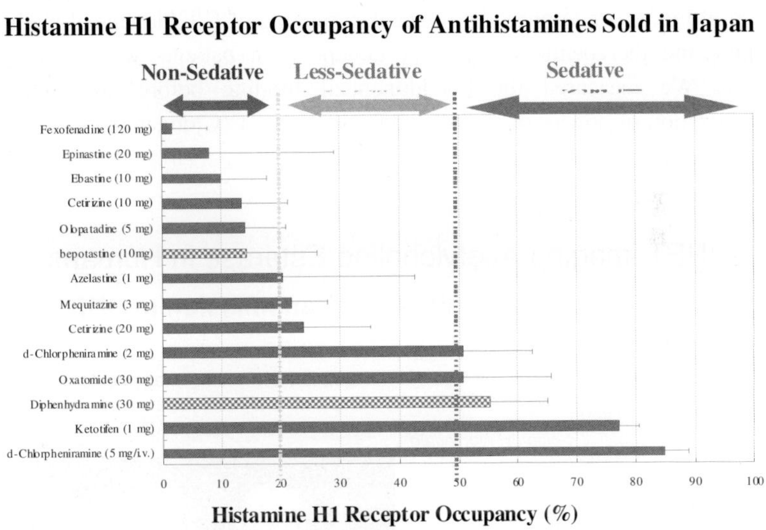

Yanai K & Tashiro M. Pharmacol. Ther. (2007)

Figure 5. Histamine H1 receptor occupancy by orally administered antihistamines in human brain.

4. Molecular Imaging in Alzheimer's Disease

Alzheimer's disease is an age-related and progressive neurodegenerative disease of the central nervous system. At the cellular levels, there are marked neurodegenerative changes accompanied with a reduction in many neurotransmitters and neuropeptides. The pathological hallmark of Alzheimer's disease is the accumulation of amyloid-forming proteins both extracellularly (amyloid-beta protein) and intracellularly (tau protein) in the brain. These protein accumulations are considered to start before the appearance of clinical symptom of dementia and regarded as a target for treatment of Alzheimer's disease. A remarkable dysfunction of the cholinergic and histaminergic system, characterized by deficits in memory and cognitive functions, has been observed in several brain regions of patients suffering from Alzheimer's disease.

We have labeled a series of carobon-11 labeled molecular imaging probes that target acetylcholinesterase (AChE) and amyloid deposits. We have performed molecular PET imaging studies using [^{11}C]donepezil, a molecular probe for imaging AChE and [^{11}C]]BF-227, a benzoxazole derivative for amyloid Aβ proteins. The clinical PET studies using [^{11}C]donepezil [3] and [^{11}C]BF-227 successfully differentiate AD patients from normal subjects. As shown in Fig. 6, we applied the molecular imaging of [^{11}C]donepezil method to examine the therapeutic efficacy of donepezil in patients with Alzheimer's disease. We measured the [^{11}C]donepezil binding before and after oral administration of donepezil in order to compare the [^{11}C]donepezil binding with the therapeutic efficacy.

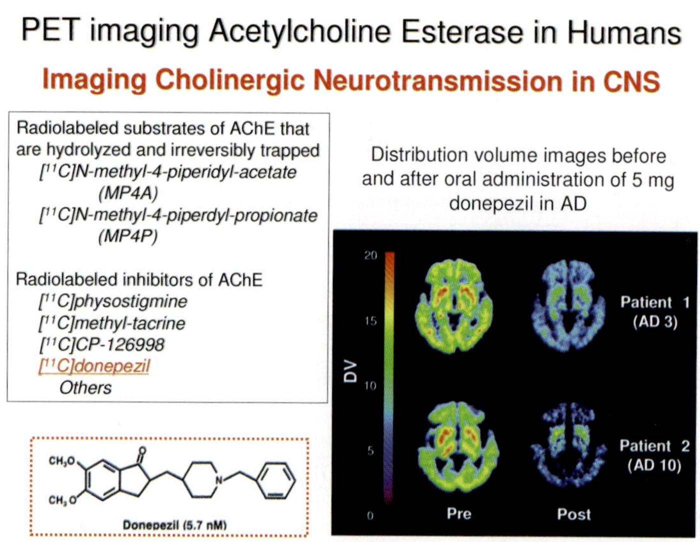

Figure 6. Molecular Imaging of Acetylcholine Esterase using [^{11}C]donepezil in Alzheimer disease.

Figure 7 illustrates the strategy of amyloid imaging at Tohoku University. Using our patented compounds, we have successfully imaged amyloid deposits in the living brains of Alzheimer's disease. Our multicenter trials demonstrate that amyloid imaging is the most sensitive method to diagnose the very early stages of Alzheimer's disease.

Molecular imaging can offer a new insight for examining pathophysiology and therapeutic efficacy of AD. In particular, amyloid imaging is a powerful tool for evaluating efficacy of anti-amyloid therapy, which is now intensively under development by pharmaceutical industry [4-7].

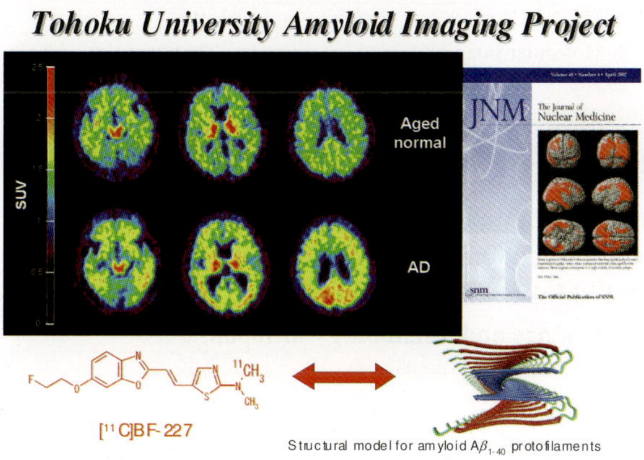

Figure 7. Amyloid imaging using benzoxazole derivatives. The used compounds are originally developed at Tohoku University School of Medicine.

5. Conclusion

Molecular imaging is a powerful and innovative tool for studying the pathology of psychiatric and neurological diseases in addition to malignancy and, more recently, for assessing the efficacy of drug treatment [8] as shown in Fig. 8. Molecular PET imaging as a new tool in therapeutic drug development can be used as follows: 1) to determine dose and dosing interval, 2) to identify homogeneous group, and 3) biomarker for drug efficacy. Technological advances in molecular imaging have made it possible to noninvasively extract information from humans regarding a drug's mechanism and sites of action. The advent of molecular imaging allows researchers to focus directly on human pharmacology.

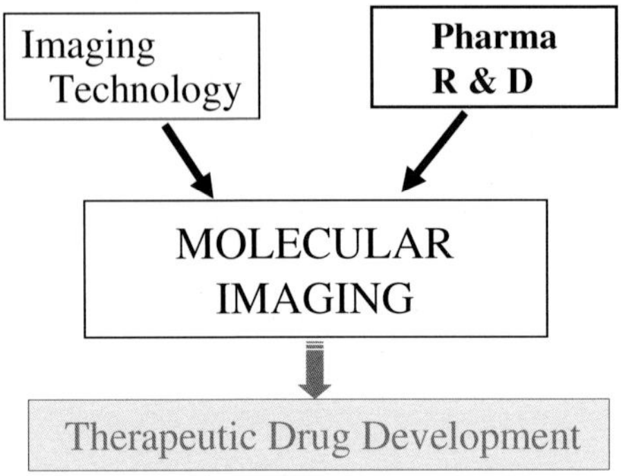

Figure 8. Molecular Imaging as an innovative tool for future drug development.

Acknowledgments

These works are supported by Grants-in Aid for Scientific Research from the Japan Society of Promotion Science (JSPS) and from Ministry of Health as well as by a grant of "Molecular Imaging" projects from the Ministry of Education, Culture, Sports, Science and Technology in Japan.

References

1. K. Yanai and M. Tashiro, The physiological and pathophysiological roles of neuronal histamine: An insight from human positron emission tomography studies. *Pharmacol Ther.* **113**, 1–15 (2007).
2. M. Tashiro, X. Duan, M. Kato, M. Miyake, S. Watanuki, Y. Ishikawa, Y. Funaki, R. Iwata, M. Itoh and K. Yanai, Brain histamine H1 receptor occupancy of orally administered antihistamines, bepotastine and diphenhydramine, measured by PET with 11C-doxepin. *Br J Clin Pharmacol.* **65**, 811–821 (2008).
3. N. Okamura, Y. Funaki, M. Tashiro, M. Kato, Y. Ishikawa, M. Maruyama, H. Ishikawa, K. Meguro, R. Iwata and K. Yanai, *In vivo* visualization of donepezil binding in the brain of patients with Alzheimer's disease. *Br J Clin Pharmacol.* **65**, 472–479 (2008).
4. N. Okamura, T. Suemoto, H. Shimadzu, M. Suzuki, T. Shiomitsu, H. Akatsu, T. Yamamoto, M. Staufenbiel, K. Yanai, H. Arai, H. Sasaki, Y. Kudo and T. Sawada, Styrylbenzoxazole derivatives for *in vivo* imaging of amyloid plaques in the brain. *J Neurosci.* **24**, 2535–2541 (2004).
5. N. Okamura, S. Furumoto, Y. Funaki, T. Suemoto, M. Kato, Y. Ishikawa, S. Ito, H. Akatsu, T. Yamamoto, T. Sawada, H. Arai, Y. Kudo and

K. Yanai, Binding and safety profile of novel benzoxazole derivative for *in vivo* imaging of amyloid deposits in Alzheimer's disease. *Geriatr Gerontol Int.* **7**, 393–400 (2007).

6. Y. Kudo, N. Okamura, S. Furumoto, M. Tashiro, K. Furukawa, M. Maruyama, M. Itoh, R. Iwata, K. Yanai and H. Arai, 2-(2-[2-Dimethylaminothiazol-5-yl]ethenyl)-6- (2-[fluoro]ethoxy) benzoxazole: A novel PET agent for *in vivo* detection of dense amyloid plaques in Alzheimer's disease patients. *J. Nucl. Med.* **48**, 553–561 (2007).
7. N. Okamura, S. Furumoto, H. Arai, R. Iwata, K. Yanai and Y. Kudo, Imaging amyloid pathology in the living brain. *Curr Med Imaging Rev.* **4**, 56–62 (2008).
8. M. Bauer, C. C. Wagner and O. Langer, Microdosing studies in humans: the role of positron emission tomography. *Drugs R D.* **9**, 73–81 (2008).

MOLECULAR IMAGING OF LYMPH NODE MICROMETASTASIS USING CONTRAST-ENHANCED HIGH-FREQUENCY ULTRASONAGRAPHY

LI LI [*], KEI TAKASE, SHOKI TAKAHASHI [†]

Radiology Department, Tohoku University Hospital,
2-1, Seiryo-cho, Aoba-ku, Sendai, Miyagi, 980-8575, Japan

Objective: We aimed to evaluate the possibility of early diagnosis of lymph node metastasis by using a 3D contrast-enhanced high-frequency ultrasonography system in a murine model of lymph node metastasis. *Methods*: A model of axillary lymph node metastasis was established by injecting a nonmetastatic fibroblastoid cell line, MRL/N-1, which stably expresses the firefly luciferase gene, into the inguinal lymph nodes of MRL/MpJ-*lpr/lpr* mice. Bioluminescence observed 2 d after implantation by using *in vivo* bioluminescence imaging system confirmed tumor establishment and metastasis. Subsequently, 3D images of normal and neoplastic lymph nodes, with and without injecting nanobubbles in the caudal vein of the mice, were constructed using a high-frequency ultrasound imaging system consisting of a scanner with a 35-MHz probe. The shapes of microvessels in the lymph nodes were reconstructed by analyzing the echogenicity of nanobubbles circulating in the lymph node blood vessels. *Results*: By using the abovementioned ultrasound imaging system, we succeeded in characterizing 3D vessel structures in the lymph nodes with the minimum axial diameter of less than 1 cm. Vascular density in the neoplastic lymph nodes was higher than that in the normal lymph nodes on day 14 ($P < 0.05$). Thus, the system established in this study may be suitable for early diagnosis of lymph node metastases. *Keyword*: nanobubble**,** high-frequency ultrasound**,** angiogenesis, lymph node micrometastasis, microvascular density (MVD), minimum axial diameter (short-axis diameter).

1. Introduction

The presence or absence of lymph node metastasis is an important prognostic factor in malignant tumor: early diagnosis of lymph node metastasis is crucial for improving immediate treatment effects as well as for the long-term survival of cancer patients [1]. Current diagnosis of regional lymph node metastasis mainly relies on noninvasive procedures such as computed tomographic (CT), and magnetic resonance (MR) imaging. Identification of metastatic lymph nodes with CT and MR imaging is mainly based on nodal size, *i.e.*, nodes with a short-axis diameter of more than 1 cm are usually regarded as being involved in

---

[*] Li Li is a Tohoku University Global COE Research Assistant.

[†] Shoki Takahashi is a Tohoku University Global COE Member.

cancer, and vice versa [2-7]. However, this size-based characterization of lymph node metastasis frequently leads to an erroneous diagnosis; for example, normal-sized lymph nodes may be metastatic, or conversely, abnormally enlarged lymph nodes may represent only reactive swelling [8-13]. Thus, although early detection of lymph node micrometastasis is desirable for treatment in the early stages of disease, it cannot be achieved by using morphological imaging modalities such as CT and or MR imaging.

Molecular imaging has been advocated for detecting the earliest pathological processes and is a rapidly expanding field, which would provide a noninvasive means for visualization of molecular events or processes. How can molecular imaging techniques visualize the molecular events occurring in the development of a tumor involving the lymph node? Angiogenesis appears to be a critical determinant of a tumor's growth, invasion, and metastatic potential [14-15]. Thus, angiogenesis due to tumor involvement may be utilized for detection of lymph node metastasis at an early stage by using molecular imaging techniques [16].

Although molecular imaging techniques have depended on the use of radiolabeled agents in the past, we consider the use of ultrasound (US) for early detection of pathological angiogenesis. Ultrasonography has emerged as a potential modality for molecular imaging owing to recent technological advancements. High-frequency US, also known as US biomicroscopy (UBM), utilizes a US frequency of 20 to 55 MHz to yield high-resolution images [17], while real-time contrast-enhanced ultrasonography (CEUS) provides a significant opportunity for visualizing microcirculation [18]. Three-dimensional (3D) sonography is the next logical step in diagnostic US examination.

In this study, to establish a system for early diagnosis of lymph node metastasis, we assessed angiogenesis in a lymph node metastatic model by quantitative 3D visualization, using high-frequency US and nanobubbles.

2. Materials and Methods

All *in vivo* studies were performed in accordance with the ethical guidelines of the Tohoku University.

2.1. *Animal Model*

Six adult male MRL/MpJ-*lpr*/*lpr* mice (35-40g, 14-20 weeks of age) were purchased from the Jackson Laboratory (Bar Harbor, Maine, USA). In MRL-*lpr*/*lpr* mice, idiopathic lymphadenopathy gradually developed with age. The inguinal lymph nodes of the mice were measured using calipers, and the mice having nodes with the maximum diameter of approximately 1 cm were selected.

2.2. *Mycoplasma-free Cell Culture and Preparation of Lymph Node Metastatic Model*

A fibroblastoid cell line, MRL/N-1, was obtained from Prof. Ono, Tohoku University, Japan. The cell line (MRL/N-1-luc) stably expressing the firefly luciferase gene were prepared by transfected pEGFPLuc and Lipofectin Transfer Reagent. MRL/N-1-luc cells were incubated at 37°C. A check for the absence of mycoplasma was performed after 1 week of culture. Approximately 1×10^5 cells (grown to 70% confluence) were suspended in 30-uL of phosphate-bufferd saline (PBS) and percutaneously injected into the inguinal lymph node of each MRL/MpJ-*lpr*/*lpr* mouse. We considered their axillary lymph node to contain metastases when the nodes's biologic fluorescence was observed by *In Vivo* Bioluminometer Imaging System (IVIS). Phosphate-buffered saline (PBS; 30 μL) was used as a negative control for the detection of neovascularities, which are known to be abundant in a tumor.

2.3. In vivo *Bioluminescence Imaging*

On days 2,6,10, and 14 after MRL/N-1-Luc cells implantation, each mouse was anesthetized with pentobarbital and injected intraperitoneally with luciferin (150-mg/kg body weight; Promega). After 10min of luciferin administration, luciferase bioluminescence was measured for 30 seconds by *in vivo* imaging system (IVIS Lumina; Xenogen Corp., Alameda, CA, USA). Then the animal was sacrificed by excessive ether for the next step.

2.4. Ex vivo *Detection of Establishment and Metastasis of MRL-N-1-Luc Cells by IVIS*

On day 14 after implantation, establishment of tumor in the inguinal lymph node and its metastasis to the axillary lymph node were detected by imaging luciferase bioluminescence 5 min after luciferin injection into the metastasis group mice, using an *in vivo* imaging system (IVIS). The inguinal lymph nodes and the axillary lymph nodes were then dissected and placed on a 6-well plate completely coverd with 300-μg luciferin in 1-mL PBS. Luciferase bioluminescence of the removed lymph nodes were then immediately quantified using the IVIS for 30 seconds.

2.5. *Nanobubble Preparation and Injection*

Acoustic liposomes (ALs) were prepared according to the method of Suzuki *et al with* some modifications [19]. The peak diameters and the zeta potential of acoustic liposome were 198 ± 30 nm (n = 3) and –4.1 ± 0.85 mV (n = 3), respectively.

Each mouse was injected a 200-μL bolus of an ultrasonic contrast agent (UCA; containing about 6.7×10^7 nanobubbles) in the caudal vein.

2.6. *Ultrasonographic Image Acquisition and Quantification*

During imaging, the mice were kept under anesthesia with 2% isoflurane in oxygen and restrained on a heated stage. All images were acquired with a Vevo770 high-frequency ultrasound system (Visual Sonics, Inc) at a 30-Hz frame rate. The transmit power of the ultrasonography system was set at 100%, and the system was equipped with a 35-MHz center frequency transducer and a mechanical scanner. At 10-mm focal depth, the image resolution was 50 × 110 μm.

The axillary lymph nodes of the 6 mice were imaged, with nanobubble administration (200-μL per mouse) and without nanobubble administration, in rendering mode by ultrasonography. Respiration gating is available in 3D mode.

Image processing and quantification were performed with the software installed in the US scanner. Image processing by the Vevo 770 system relied on 2 sets of images: a pre-nanobubble administration set (reference) and a post-nanobubble administration set. Once the image pairs were determined, the change in the brightness of the bubble flowing in the blood vessel was extracted. Reconstruction image was generated and displayed in shades of green on top of the 3D image by a blending algorithm to produce a microvascular construction image. Further, minimum axial diameters of the lymph nodes of mice in both the groups were measured on day 14.

2.7. *Statistical Analysis*

All measurements are expressed as mean ± S.E.M. (standard error of the mean). An overall difference between the groups was determined by one-way analysis of variance (one-way ANOVA). When the one-way ANOVA results were significant for 3 samples, the differences between each group were estimated using the Tukey-Kramer test. Simple comparisons of the mean and S.E.M. of the data were performed using Student's t-test. The differences were considered to be significant at $P < 0.05$.

3. Results

All the mice survived after undergoing the experimental procedure.

3.1. *Axillary Lymph Node Metastasis was Confirmed*

Bioluminescence observed 2 d after implantation by using *in vivo* bioluminescence imaging system confirmed tumor establishment and metastasis (Fig. 1). Luciferase activity in the inguinal and axillary lymph nodes, observed on days 2, 6, 10, and 14 post-implantation, was found to increase with time (Fig. 1A).

Ex vivo bioluminescence was visualized in the inguinal as well as axillary lymph nodes, which confirmed axillary lymph node metastasis (Fig. 1C)

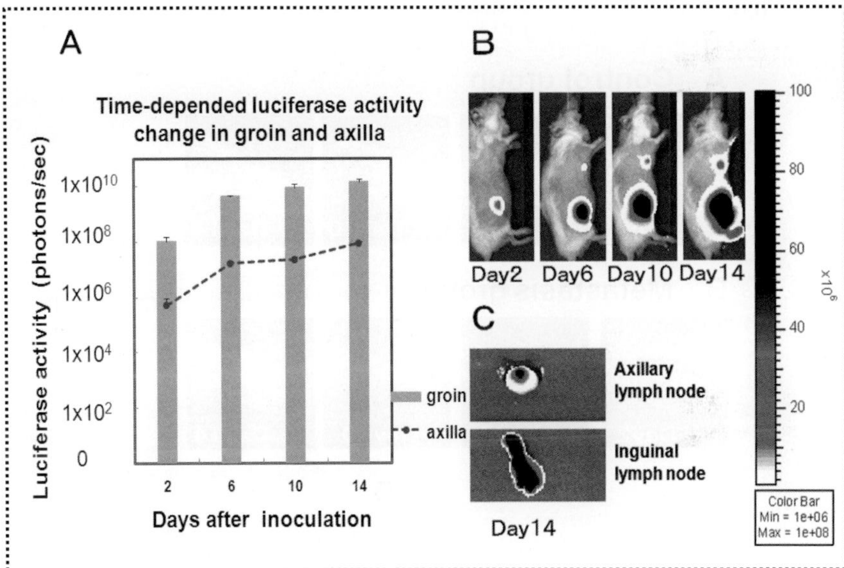

Figure 1. Visualization of tumor establishment and metastasis in lymph nodes by using *in vivo* imaging system (IVIS).

3.2. *Volume of the Metastatic Lymph Node was Constructed by Using a High-frequency Ultrasonograph*

Figure 2A and B shows the representative 3D volume construction image of axillary lymph node with elapsed time; the image was obtained using a 3D visualization high-frequency ultrasonograph and nanobubbles. The mice were divided into 2 groups: control group (n = 3) and metastasis group (n = 3).

Quantitative analysis of the construction sonograms revealed that the volume-to-growth ratios in the control and metastasis groups as well as within the metastasis group did not differ significantly between days 0 and 14 (Fig. 2C).

3.3. *Minimum Axial Diameters of Lymph Nodes were Measured*

On day 14, the minimum axial diameters of lymph nodes of the control and metastasis group mice were measured (Fig. 2D). The metastasis group mice showed the mean diameter of 4.37 ± 0.62 mm (n = 3), while the control group mice showed the mean diameter of 3.54 ± 0.60 mm (n = 3). Thus, the mean minimum axial diameter of lymph nodes of the metastasis group mice was greater than that of the control group mice; however, the difference was not significant ($P > 0.05$).

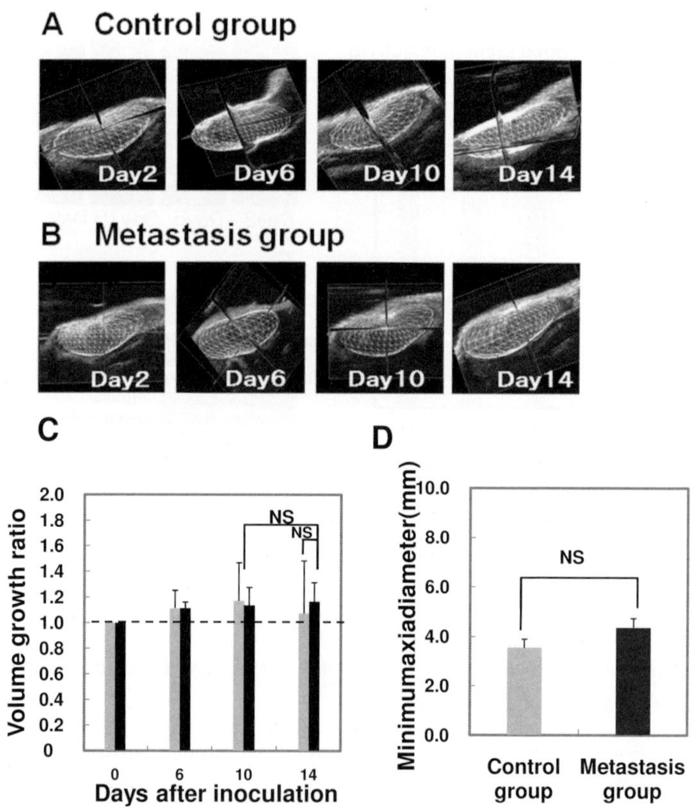

Figure 2. Analysis of volume construction of metastatic axillary lymph node by 3D high-frequency ultrasound Imaging.

3.4. *Vascular Density of the Metastatic Lymph Node was Constructed by Using a High-frequency Ultrasonograph*

The mice were divided into 2 groups as previously described. Microvascular construction image of the axillary lymph nodes was obtained using a 3D visualization high-frequency ultrasonograph and nanobubbles. Quantitative analysis of the construction sonograms revealed that the normalized vascular density in the control and metastasis groups were significantly different on day 14 ($P < 0.05$). Further, the vascular density within the metastasis group continued to mount on days 6, 10, and 14 post-inplantation, and were significantly different on days 6 and 14 ($P < 0.05$) (Fig. 3C) .

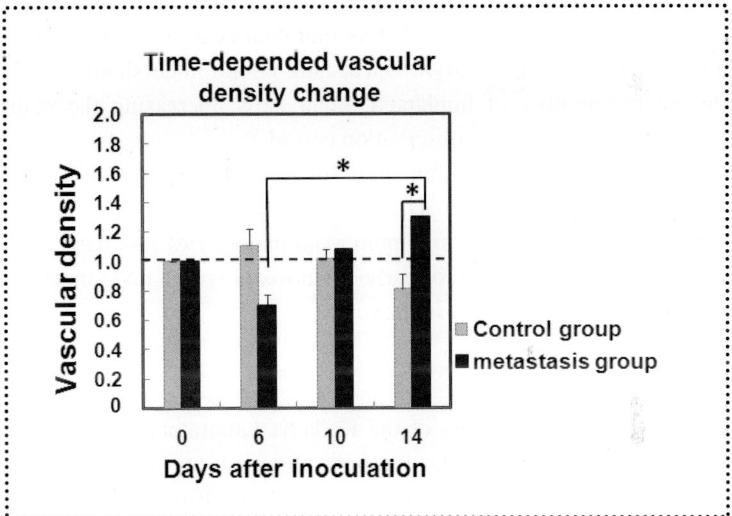

Figure 3. Quantitative analysis of microvascular construction sonograms.

4. Discussion

In this study, we demonstrated for the first time that microvascular density can be assessed in a murine model of lymph node metastasis by 3D high-frequency US imaging of angiogenesis and use of nanobubbles.

First, we established a lymph node metastatic model in MRL-*lpr*/*lpr* mice. In our model, the establishment of tumor in inguinal lymph node and its metastasis to ipsilateral axillary lymph node were confirmed *in vivo* by using a bioluminescence imaging system as well as *ex vivo*; this makes the model more reliable.

Next, we attempted to detect lymph node micrometastasis by using 3D visualization high-frequency ultrasonograph and nanobubbles. The axillary lymph nodes of 6 mice were imaged in 3D render mode by ultrasonograph with and without nanobubble administration. Quantitative analysis of the construction sonograms revealed that normalized vascular density in the control and metastasis groups were significantly different on day 14 ($P < 0.05$). In our study, angiogenesis relative to metastatic lymph node growth was visualized by ultrasonography. In addition, on the last day of the experiment, the minimum axial diameters of axillary lymph nodes of the mice in both the groups measured less than 5 mm. Taken together, we consider that early lymph node metastasis can be diagnosed by assessing the vascular density with elapsed time.

This study had some limitations. Although significant, the number of animals used in this study was small. Vascular density decreased temporarily. In the future, microvascular density of metastatic lymph node should be assessed by reducing the number of implanted tumor cells, increasing the number of study animals, and prolonging observation period.

In conclusion, the results of our study suggest that US imaging with contrast nanobubbles allows noninvasive 3D spatial visualization of volume and vascular density in lymph nodes with a minimum axial diameter of less than 1 cm. This should serve as direct evidence for early diagnosis of lymph node metastasis and monitoring of antitumor therapy in the future.

Acknowledgments

We thank all the staff members of the Kodama Laboratory for their technical assistance and acknowledge the support provided by a grant of Tohoku University Global COE Program "Global Nano-Biomedical Engineering Education and Research Network Centre".

Reference

1. H. Nakanishi, S. Ito, Y. Mochizuki *et al.*, *Methods Mol Med.* **111**, 351–362 (2005).
2. H. Hricak, C. G. Lacey, L. G. Sandles *et al.*, *Radiology* **166**, 623–631 (1998).
3. K. Togashi, K. Nishimura, T. Sagoh *et al.*, *Radiology* **171**, 245–251 (1989).
4. K. Matsukuma, N. Tsukamoto *et al.*, *Gynecol Oncol.* **33**, 168–171 (1989).
5. S. H. Kim, B. I. Choi, H. P. Lee *et al.*, *Radiology* **175**, 45–51 (1990).
6. S. H. Kim, B. I. Choi, J. K. Han *et al.*, *J Comput Assist Tomogr.* **17**, 633-640 (1993).

7. S. H. Kim, S. C. Kim, B. I. Choi and M. C. Han, *Radiology* **190**, 807–811(1994).
8. H. Hricak, C. B. Powell, K. K. Yu *et al.*, *Radiology* **198**, 403–409 (1996).
9. J. Scheidler, H. Hricak *et al.*, *JAMA*. **278**, 1096–1101 (1997).
10. H. Hawighorst, S. O. Schoenberg, P. G. Knapstein *et al.*, *J Comput Assist Tomogr.* **22**, 75–81 (1998).
11. H. Kaur, P. M. Silverman, R. B. Iyer *et al.*, Diagnosis, staging, and surveillance of cervical carcinoma. *AJR Am J Roentgenol* **180**, 1621–1631 (2003).
12. W. T. Yang, W. M. Lam, M. Y. Yu, T. H. Cheung and C. Metreweli, *AJR Am J Roentgenol* **175**, 759–766 (2000).
13. A. D. Williams, C. Cousins, W. P. Soutter *et al.*, *AJR Am J Roentgenol.* **177**, 343–348 (2001).
14. N. Weidner, J. P. Semple, W. R. Welch *et al.*, *Annu Rev Med.* **49**, 407–424 (1998).
15. R. Kerbel and J. Folkman, *Nat Rev Cancer.* **2**, 727–739 (2002).
16. J. K. Willmann, R. Paulmurugan, K. Chen *et al.*, *Radiology* **246**, 508–518 (2008).
17. A. M. Y. Cheung, A. S. Brown, L. A. Hastie, *Ultrasound in Med. & Bo.* **31**, 865–870 (2005).
18. J. Du, F. H. Li, H. Fang, *J Ultrasound Med.* **27**, 821–831 (2008).
19. R. Suzuki, T. Takizawa, Y. Negishi, K. Hagisawa, *J Control Release.* **117**, 130–136 (2007).

EVALUATION OF EXERCISE INDUCED ORGAN ENERGY METABOLISM USING TWO ANALYTICAL APPROACHES: A PET STUDY

MEHEDI MASUD [1], TOSHIHIKO FUJIMOTO [2], MASAYASU MIYAKE [1], SHOIICHI WATANUKI [1], MASATOSHI ITOH [1], MANABU TASHIRO [1]*

1) Divisoin of Nuclear medicine, Cyclotron and Radioisotope Center, Tohoku University, Sendai, Japan

2) Center for the Advancement of Higher Education, Tohoku University, Sendai, Japan

Our purpose was to evaluate glucose metabolism of skeletal muscles and viscera after two exercise loads using ^{18}FDG and 3D-PET technique, comparing two analytical procedures. Eleven healthy male subjects were studied as controls (n=6) and exercise group (n=5). Exercise group subjects performed ergometer bicycle exercise for 40 minutes at 40% and 70% of $\dot{V}O_{2max}$. ^{18}FDG (39.2 ± 2.6 MBq) was injected 10 min after exercise task. A whole body 3D-PET scan was performed immediately after exercise using PET scanner (SET-2400W, Shimadzu Co. Kyoto, Japan). PET image data was reconstructed by filtered 3D back projection algorithm with supercomputer (SX-4/128H4, synergy Center, Tohoku Univ). Controls were studied using identical study criteria with exercise group. Two analytical procedures, semiquantitative (SUV) and quantitative (rMRGlc) [autoradiographic method (Phelps *et al.*)] were applied to assess organ glucose metabolism. ROIs were drawn on lower limb muscles (*e.g.*, Thigh and lumbar/gluteal muscles) and viscera (*e.g.*, Liver, heart and brain). To compare analytical procedures, correlation coefficient analysis was done between SUV and rMRGlc data of lower limb muscles, brain, and heart. Quantitative analysis revealed that rMRGlc was increased ($p<0.05$) in skeletal muscles of thigh and lumbar/gluteal region, and decreased in brain ($p<0.05$) at mild or moderate exercise loads. Semiquantitative analysis revealed the identical results except in lumbar/gluteal muscles. It was found a correlation between SUV and MRGlc at thigh, brain and heart, however; correlation was not suggestive at lumbar/gluteal muscles. We validated the semiquantitative procedure (SUV) taking absolute quantification method of glucose metabolic rate as a standard. It was observed that the MRGlc was increased in thigh muscles and decreased in brain at mild or moderate workloads. However, discrepancies between SUV and rMRGlc in some organs, demonstrate that semiquantitative approach needs a great care when metabolic rate of glucose utilization changes at whole-body level.

Keywords: [^{18}F]FDG-3DPET, Exercise, Glucose-Metabolism, Semiquantification, Autoradiography.

* Manabu Tashiro is a Tohoku University Global COE Member.

1. Introduction

Our previous reports [1,2] employed semiquantification method (standardized uptake values: SUV) to reveal organ glucose metabolic changes after exercise task, using ^{18}F-2-fluoro-2-deoxyglucose and three-dimensional positron emission tomography technique ([^{18}F]FDG and 3D-PET). Recently, Kemppainen J. and co-workers assessed the glucose metabolism of lower limb skeletal muscle and myocardium using absolute quantification method (rMRGlc) [3]. However, no reports have evaluated whole-body glucose metabolic changes after exercise using 3D-PET technique, comparing semiquantitative and quantitative analytical methods.

Our purpose was to evaluate workloads-induced whole-body organ glucose metabolism using two analytical methods (semiquantitative and quantitative), as to establish a relationship between two PET quantification procedures.

2. Research Design

2.1. *Subjects*

Eleven healthy male volunteers collaborated with this investigation. All subjects abstained from eating and drinking for at least 5 hours before the experiment. They were asked not to perform any kind of physical exercise from one day before investigation. 5 subjects served as exercise group whose ages ranged from 21 to 23 y (21.80 ± 0.84 y; mean ± S.D.). Another 6 subjects, aged mean 24 ± 5.34 y (range; 19 ~ 33 y) were studied as resting control maintaining the same study protocol without exercise. [^{18}F]FDG dose for control was in average 42.48 ± 6.63 MBq (mean ± S.D.). A written fully informed consent was obtained from each subject before the study. This study protocol was approved by the Clinical Committee for Radioisotope Studies of Tohoku University.

2.2. *Study Protocol*

Ergometer bicycle exercise was arranged at 40% and 70% $\dot{V}O_{2max}$ workloads. $\dot{V}O_{2max}$ was measured by intermittent exercise on an ergometer bicycle (Monark 818E, Sweden), and oxygen consumption rate was determined by an automated metabolic unit machine (AE280-S, Minato Co. Ltd. Osaka, Japan). Before the experiment, subjects rested for 20 minutes in a dim lit quiet room. One teflon catheter was inserted to their antecubital veins of the left hand for blood sampling to measure plasma glucose, lactate and insulin. Another teflon catheter was inserted to opposite antecubital veins for [^{18}F]FDG administration. Then, they started ergometer bicycle riding at the speed of 60 revolution/min (Monark 818E, Sweden) at both workloads (40% and 70% $\dot{V}O_{2max}$). [^{18}F]FDG was injected through a catheter at 10 min later following exercise task. The radioactivity dose for the exercise group was 38.37 ± 2.15 MBq (mean ± S.D.). After injection, subjects continued to pedal the bicycle for another 30 min, completing a total of 40 min task. Immediately after intravenous administration of [^{18}F]FDG, heated arterialized venous blood was sampled from cubital vein opposite to the injection site. Plasma [^{18}F]FDG concentrations were measured both during exercise and PET scan for 24 times. Plasma metabolite concentrations (*i.e.*, Glucose, lactate and insulin) were measured at two points such as pre and post exercise states.

Subjects lay down in supine position on PET table with eyes open following exercise task. The PET room was kept dimmed and quiet. The scan protocol was as follows: a 3 dimensional (3D) whole-body emission scan (3 min × 9F) was performed from knee to the vertex followed by transmission scan (3 min × 9F) using a PET apparatus (SET2400W, Shimadzu, Kyoto, Japan). The transmission scan (post-injection mode) was performed with a ^{68}Ge/^{68}Ga external rotating line source (370 MBq at purchase).

2.3. *Quantitative Methods*

Regions of interest (ROIs) were set on the skeletal muscles of thigh, lumbar/gluteal regions, and visceral organs such as liver, heart and brain *etc.* (Fig. 1). To evaluate the rate of glucose utilization, an autoradiographic method [4] was applied using the following equation:

$$rMRGlc = \frac{Cp}{LC}\left[\frac{K^*_1\,k^*_3}{k^*_2+k^*_3}\right]\left[\frac{C^*i(T)-C^*e(T)}{C^*m(T)}\right]$$

In another, semiquantitative analysis (Standard uptake value; SUV) was done by using the following equation:

$$\frac{Mean\ ROIcts\ (cps/pxls)\times Body\ weight(g)}{Injected\ dose\ (\mu Ci)\times Calibration\ factor\ (cps/\mu Ci)}$$

2.4. *Statistical Analysis*

Group comparisons were done by using one-way analysis of variance (ANOVA) and Tukey's test (post-hoc) analysis. The significant differences were set at

$p<0.05$. Correlation was calculated using Pearson's correlation coefficient analysis.

3. Results

[^{18}F]FDG uptake was only remarkable in the brain, heart and urinary bladder in the resting subject, while high uptake was visualized in skeletal muscles at exercise state (Fig. 1). Glucose metabolism (SUV and rMRGlc) was increased in the skeletal muscles of thigh and lumbar/gluteal regions ($p<0.05$), and was decreased in the brain ($p<0.05$) after exercise task (40% and 70% $\dot{V}O_{2max}$ workloads). A correlation between SUV and rMRGlc was found among organs (*i.e.*, Thigh, liver, heart and brain), except in the lumbar/gluteal muscles. Figure 2 clearly depicted a good correlation between SUV and rMRGlc in the brain (a) and heart (b); however, a non-suggestive correlation was found in the lumbar/gluteal skeletal muscle (c). The changes in plasma metabolites were as follows: stable plasma glucose concentrations, an increase ($p<0.05$) plasma lactate concentration at post-exercise condition of 70% $\dot{V}O_{2max}$ (5.3 ± 2.4 mmol/liter) to compare with pre-exercise condition (0.9 ± 0.2 mmol/liter). The plasma insulin concentration was decreased ($p<0.05$) only at post-exercise workload of 70% $\dot{V}O_{2max}$ (2.0 ± 0.7 μU/mol) than pre-exercise condition (4.6 ± 1.5 μU/mol).

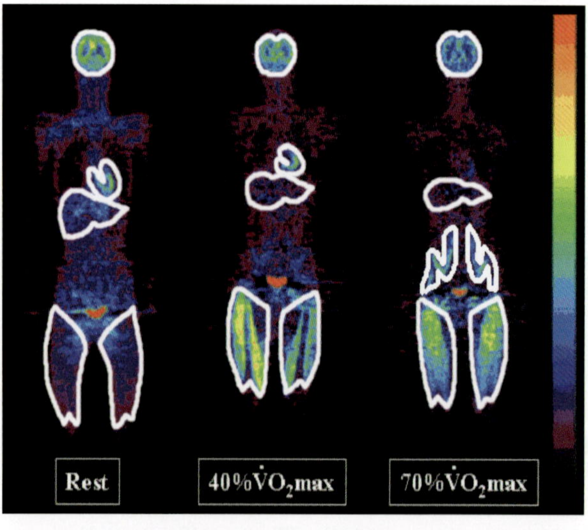

Figure 1. ROIs procedure and [^{18}F]FDG uptake of individual organs at rest (left) and exercise loads; 40% and 70% $\dot{V}O_{2max}$ (right).

4. Discussions and Conclusion

Organ glucose uptake either increased or decreased almost linearly with exercise loads up to moderate workload (70% $\dot{V}O_{2max}$). In spite of complexity of energy metabolic controls such as glucose-fatty acid metabolic interaction, aerobic-anaerobic interaction, and involvement of glycogenolysis [5], exercise-induced organ glucose metabolism were successfully assessed with [^{18}F]FDG-3D-PET technique and two analytical approaches. Organ glucose uptake either increased or decreased almost linearly with exercise loads up to moderate workload (70% $\dot{V}O_{2max}$), suggesting of homeostatic metabolic control. Semiquantitative method without blood samplings was found useful to estimate a rough trend of glucose consumptions. However, one organ failed to have good correlations between SUV and rMRGlc, the lumbar and gluteal muscles for example, which demonstrates that semiquantitative approach needs a great care when metabolic rate of glucose utilization changes at whole-body level.

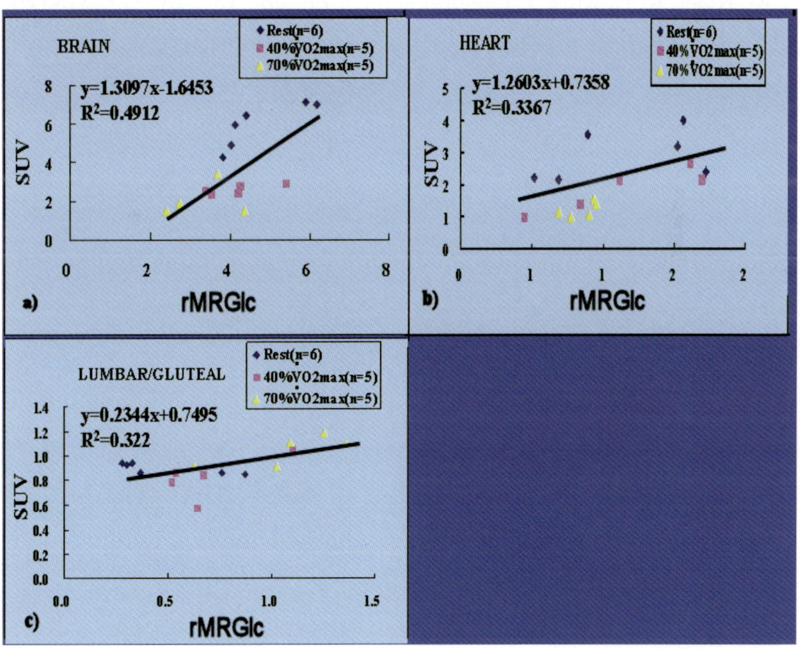

Figure 2. Correlation between SUV and rMRGlc in the viscera (a, b) and skeletal muscle (c), showing good correlation in the brain (a) and heart (b), and non-suggestive correlation in the lumbar/gluteal muscle (c).

Acknowledgments

This work was supported by grants-in-aid (No. 14704059 to T Fujimoto and No. 16790308 to M Tashiro) from the Japan Society of Promotion of Science (JSPS) and the Ministry of Education, Culture, Sports, Science and Technology in Japan, by a grant on research and education in "molecular imaging" from the Japan Society of Technology (JST), as well as by the 13th and 23rd Research-aid by Meiji Life Foundation and Welfare.

References

1. T. Fujimoto, M. Itoh, H. Kumano *et al.*, *Lancet* **48**, 266 (1996).
2. M. Tashiro, T. Fujimoto, M. Itoh *et al.*, *J Nucl Med.* **40**, 70–76 (1999).
3. J. Kemppainen, T. Fujimoto, K. K. Kari *et al.*, *J Physiol.* **542**, 403–412 (2002).
4. M. E. Phelps, S. C. Huang, E. J. Hoffman *et al.*, *Ann Neurol.* **6**, 371–388 (2005).
5. M. Kjaer, K. Howlett, J. Langfort *et al.*, *J Physiol.* **15**, **528**, 371–378 (2000).

DEVELOPMENT OF A NEW POSITRON EMISSION MAMMOGRAPHY (PEM)

MASAYASU MIYAKE [1)], SEIICHI YAMAMOTO [3)], MASATOSHI ITOH [1,4)], KAZUAKI KUMAGAI [1)], TAKEHISA SASAKI [1)], TARGINO RODRIGUES DOS SANTOS [1,5)], MANABU TASHIRO [1)*], MAMORU BABA [2)]

1) Divisions of Cyclotron Nuclear Medicine and Tohoku University Cyclotron and Radioisotope Center, Tohoku University,
6-3 Aoba, Aramaki, Aoba-ku, Sendai, Miyagi 980-8578, Japan

2) Divisions of Radiation Protection and Safety Control, Tohoku University Cyclotron and Radioisotope Center, Tohoku University,
6-3 Aoba, Aramaki, Aoba-ku, Sendai, Miyagi 980-8578, Japan

3) Department of Electrical Engineering, Kobe City College of Technology,
8-3 Gakuen-higashi-machi, Nishi-ku, Kobe, Hyogo 651-2194, Japan

4) Sendai Medical Imaging Clinic,
2-1-25 Itsutsubashi, Aoba-ku, Sendai, Miyagi 980-0022, Japan

5) CMI, Inc., Suite5114, Yebisu Garden Place Tower 18F, 4-20-3 Ebisu, Shibuya-ku, Tokyo 150-6018, Japan

Breast cancer is one of the leading causes of death among Japanese women. This malignancy can be treated relatively easily when detected early. However, it is now clear that breast cancer detection is difficult without using imaging modalities such as X-ray mammography and/or ultrasonic echography. While these imaging techniques have been proven useful, morphological information alone provided by these techniques is not sufficient for differentiating cancer from other benign diseases. Positron emission tomography for whole-body diagnosis can provide functional information regarding the degree of malignancy. Nevertheless, various problems still remain such as limited spatial resolution and motion artifacts due to respiratory movement, as well as high costs for device installment and maintenance. To compensate these weak points, we have started to develop positron emission mammography (PEM) dedicated for the local diagnosis of breast cancer.

Last year, we developed a test device we called the "PEM simulator". Moreover, we developed a prototype PEM scanner. Here, we introduce the basic features of this scanner in terms of the gantry, chair, detector unit and measurement system of the PEM scanner, as well as our first results obtained through clinical use.

---

[*] Manabu Tashiro is a Tohoku University Global COE Member.

1. Introduction

Breast cancer is one of the leading causes of death among Japanese women. This malignancy can be treated relatively easily when detected early because it can be easily identified and grows on the surface of the body. The highest incidence of breast cancer is observed among those in their thirties and forties. The high incidence among these age groups considerably affects our society because women in these generations usually play important roles in their families and working places.

Recently, imaging modalities such as X-ray mammography and/or ultrasonic echography have proven useful for the detection of breast cancer. However, these modalities provide only morphological information. On the other hand, younger women or women without a history of breastfeeding tend to develop mammary gland tumor. In some cases of X-ray mammography, a developing mammary gland produces shadows on mammography images, making detection of breast cancer difficult.

Positron emission tomography (PET) for whole-body diagnosis can provide functional information on cancerous tissues including the degree of malignancy, as reflected by the level of accumulation of a tracer, [^{18}F]fluoro-deoxyglucose. However, various problems still remain such as limited spatial resolution and motion artifacts due to respiration movement, as well as high costs for device installment and maintenance. In measuring events of the coincidences of positrons, a larger distance between detectors, as seen with PET scanners, for whole-body imaging limits spatial resolution.

To compensate these weak points, we have started to develop a positron emission mammography (PEM) scanner dedicated for the local diagnosis of breast cancer. This PEM scanner is much smaller than the conventional PET scanner. A smaller scanner presents advantages of improved spatial resolution and lower costs.

2. PEM Development and Applications

In recent years, a number of PET systems for detecting breast cancer have been developed. Moses has proposed a PEM camera that had parallel planes of detector modules. The modules had a larger solid angle around the breast than the conventional PET scanner, and thus it offered advantages of both higher efficiency and lower cost [1]. Freifelder *et al.* have employed computer simulations to compare two designs, namely, a ring scanner surrounding the breast and another scanner consisting of two planar detectors placed on opposite sides of the breast [2] Doshi *et al.* have designed and evaluated a PET scanner using LSO, and optical tapered light guides have been used to reduce the gaps of scintillator arrays [3] Murthy *et al.* [4] and Murthy *et al.* [5] have combined PEM with X-ray mammography to achieve overlaying of images. Lecoqa *et al.* have used LYSO as a scintillator and an avalanche photodiode as a scintillation detector [6]. On the other hand, Raylmana *et al.* have developed a dedicated system for biopsy of breast cancer [7,8].

3. Development Prototype PEM Scanner

3.1. *PEM Simulator*

Presently, we are in the process of developing a PEM scanner, and we will also fabricate a prototype PEM scanner for commercial use. This scanner will be used for detecting smaller tumors, and it will be cheaper and smaller than the conventional PET scanner. We are aiming for a spatial resolution of <1-2 mm, and the price will be 20% lower than that of the conventional PET scanner.

Last year, we developed several test devices and a PEM simulator. One of these devices is a block detector device. This device is constructed from a flat-panel photomultiplier tube (FP-PMT), an inorganic scintillator crystal array, an aluminum case and a signal processing circuit. The FP-PMT is a multianode PMT with 16x16 anodes and has positional sensitivity.

The PEM simulator is a testing and an evaluating system of the PEM scanner. The simulator has four block detector devices, high voltage supply units for PMTs and signal processing units. The latter consists of an analog amplifier, an analog to digital converter, FPGA devices and communication interface for PC.

We used this device for evaluation of the basic property of the PEM scanner we are developing. We measured point sources and a line source, and took some distribution images of positron sources.

We show below the distribution images of a positron line source. A crystal pair is set parallel, and then a curved tube filled with the liquid of a positron source (18-F) is placed in the center of the crystals (Fig. 1, left). Images shown in the right side of Fig. 1 are the reconstructed distribution images of a positron source. These images are parallel to the crystals and ordered from one crystal to another. Thus, the images show a shape of the line source.

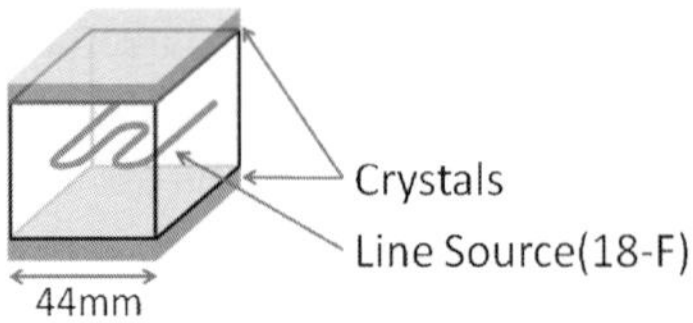

Figure 1. Geometry of the measurement of a positron line source (left) and the measured images of this line source (right).

3.2. *Large Area PEM Detector Unit*

We have developed a large area PEM detector unit (156 x 208 mm^2) consisting of three 140.8 x 44 mm^2 inorganic scintillator crystals arrays (Fig. 2) and 3 x 4 FP-PMTs (Fig. 3). These are stored in an aluminum case shielded from light (Fig. 4). Two PEM detector units are used for the prototype PEM scanner and are set up in parallel. Presently, we are able to measure some radioactive phantoms using these detector units.

Moreover, we are now developing control and acquisition software, a gantry and a chair. Soon, we will finally assemble the detector unit and these parts and construct the PEM scanner. Importantly, we will check the safety of the scanner and assess its applicability in clinical use.

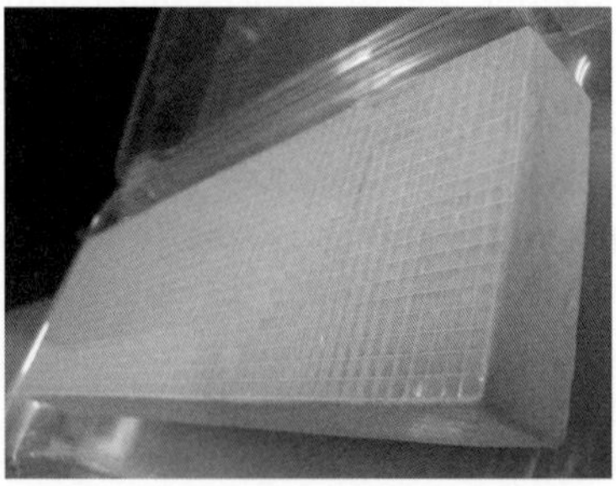

Figure 2. Inorganic scintillator crystals arrays. The size of the crystal is 2.2 x 2.2 x 15 mm^3 and that of the arrays is 140.8 x 44 mm^2.

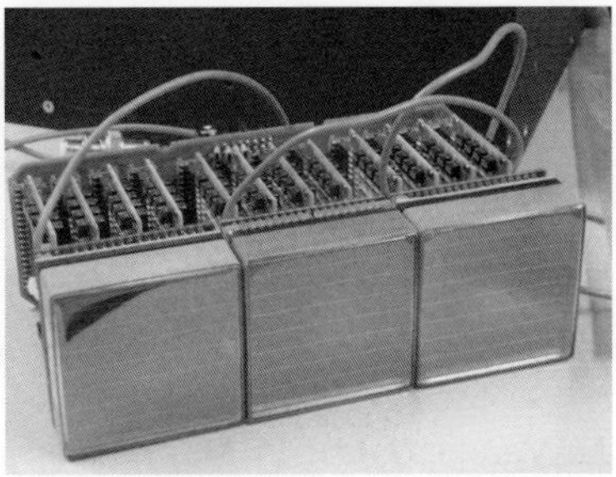

Figure 3. Parts of the FP-PMTs of the PEM detector unit.

Figure 4. Opened cover of the PEM detector unit shows an aluminum case, an inorganic scintillator crystal array and electronic circuits for FP-PMTs.

Acknowledgments

This project named, "Development of high-resolution PEM equipment for the next-generation breast cancer diagnosis" was supported by a grant from the Regional Research and Development Resources Utilization Program of the Japan Science and Technology Agency (JST).

References

1. W. W. Moses, Positron emission mammography imaging. *Nucl Instrum Methods Phys Res A* **525**, 249–252 (2004).
2. R. Freifelder and J. S. Karp, Dedicated PET scanners for breast imaging. *Phys Med Biol* **42**, 2463–2480 (1997).
3. N. K. Doshi, Y. Shao, R. W. Silverman and S. R. Cherry, Design and evaluation of an LSO PET detector for breast cancer imaging. *Med Phys* **27**, 1535–1543 (2000).

4. K. Murthy, M. Aznar, A. M. Bergman, C. J. Thompson, J. L. Robar, R. Lisbona, A. Loutfi and J. H. Gagnon, Positron emission mammographic instrument: initial results. *Radiology* **215**, 280–285 (2000).
5. K. Murthy, M. Aznar, C. J. Thompson, A. Loutfi, R. Lisbona and J. H. Gagnon. Results of preliminary clinical trials of the positron emission mammography system PEM-I: A dedicated breast imaging system producing glucose metabolic images using FDG. *J Nucl Med* **41**, 1851–1858 (2000).
6. P. Lecoqa and J. Varela, Clear-PEM, a dedicated PET camera for mammography. *Nucl Instrum Methods Phys Res A* **486**, 1–6 (2002).
7. R. R. Raylman, S. Majewski, A. G. Weisenberger, V. Popov, R. Wojcik, B. Kross, J. S. Schreiman and H. A. Bishop, Positron emission mammography–guided breast biopsy. *J Nucl Med* **42**, 960–966 (2001).
8. R. R. Raylmana, S. Majewskib, B. Krossb, V. Popovb, J. Proffittb, M. F Smithb, A. G. Weisenbergerb and R. Wojcikb, Development of a dedicated positron emission tomography system for the detection and biopsy of breast cancer. *Nucl Instrum Methods Phys Res A* **569**, 291–295 (2006).

ANALYSIS OF ANATOMICAL NETWORKS USING REGIONAL GRAY MATTER VOLUME WITH JAPANESE BRAIN MRI DATABASE

KAI WU [*], HIROSHI FUKUDA [†]

Department of Nuclear Medicine and Radiology,
Institute of Development, Aging and Cancer, Tohoku University, Sendai, Japan

Recent research has revealed that human brain networks that are functional during activity or even at the resting state exhibit a small-world architecture. This study attempted to examine the network of the anatomical connections in the human brain. We investigated large-scale anatomical connection patterns of the human brain using regional gray matter volume from a Japanese MR image database. The regional gray matter volume network examined in this study exhibited 'small-world' attributes, characterized by the presence of abundant clustering of connections combined with short average distances between the regions. More importantly, our results demonstrated that a significant statistical difference existed in the small-world properties between the sexes.

1. Introduction

In recent years, significant progresses have been made in the study of human brain aging, using MRI, fMRI, DTI, PET, and SPECT [1-3]. Neuroimaging parameters have made interesting contributions to understanding the structural and functional aging of the brain. Our previous studies using a Japanese brain MR image database (1637 subjects, aged between 12 and 87 years) have shown that the gray matter volume (GMV) showed a significant correlation with age in both sexes, while the white matter volume did not exhibit such a correlation [4,5]. Gray matter ratio (GMR) and age also showed a significant negative correlation in both sexes. A graph theoretical analysis of the influence of aging on complex brain networks investigated the functional connectivity of different EEG bands at the baseline situation (rest) and during mathematical thinking in children and young adults to study the maturation effect on brain networks both at rest and during a cognitive task. Results have indicated that (i) enhanced synchronization for the theta band during math was more prominent in adults,

---

[*] Kai Wu is a Tohoku University Global COE Research Assistant.
[†] Hiroshi Fukuda is a Tohoku University Global COE Member.

(ii) decrease of the optimal small-world network (SWN) organization of the alpha 2 band during math, (iii) the beta and, particularly, the gamma bands showed lower synchronization and signs of lower SWN organization in both situations in adults [6]. Another recent study using resting fMRI to analyze the efficiency and cost of brain functional networks demonstrated that brain functional networks have economical small-world properties supporting efficient parallel information transfer at a relatively low cost, which is differently impaired by normal aging [7]. Efficiency was reduced at a rate disproportionate to cost in older people, and the detrimental effects of age on efficiency were localised to the frontal and temporal cortical and subcortical regions.

The graph theory is of special interest as it applies to the structural and functional brain connectivity. Recent study on large-scale anatomical connectivity patterns of the human brain using cortical thickness measurement from in vivo MRI indicated that the human brain anatomical network had robust small-world properties with cohesive neighborhoods and short mean distances between regions that were insensitive to the selection of correlation thresholds [8,9].

Despite these studies, the influence of aging on human brain connectivity and the differences between the sexes remain unclear. In the present study, we investigated the large-scale anatomical connectivity patterns in the human brain consisting of 56 regions, using the regional gray matter volume (RGMV) from the Japanese brain MRI database and also examined the differences in the small-worldness between the sexes, and among decades from 20 to 70 years.

2. Materials and Methods

2.1. *Subjects and Image Acquisition*

The subjects were all community-dwelling normal Japanese subjects recruited by 2 projects: the Aoba Brain Imaging Project, Sendai, Japan and the Tsurugaya Project, Sendai, Japan. Then, we collected the brain images of 1496 subjects and the subjects of this study consisted of 754 men (mean ± SD; age, 44.89 ± 18.33 years; range: 18–79 years) and 742 women (mean ± SD; age, 48.16 ± 15.49 years; range: 18–79 years). Characteristics of the subjects are shown in Table 1. Brain images were obtained from each subject using an 0.5 T MR scanner (Signa contour, GE-Yokogawa Medical Systems, Tokyo).

Table 1. Characteristics of the subjects

Decade	Female			Male		
	Number of subjects	Age Mean ± SD	Group ID	Number of subjects	Age Mean ± SD	Group ID
20 *	127	24.23 ± 3.11	F20	215	21.88 ± 3.14	M20
30	99	34.60 ± 2.73	F30	96	34.78 ± 2.84	M30
40	139	45.76 ± 2.70	F40	111	45.02 ± 3.04	M40
50	186	53.63 ± 2.91	F50	135	54.61 ± 2.86	M50
60	107	63.88 ± 2.66	F60	99	64.11 ± 2.99	M60
70	84	72.20 ± 1.89	F70	98	72.31 ± 2.06	M70

*Decade of 20 included the subjects which ages are from 18–29 years.

2.2. *Image Analysis*

After the image acquisition, all the T1-weighted MR images were analyzed using statistical parametric mapping 2 (SPM2) (Wellcome Department of Cognitive Neurology, London, UK) in Matlab. First, T1-weighted MR images were registered into the same stereotactic space and transformed using linear and non-linear parameters into a template brain image. The template image used was the ICBM 152 template (Montreal Neurological Institute, Montreal, Canada), which approximates the Talairach space, derived from 152 normal subjects. Then, tissue segmentation from the raw images into the gray matter, white matter, cerebrospinal fluid (CSF) space, and non-brain tissue segments was performed using the SPM2 default segmentation procedure. We applied these processes using the Matlab file 'cg_vbm_optimized' (http://dbm.neuro.uni-jena.de/vbm.html).

WFU_pickatlas software was employed to label the regions in the gray matter images. We chose 28 separate regions for each hemisphere to calculate the RGMV for each subject. RGMV was calculated as the percentage fidelity, which is computed as the (total number of probed voxels – total number of mismatched voxels)/(the total number of probed voxels).

The main objective of this study is to investigate whether the properties of the anatomical network vary significantly with the sexes and the age decades. We divided the data of RGMV obtained from all the subjects into 12 groups according to sex and age decades between 20 and 70 years, the Group IDs of which were as shown in Table 1.

2.3. *Graph-theoretical 'Small-world' Analysis of the Age Decades in Each Gender*

To study the properties of the anatomical network between the sexes and among the age decades between 20 and 70 years, we constructed a correlation matrix using F20, F30, F40, F50, F60, F70, M20, M30, M40, M50, M60, and M70 and calculated the network parameter based on the graph-theoretical analysis.

To analyze the anatomical network by using the RGMV, we first applied the method introduced by Yong He to construct the anatomical connection matrix [8]. We computed the Pearson correlation coefficient (PRC) between the RGMV across the subjects in each group to construct the interregional correlation matrix ($N \times N$, where N is the number of gray matter regions, here $N = 56$). Each connection matrix (12 in total) can be converted to a binarized and undirected graph G by considering a threshold T (range from 0 to 1). For increasing values of the threshold, more and more edges will be lost and the resulting graphs will become sparse, leading to a decrease of the mean degree. As the correlation threshold reaches the maximum, *i.e.*, the T_{max}, the mean degree of the resulting network will be less than the log of the number of nodes (here $K_{min} < \log(N) = 4.025$) and the properties of small-world will become unestimable. Since the structure of the graph is generally biased by the number of existing edges, statistical measures should be calculated on graphs of equal degree K. Therefore, the threshold T was chosen for each connection matrix (12 in total) in such a manner that all the produced graphs had a fixed mean degree. In this study, the best choice for the fixed mean degree was $K = 6$ with the total number of edges $N = 168$, equivalent to 10.9% of the maximum number of edges possible (1540) in a network of 56 nodes [7]. Thresholding each connection matrix with $T_{K=6}$, we obtained the anatomical connection matrix described as the binarized and undirected graph G.

We calculated the 3 important metrics of the graph G, namely, the mean degree K, the clustering coefficient C, and the characteristic path length L [9]. Network topology was said to correspond to a "small world" if the network's clustering coefficient is much greater than that of equivalent random controls $C > C_{rand}$ or $\gamma = C/C_{rand} > 1$, while their path lengths are comparable $L \approx L_{rand}$ or $\lambda = L/L_{rand} \approx 1$. The small-worldness σ_{sw} is defined as $\sigma_{sw} = \gamma/\lambda > 1$. Comparisons are carried out against populations of $n = 1000$ degree-matched random networks [8,9].

2.4. *Statistical Analysis*

For statistics, we used the SPSS 14.0 for Windows. We used a T-Test to analyze the differences in the graph-theoretical properties of anatomical network between the sexes, using the fixed mean degree K = 6.

3. Results

With the fixed mean degree K = 6, the results by the different gender groups showed that significant differences exist in C, gamma, sigma between females and males at level of $P < 0.05$, with the probability of 0.011, 0.012, and 0.034, respectively (Table 2).

Table 2. Small-worldness in each gender▲

	Female		Male			
	Mean	S. D.	Mean	S. D.	F	Sig.
C	0.5136	0.0365	0.6278	0.0818	9.741	0.011 *
L	2.6818	0.2119	3.0549	0.4526	3.345	0.097
Gamma	2.4137	0.2300	3,7912	1.0722	9.468	0.012 *
Lamda	1.0885	0.0880	1.2540	0.1882	3.809	0.080
Sigma	2.2353	0.3256	3.0444	0.7386	6.030	0.034 *

▲ K = 6;
* Correlation is significant at the 0.05 level (2-tailed).

4. Conclusion

To our knowledge, the present study provides the first evidence of gender and generational differences in the properties of anatomical network, using RGMV. The results indicated that significant differences exist in the properties of the anatomical network between the sexes although its biological significance still remains unknown.

Acknowledgments

The authors wish to acknowledge the important contribution from Y. Taki, S. Kinomura, K. Sato, R. Goto, K. Inoue, and K. Okada. We would like to thank Y. He and A. Evans for their insightful comments. This study was supported by the 2007 Tohoku University Global Century Centre of Excellence (GCOE) Programme (Ministry of Education, Culture, Sports, Science and Technology; MEXT), entitled "Global Nano-Biomedical Engineering Education and Research Network Centre". The work in the present study had been partially reported in the 7th International Symposium on Nano-Biomedical Engineering, October 16–17, National Cheng Kung University, Tainan, Taiwan.

References

1. C. Smith, H. Chebrolu, D. Wekstein, F. Schmitt and W. Markesbery, Age and gender effects on human brain anatomy: A voxel-based morphometric study in healthy elderly. *Neurobiology of Aging* **28**, 1075–1087 (2007).
2. H. Lemaitre, F. Crivello, B. Grassiot, A. Alperovitch, C. Tzourio and B. Mazoyer, Age- and sex-related effects on the neuroanatomy of healthy elderly. *NeuroImage* **26**, 900–911 (2005).
3. E. Sullivan and A. Pfefferbaum, Diffusion tensor imaging in normal aging and neuropsychiatric disorders. *European Journal of Radiology* **45**, 244–255 (2003).
4. Y. Taki, R. Goto, A. Evans, A. Zijdenbos, P. Neelin, J. Lerch, K. Sato, S. Ono, S. Kinomura, M. Nakagawa, M. Sugiura, J. Watanabe, R. Kawashima and H. Fukuda, Voxel-based morphometry of human brain with age and cerebrovascular risk factors. *Neurobiol Aging* **25**, 455–463 (2004).
5. Y. Taki, S. Kinomura, S. Awata, K. Inoue, K. Sato, H. Ito, R. Goto, S. Uchida, I. Tsuji, H. Arai, R. Kawashima and H. Fukuda, Male elderly subthreshold depression patients have smaller volume of medial part of prefrontal cortex and precentral gyrus compared with age-matched normal subjects: A voxel-based morphometry. *Journal of Affective Disorders* **88**, 313–320 (2005).
6. S. Micheloyannis, M. Vourkas, V. Tsirka, E. Karakonstantaki, K. Kanatsouli and C. Stam, The influence of ageing on complex brain networks: A graph theoretical analysis. *Human Brain Mapping*, doi:10.1002/hbm.20492 (2007).
7. S. Achard and E. Bullmore, Efficiency and Cost of Economical Brain Functional Networks. *PLoS Computational Biology.* **3**, e17 doi:10.1371/journal.pcbi.0030017 (2007).
8. Y. He, Z. Chen and A. Evans, Small-world anatomical networks in the human brain revealed by cortical thickness from MRI. *Cerebral Cortex.* **17**, 2407 (2007).
9. O. Sporns, C. Honey and R. Kötter, *PLoS ONE.* **2**, e1049 doi:10.1371/journal.pone.0001049 (2007).

Section 3

Nano-Biodevices

PASSIVE INTELLIGENT WALKER CONTROLLED BY SERVO BREAKS

YASUHISA HIRATA

Department of Bioengineering and robotics, Tohoku University, 6-6-01, Aoba, Aramaki, Aoba-ku, Sendai 980-8579, Japan

KAZUHIRO KOSUGE *

Department of Bioengineering and robotics, Tohoku University, 6-6-01, Aoba, Aramaki, Aoba-ku, Sendai 980-8579, Japan

We propose a new intelligent walker based on passive robotics that assists the elderly, handicapped people, and the blind who have difficulty walking. We developed a prototype of the RT Walker (Robot Technology Walker), a passive intelligent walker that uses servo brakes. The RT Walker consists of a support frame, two casters, two wheels equipped with servo brakes, and it has passive dynamics that change with respect to applied force/moment. This system is intrinsically safe for humans, as it cannot move unintentionally, *i.e.*, it has no driving actuators. In addition, the RT Walker provides a number of navigational features, including good maneuverability, by appropriately controlling the torque of servo brakes based on robot technology. We propose an environmentally-adaptive motion control algorithm, which incorporates environmental information to provide obstacle/step avoidance and gravity compensation functions. The proposed control algorithms are experimentally applied to the RT Walker to test their validity.

1. Introduction

As societies age and experience a shortage of people for nursing care, handicapped people, including the elderly and blind, find it increasingly necessary to be self-supporting. However, many such people suffer from injuries, poor eyesight, or a general lack of muscular strength, and need the support of other people in daily activities. In the living environment, the ability to walk is one of the most important and fundamental functions for humans, and enables them to realize high-quality lives.

This paper focuses on a walker-type support system, which works on the basis of the physical interaction between the system and the user. Walkers are widely used by the handicapped because they are simple and easy to use. Many

---

* Kazuhiro Kosuge is a Tohoku University Global COE Member.

robotics researchers have considered improving their functionality by adding wheels with actuators and controlling them based on robot technology (RT), such as motion control technology, sensing technology, vision technology, and computational intelligence.

Fujie *et al.* developed a power-assisted walker for physical support during walking [1]. We developed a motion control algorithm for an intelligent walker with an omni-directional mobile base, in which the system is moved based on the user's intentional force/moment [2]. Manuel *et al.* proposed a non-holonomic navigation system for a walking-aid robot named Care-O-bot II [3]. Savaniti *et al.* developed a motorized rollator [4]. Yu *et al.* proposed the PAMM system to provide mobility assistance and user health status monitoring [5]. Kotani *et al.* proposed the HITOMI system, which helps the blind navigate outdoors [6]. Various assisting technologies are also surveyed and some electric intelligent walkers are introduced [7].

Many intelligent walkers based on RT consist of servo motors mounted on a mobile base and sensors such as force/torque and ultrasonic sensors. Information from many types of sensors controls the servo motors. By appropriately controlling the servo motors, these intelligent walkers provide many functions, such as variable motion, obstacle avoidance, and navigation; thus, they provide a maneuverable system that supports walking.

On the other hand, simple walkers without servo motors consisting of a frame, wheels or casters, and handbrakes, are well known and commercially available. People can use these walkers intuitively, because they move these systems passively using intentional force/moment. However, unlike active intelligent walkers that employ servo motors, it is impossible to change the apparent dynamics of these systems; they are completely passive and cannot adapt to differing user difficulties. In addition, these systems cannot move properly in an environment with obstacles or slopes, because they offer no environment-based control over their motion.

In this paper, we consider a passive intelligent walker, which is not only simple and safe but also offers many functions similar to those found in active walkers. We develop a passive intelligent walker called the RT Walker that uses servo brakes and incorporates passive robotics. First, we explain the concept of passive robotics, and introduce the passive intelligent walker. Then, we introduce several functions that allow the walker to move based on information about its environment, and evaluate the validity of these motion control algorithms through several experiments.

2. Passive Robotics

For practical use of intelligent systems in the real world, we need to consider two main points: achieving high performance and user safety. Most conventional intelligent systems have servo motors that are controlled based on sensory information from sensors such as force/torque and ultrasonic sensors. The high performance of intelligent systems is realized in the form of functions such as power assistance, collision avoidance, navigation, and variable motion.

However, if we cannot appropriately control the servo motors, they can move unintentionally and might be dangerous for a human being. In particular, in Japan, legislation must be formulated for using them in a living environment. In addition, active intelligent systems tend to be heavy and complex because they require servo motors, reduction gears, sensors, a controller, and rechargeable batteries. Batteries present a significant problem for long-term use because servo motors require a lot of electricity.

Goswami *et al.* proposed the concept of passive robotics [8], in which a system moves passively based on external force/moment without the use of actuators, and used a passive wrist comprising springs, hydraulic cylinders, and dampers. The passive wrist responds to an applied force by computing a particular motion and changing the physical parameters of the components to realize the desired motion. Peshkin *et al.* also developed an object handling system referred to as Cobot [9] consisting of a caster and a servo motor for steering the caster based on passive robotics.

Wasson *et al.* [10,11] and Rentschler *et al.* [12] proposed passive intelligent walkers. In most of these walkers, a servo motor is attached to the steering wheel, similar to the Cobot system, and the steering angle is controlled depending on environmental information. The RT Walker proposed in this paper also has passive dynamics with respect to the force/moment applied. It differs from other passive walkers in that it controls servo brakes appropriately without using any servo motors. These passive systems are intrinsically safe because they cannot move unintentionally. Thus, passive robotics will prove useful in many types of intelligent systems through physical interaction between the systems and humans.

3. RT Walker

Conventional passive intelligent walkers consist of a frame for supporting the user's weight, a steering wheel, manual or automatic brakes, and sensors for detecting environmental information. A steering-wheel actuator is used for navigation in the environment. Brake systems such as hand brakes or automatic brakes limit the speed of the walker and prevent the user from falling. The brake function, in particular, is important and essential for the safety of users.

In this research, we pay special attention to the braking system, and propose a new passive intelligent walker (RT Walker), which uses servo brake control. It differs from conventional passive intelligent walkers in that it does not have servo motors for steering. However, the servo brakes can navigate the RT Walker, and its maneuverability can change based on environmental information or the difficulties and conditions faced by the user.

The developed RT Walker is shown in Fig. 1(a). This prototype consists of a support frame, two passive casters, two wheels with servo brakes (referred to as powder brakes), a laser range finder, tilt angle sensors, and a controller. The part of the rear wheel with the powder brake is shown in Fig. 1(b); the brake torque is transferred directly to the axle. For controlling the brake torque, we investigate the actual torque of the powder brake by removing the brake from the wheel and measuring its actual torque with a spring scale, as shown in Fig. 1(c). As shown in Fig. 1(c), the brakes change the torque almost in proportion to the input current. We use values derived from the experimental results by least squares estimation.

RT Walker is lightweight because its structure is relatively simple compared to active intelligent walkers, and it needs little electricity to operate the servo brakes. The driving force of the RT Walker is the actual force/moment applied by the user, and therefore, he/she can move it passively without using the force/torque sensor. By changing the torque of the two rear wheels appropriately and independently, we can control the motion of the RT Walker, which receives environmental information from its laser range finder and tilt angle sensors. Based on this information, the RT Walker can realize the collision avoidance, gravity compensation, and other functions.

The servo brake and laser range finder are relatively expensive; however, they are high-performance components intended for industrial fields. In the welfare fields, human-assistance devices do not require high precision motions and the part costs could be reduced if they are designed for the human support systems. Therefore, from the viewpoint of safety and cost effectiveness, the RT Walker will be used in the near future.

4. Experiments for Environmentally-adaptive Functions

We experimented with the RT Walker to illustrate the validity of the performance of a passive intelligent walker using servo brakes. First, we experimented with collision avoidance and preventing falling due to steps. In these experiments, a human pushed the RT Walker just enough to move it. This allowed us to determine the validity of the proposed control algorithm easily and intuitively. The experiments are shown in Fig. 2, and the paths of the representative point, which is a point attached to the RT Walker for detecting the obstacle/step, and the center of the walker's axle are shown in Fig. 3. These results show that the RT Walker detects a wall and a step, and avoids them using only brake control.

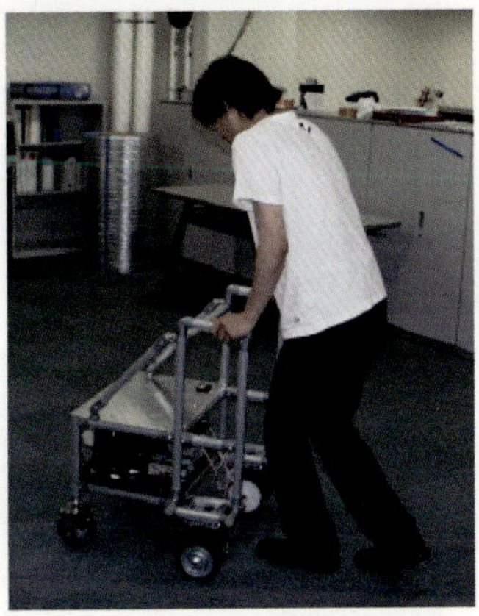

(a) RT Walker

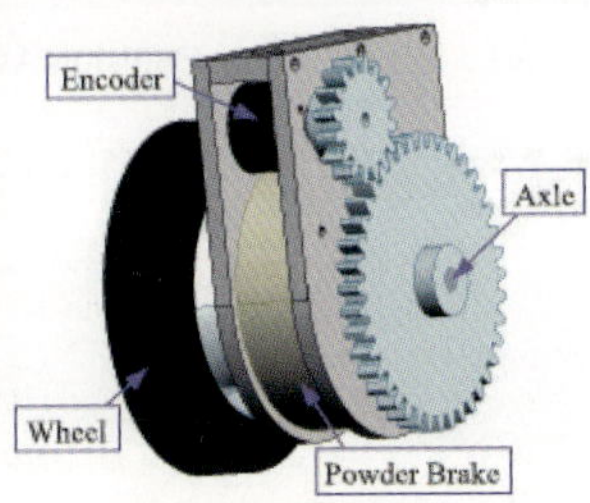

(b) Wheel with Servo Brake

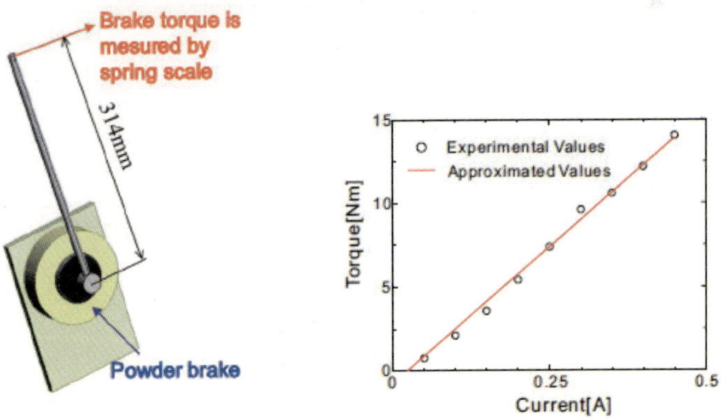

(c) Characteristic of Servo Brake

Figure 1. Passive Intelligent Walker

In the next experiment, we used the RT Walker in the environment shown in Fig. 4. We generated an artificial potential field. In this experiment, four university students used the RT Walker and each wore a blindfold to prevent visual feedback. The experimental results, shown in Fig. 4, illustrate the path of the RT Walker detected by the encoder system. Four people successfully used the RT Walker without colliding with the wall.

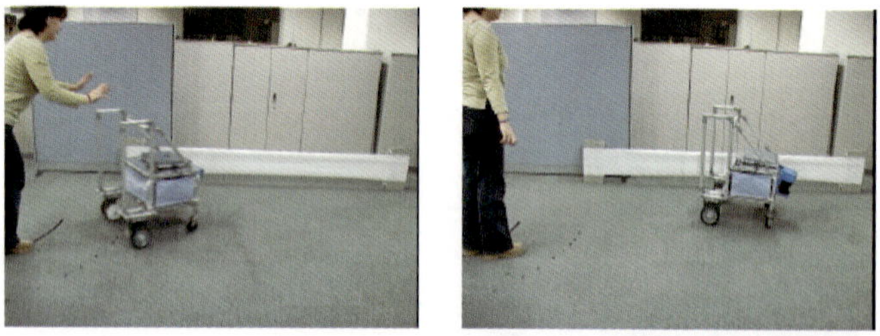

(a) Collision Avoidance with Obstacle

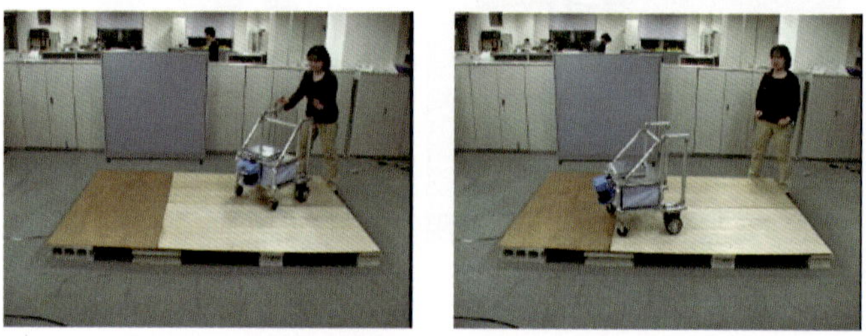

(b) Step Avoidance Motion

Figure 2. Collision/Step Avoidance Functions

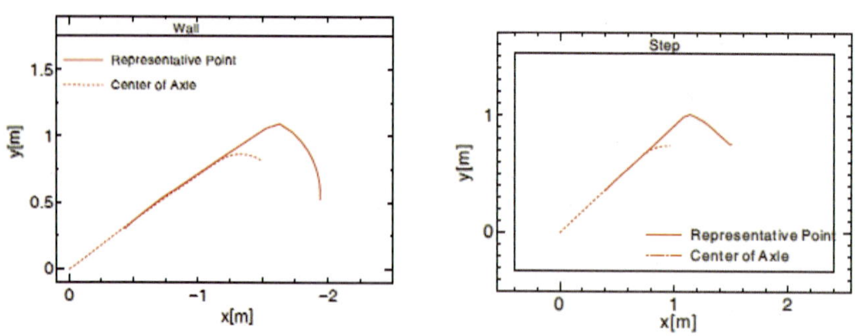

Figure 3. Experimental Results of Collision/Step Avoidance Functions

We also experimented with a path leading from a start point to a destination. In this experiment, we generated an S-line path for the RT Walker using the artificial potential field, which can generate the potential with a steep gradient. In this experiment, five university students with blindfolds operated the RT Walker. The results are shown in Fig. 5, and the differences between the desired and actual paths were almost zero; path following was successfully achieved.

Next, we experimented with gravity compensation. We tested the RT Walker in a real-world environment, as shown in Fig. 6. Although the user did not touch the walker, which was either on a downhill or uphill road, it did not move because of the force of gravity. That is, it only moved when the user intentionally applied force to it, just as if it was always on a horizontal road. Note that the RT Walker cannot pull the user against gravity when walking uphill, which makes it different from active walkers. This is the disadvantage of the passive walker. However, when walking uphill, the user's load is less, because the passive walker is relatively lightweight and is not pulled downward by gravity.

In a final experiment, we used the RT Walker in a real-world environment with stairs, obstacles, and a down slope. A user wore a blindfold to prevent visual feedback, as shown in Fig. 7. From this experimental result, you can see that the user can safely use the RT Walker, and can depend on it for navigation even in a complicated environment.

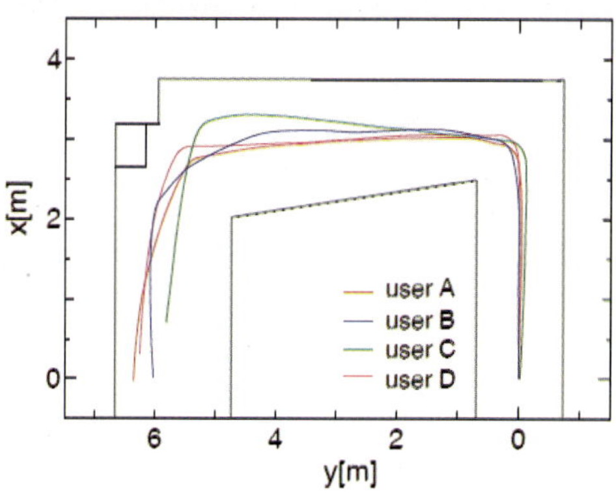

Figure 4. Motions based on Environment Information

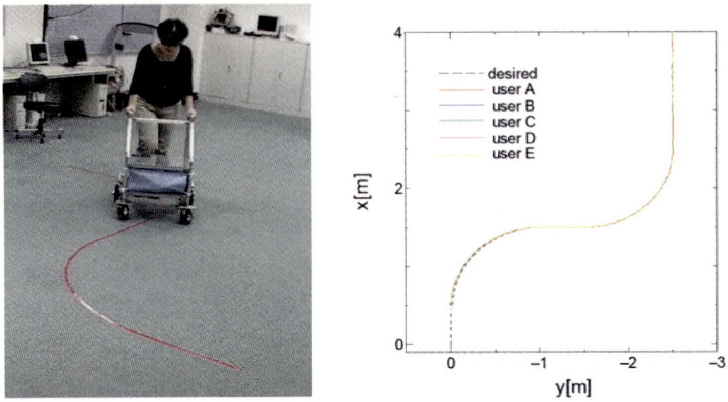

Figure 5. Path Following Function

(a)

(b)

Figure 6. Gravity Compensation Function. (a) Experiments Going Downhill. (b) Experiments Going Uphill

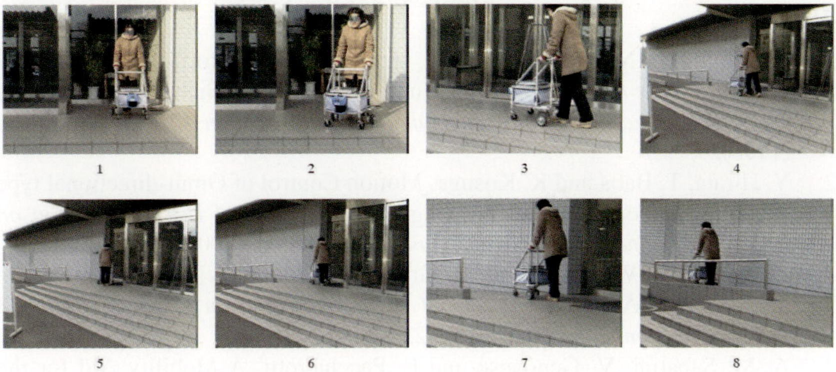

Figure 7. Experiments in Real World, with Stairs, Obstacles, and Downward Slope

5. Conclusion

In this paper, we proposed a new intelligent walker, based on passive robotics, to help the elderly, handicapped, or blind people who have difficulty walking. We developed a prototype walker, the RT Walker, which employs servo brakes. We also introduced several functions, which change the motion of the passive intelligent walker, to adapt to the environment.

Although this paper focused on realizing several important functions of intelligent walkers, such as environmentally-adaptive controls, the estimation of the human situation and selection of the appropriate functions are critical in creating dependable walkers. If an intelligent walker can recognize the user's situation, it can select from several functions, such as obstacle/step avoidance, gravity compensation, variable motion characteristics, fall-prevention, etc. Thus, users who are not professionals of the intelligent systems, can use them dependably and practically, without considering the functions of an intelligent walker. Towards this direction, we also proposed a method for selecting the appropriate functions [13], in which the user situation is estimated based on the distance between the user and the walker.

In future work, we will consider the human adaptive and environmentally-adaptive motion control algorithms in more detail to improve the maneuverability of the RT Walker. In addition, we will evaluate the validity of the RT Walker through comments from many people, including the elderly, handicapped, blind people, physical therapists, and medical doctors. We will improve the mechanisms and control algorithms of the RT Walker.

References

1. M. Fujie, Y. Nemoto, S. Egawa, A. Sakai, S. Hattori, A. Koseki and T. Ishii, Power Assisted Walking Support and Walk Rehabilitation. *Proc. of 1st International Workshop on Humanoid and Human Friendly Robotics*, (1998).
2. Y. Hirata, T. Baba and K. Kosuge, Motion Control of Omni-directional type Walking Support System "Walking Helper". *Proc. of IEEE Workshop on Robot and Human Interactive Communication*, 2A5 (2003).
3. J. Manuel, H. Wandosell and B. Graf, Non-Holonomic Navigation System of a Walking-Aid Robot. *Proc. of IEEE Workshop on Robot and Human Interactive Communication*, 518–523 (2002).
4. A. M. Sabatini, V. Genovese and E. Pacchierotti, A Mobility Aid for the Support to Walking and Object Transportation of People with Motor Impairments. *Proc. of IEEE/RSJ Intl. Conf. on Intelligent Robots and Systems*, (2002).
5. H. Yu, M. Spenko and S. Dubowsky, An Adaptive Shared Control System for an Intelligent Mobility Aid for the Elderly. *Auton. Robots* **15**, 53–66 (2003).
6. S. Kotani, H. Mori and N. Kyohiro, Development of the robotic travel aid HITOMI. *Proc. of IEEE Int. Conf. on Robotics and Automation* (1990).
7. K. Z. Haigh and H. A. Yanco, Automation as caregiver: A survey of issues and technologies. *Proc. of the AAAI-2002 Workshop on Automation as Caregiver: The Role of Intelligent Technology in Elder Care* (2002).
8. A. Goswami, M. A. Peshkin and J. Colgate, Passive robotics: an exploration of mechanical computation (invited). *Proc. of the IEEE Int. Conf. on Robotics and Automation*, 279–284 (1990).
9. M. A. Peshkin, J. E. Colgate, W. Wannasuphoprasit, C. A. Moore, R. B. Gillespie and P.Akella, Cobot Architecture. *IEEE Transactions on Robotics and Automation*, **17** (2001).
10. G. Wasson, P. Sheth, M. Alwan, K. Granata, A. Ledoux and C. Huang, User Intent in a Shared Control Framework for Pedestrian Mobility Aids. *Proc. of the 2003 IEEE/RSJ Intl. Conf. on Intelligent Robots and Systems* (2003).
11. M. Alwan, P. Rajendran, A. Ledoux, C. Huang, G. Wasson and P. Sheth, Stability Margin Monitoring in Steering-controlled Intelligent Walkers for the Elderly. *AAAI Fall 2005 Symposium* (2005).
12. A. J. Rentschler, R. A. Cooper, B. Blaschm and M. L. Boninger, Intelligent walkers for the elderly : Performance and safety testing of VA-PAMAID robotic walker. *Journal of Rehabilitation Research and Development*, **40** (2003).
13. Y. Hirata, A. Muraki and K. Kosuge, Motion Control of Intelligent Walker based on Renew of Estimation Parameters for User State. *Proc. of the 2006 IEEE/RSJ International Conference on Intelligent Robots and Systems*, 1050–1055 (2006).

NEW OBJECTIVE ASSESSMENT OF ACOUSTIC TRANSFER FUNCTION VIA PATULOUS EUSTACHIAN TUBE USING TIME-STRETCHED PULSE

TETSUAKI KAWASE *

Laboratory of Rehabilitative Auditory Science,
Tohoku University Graduate School of Biomedical Engineering,
Sendai, Miyagi, 980-8574, Japan

Department of Audiology, Tohoku University Graduate School of Medicine
Sendai, Miyagi, 980-8574, Japan

Department of Otolaryngology-Head and neck Surgery, Tohoku University Hospital
Sendai, Miyagi, 980-8574, Japan

YOKO HORI, YASUSHI BABA, TOSHIMITSU KOBAYASHI

Department of Otolaryngology - Head and Neck Surgery, Tohoku University Graduate School of Medicine
Sendai, Miyagi, 980-8574, Japan

SHUICHI SAKAMOTO, YÔICHI SUZUKI

Research Institute of Electrical Communication, Tohoku University Sendai, Miyagi, 980-8577, Japan

Patulous Eustachian tube (ET) is a condition caused by persistent opening or insufficient closing of the normally closed ET. Diagnosis of patulous ET is usually based on the verification of sound and/or pressure transmission from the pharyngeal cavity to the middle ear. In this study a new type of sonotubometry using time-stretched pulses (TSPs) was developed to evaluate the frequency characteristics of sound transmission from the pharyngeal cavity to the middle ear in the patulous ET. The obtained signal is characterized by a completely flat frequency spectrum with a very low crest factor. Using this method, the acoustic transfer function can be precisely measured very quickly (in about 10 seconds). Experimental measurements using a simple model showed that the acoustic transfer function reflects the caliber of the intermediate tube; *i.e.*, larger caliber results in greater sound transmission to the ear. Preliminary application to clinical cases suggested that the patency of the ET can be accurately assessed by TSP sonotubometry. The acoustic transfer functions were significantly increased under patulous conditions and were decreased after obstructive treatment of the patulous ET. TSP sonotubometry is an effective method for the management of patients with patulous ET.

---

* Tetsuaki Kawase is a Tohoku University Global COE Member.

1. Introduction

Patulous Eustachian tube (ET) is a condition caused by persistent opening or insufficient closing of the normally closed ET, associated with intense sound and/or pressure transmission from the pharyngeal cavity to the middle ear cavity resulting in uncomfortable symptoms such as aural fullness, autophony, hearing one's own breathing, *etc*. [1-3].

Diagnosis of patulous ET is usually based on the verification of sound and/or pressure transmission from the pharyngeal cavity to the middle ear. Pressure transmission causes synchronous movement of the tympanic membrane with respiration, which can be measured by tympanometric methods as well as tubo-tympano-aerodynamic-graphy (TTAG) based on impedance or pressure changes measured in the external auditory canal (EAC) [4-7].

On the other hand, sound transmission, which appears to be more related to autophony, can be partially assessed by conventional sonotubometry. However, conventional sonotubometry was basically developed to assess the opening function of the ET [8], so sound transmission between the pharyngeal cavity and the middle ear cannot be fully examined [9]. The recently reported methods of nasal audiometry (which investigates detection thresholds for tones presented in the nasal cavity) and nasal-noise masking audiometry (in which audiometric measurement is conducted with and without masking noise presented in the nasal cavity) are methods with potential to assess patulous ET with ease [10,11]. However, the results of these methods may be influenced by a subject's condition because a conventional audiometric procedure is used to decide the threshold.

The present paper describes a new type of sonotubometry using time-stretched pulses (TSPs) to assess the frequency characteristics of sound transmission from the pharyngeal cavity to the middle ear in patients with patulous ET.

2. Materials and Method

2.1. *System*

A acoustic signals to measure the transfer function were presented to the nostril on the test side through a special speakerphone (JK04S110; RION, Kokubunji, Tokyo, Japan) with an olive-shaped tip (JK040330; RION), originally designed for use in conventional sonotubometry [8]. The patient was asked to close the contralateral nostril by pressing the wing of the nose with the finger (Fig. 1). A microphone (ER-10C; Etymotic Research, Elk Grove Village, IL, U.S.A.) was placed in the EAC of the test side to measure the sound pressure level transmitted from the nasopharyngeal cavity to the middle ear. Earmuffs were used to minimize possible sound interference directly reaching the EAC, such as ambient noise or test sound leaked from the nostril (Fig. 1).

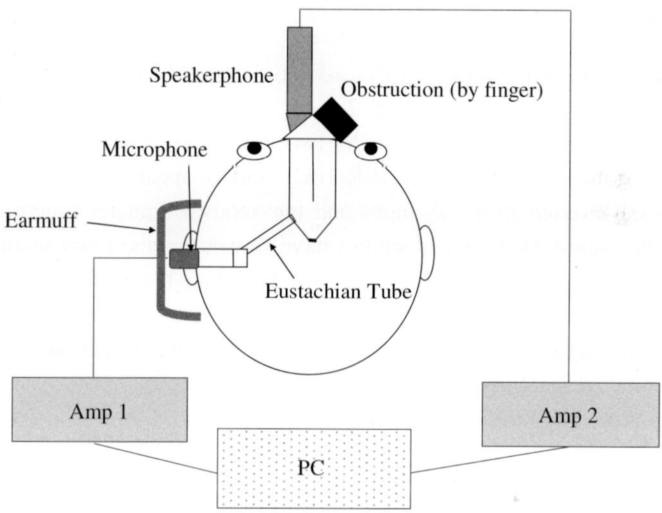

Figure 1. Schema of the measurement system. The acoustic transfer functions were analyzed using software AEIRM (Nittobo Acoustic Engineering Co., Ltd., Tokyo, Japan) and a PC board with two 24-bit AD/DA channels. The signals were presented to the nostril on the test side through a special speakerphone (JK04S110; RION, Kokubunji, Tokyo, Japan) with an olive-shaped tip (JK040330; RION). The contralateral nostril was obstructed by pressing the wing of the nose. A microphone (ER-10C; Etymotic Research, Elk Grove Village, IL, U.S.A.) was placed in the external auditory canal.

The acoustic transfer functions from the nasopharyngeal cavity to the middle ear were analyzed using software AEIRM (Nittobo Acoustic Engineering Co., Ltd., Tokyo, Japan) and a PC board with two AD/DA channels of 24-bit quantization. The acoustic signal was a time-stretched pulse signal with a duration of 0.5 sec [12]. This signal is characterized by a completely flat frequency spectrum with a very low crest factor. Therefore, high precision and high signal-to-noise ratio may be realized in the measurement of transfer functions. In the present study, a TSP designed with 2^{14} (= 16384) sample points at a sampling rate of 44100 Hz was used. The recorded responses were deconvolved by the software AEIRM to obtain the impulse responses of 1024 points long after a pure delay. To further improve the signal-to-noise ratio, 16 measurements were averaged. The averaged impulse responses were then converted to transfer functions by FFT (fast Fourier transform) analysis. The TSP signal was not compensated for the frequency response of the output transducer.

2.2. *Model Experiment*

The effect of the ET caliber on the acoustic transfer function was examined using the simple model shown in Fig. 2.

The model was constructed with two truncated syringes and a narrow connecting tube. A microphone (ER-10C) and a speakerphone (JK04S110; RION) were inserted in the syringes and the acoustic transfer function via the tube was measured. Histological studies have shown that the cross-sectional area of the narrowest portion of the ET lumen is less than 1 mm^2 [13,14]. Based on these reports, the calibers of the narrow tubes used in the present study were 0, 0.28, 0.38, 0.58, 0.76, 1.14, and 1.57 mm (=0, 0.06, 0.11, 0.26, 0.45, 1.02, and 1.93 mm^2, respectively, in area).

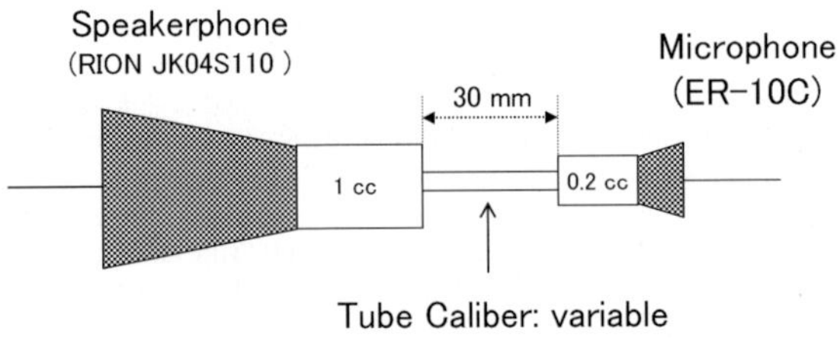

Figure 2. Schema of the model experiment. The model was constructed with two truncated syringes and a narrow connecting tube with calibers of 0, 0.28, 0.38, 0.58, 0.76, 1.14, and 1.57 mm.

2.3. *Measurement in Clinical Cases*

This new TSP sonotubometry method was applied to assess the acoustic transfer function in patients with possible patulous ET, as well as normal subjects. The measurements were conducted before and after treatment for ET obstruction in the patients. All parts of the present study were performed in accordance with the guidelines of the Declaration of Helsinki.

3. Results

3.1. *Model Study*

The effects of caliber diameter on the acoustic transfer function are shown in Fig. 3.

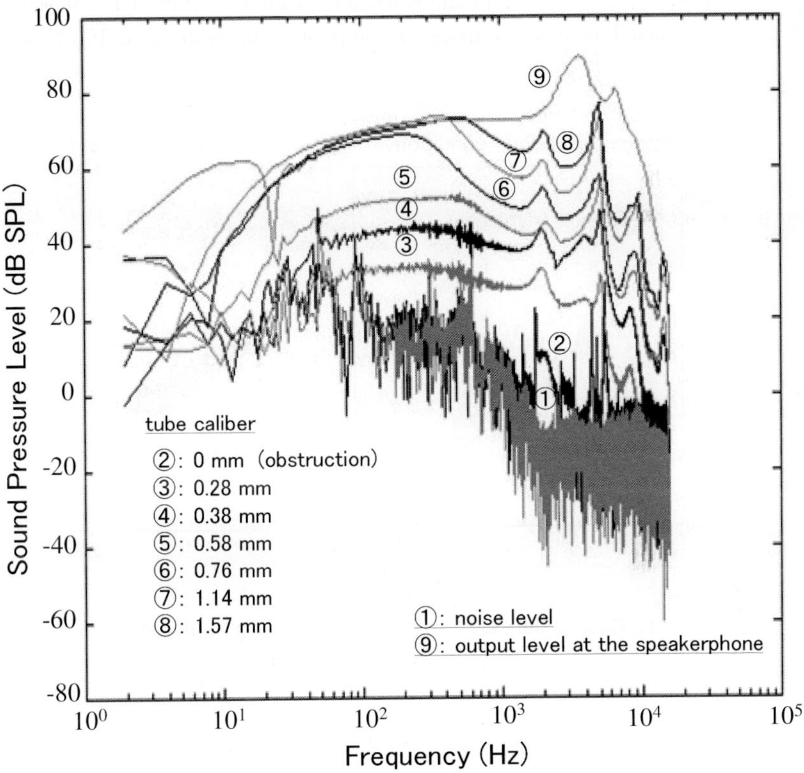

Figure 3. Effects of caliber diameter on the model acoustic transfer function. The transferred sound pressure recorded by the microphone is plotted as a function of frequency. Higher recorded sound pressure level indicates better acoustic transfer function. The transferred sound pressure was gradually reduced with smaller tube caliber. See text for further details.

The transferred sound pressure recorded by the microphone was gradually reduced with smaller tube caliber, but the acoustic transfer function varied at different frequencies. When the tube caliber was relatively large, the transferred sound pressure was reduced only at higher frequencies (Fig. 3, tube caliber: 0.76,

1.14, and 1.67 mm). On the other hand, when the caliber of the tube was smaller than 0.58 mm, the sound pressure was reduced at lower and higher frequencies.

3.2. *Clinical Cases*

Representative measurements of the acoustic transfer function in three patients complaining of patulous symptoms are shown in Fig. 4.

Case 1 was a 25-year-old female who had suffered from bilateral autophony for 2 years. Based on the findings of tympanic membrane and ET function

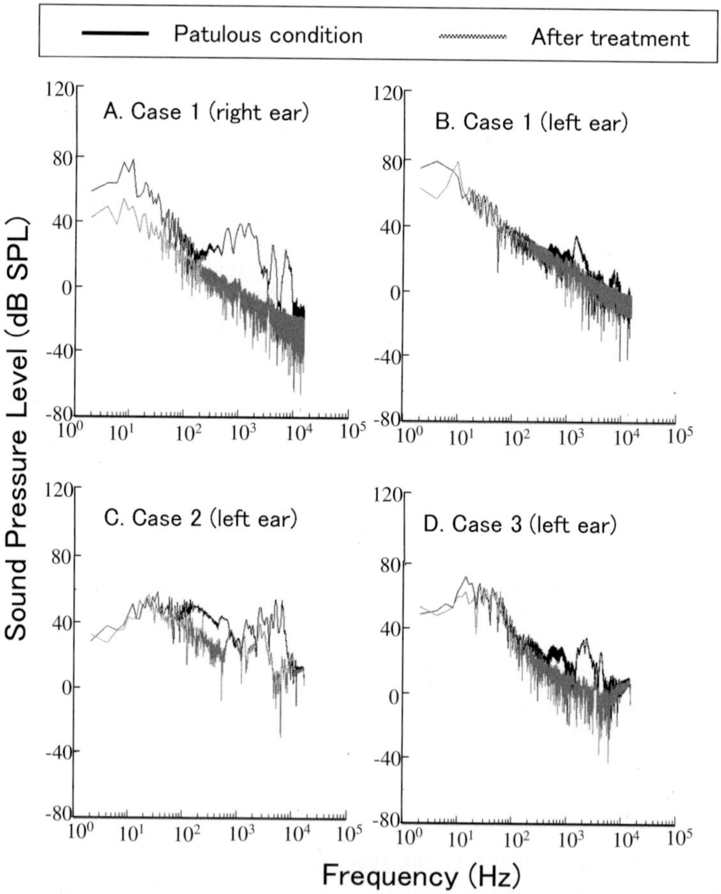

Figure 4. Representative measurements of the acoustic transfer function in three patients complaining of patulous symptoms (A-D). The transferred sound pressure recorded by the microphone is plotted as a function of frequency. See text for further details.

examinations such as TTAG, the diagnosis was bilateral patulous ETs. Her patulous symptoms were relieved by insufflation with Lugol's solution (iodine solution). On the day of examination, her subjective symptoms of autophony were somewhat more severe in the right ear. However, no apparent differences could be observed in the mobility of the ear drum between the right and left ears. The findings of TSP sonotubometry before and after the insufflation with Lugol's solution are shown in Fig. 4A and B.

The acoustic transfer functions were greater in the patulous condition and were decreased after obstruction of the ET in both ears. The acoustic transfer function in the right ear appeared to be larger than that in the left ear; *i.e.*, the sound presented in the nasal cavity was transmitted to the external ear with less attenuation in the right ear, in which the subjective symptoms (autophony) were more severe. The acoustic transfer function after the treatment in Case 1 was nearly the same as the level of background noise measured in normal subjects.

Case 2, a 68-year-old female, and Case 3, a 73-year-old female, had suffered from patulous ET in the left ear for more than 1 year. Both patients underwent treatment to relieve their patulous symptoms, consisting of insufflation with Lugol's solution (Case 2) and instillation of gelatinous iodine solution in the pharyngeal orifice of the ET (Case 3). The acoustic transfer functions were greater in the patulous condition and decreased after obstruction of the ET (Fig. 4C and D).

4. Discussion

The present method of TSP sonotubometry enables objective assessment of the patulous ET. The acoustic transfer function can be easily averaged and assessed with a good signal-to-noise ratio within a minute. The present study showed that the acoustic transfer function reflected the caliber of the tube (Fig. 3), enabling assessment of the patency of the ET by the TSP sonotubometry. Preliminary application of TSP sonotubometry to clinical patients suggested the usefulness of the present method for assessment of ET patency (Fig. 4). The acoustic transfer functions were significantly increased under the patulous condition and were decreased to the normal level after obstructive treatment of the patulous ET.

TSP sonotubometry is based on the same concept as conventional sonotubometry [8,9], *i.e.*, the sound transmission via the ET is used as an indicator of its patency. However, the main aim of conventional sonotubometry is to assess ET patency at swallowing [8]. Therefore, the actual acoustic transfer function from the nasopharyngeal cavity to the ear canal is relatively difficult to assess. In contrast, TSP sonotubometry can assess the acoustic transfer function over a wide frequency range with a good signal-to-noise ratio, achieved by averaging, although any change in the ET patency at swallowing cannot be assessed.

Nasal audiometry and/or nasal-noise masking audiometry are also useful methods to investigate the acoustic transfer function of the patulous ET [10,11]. However, these methods use an audiometric procedure to measure the detection thresholds for tones presented in the nasal cavity and require a longer time than TSP sonotubometry.

Objective information regarding the severity of the patulous condition would be helpful to decide the indications for treatment, such as infusion of absorbable gelatin sponge solution, injection of Teflon into the peritubal area, and placement of an indwelling catheter, autologous fat tissue, or a pin-shaped silicone plug [1,3,15-17]. TSP sonotubometry can provide such data for the effective management of patients with patulous ET.

Acknowledgments

This study was supported by Grant-in-Aid for Scientific Research (B) 17390456 from the Ministry of Education, Science and Culture, Japan. Authors also acknowledge the support of Tohoku University Global COE Program "Global Nano-Biomedical Engineering Education and Research Network Centre". We wish to thank Mr. Katsuo Watanabe and Ms. Fumie Nakajima for their continuous support of this work.

References

1. S. Ogawa, I. Satoh and H. Tanaka, *Arch Otolaryngol.* **102**, 276 (1976).
2. A. F. O'Connor and J. J. Shea, *Laryngoscope* **91**, 1427 (1981).
3. J. K. Doherty and W. H. Slattery, *Otolaryngol Head Neck Surg.* **128**, 88 (2003).
4. T. Kumazawa, I. Honjo and K. Honda, *Arch Otorhinolaryngol.* **208**, 147 (1974).
5. T. Kumazawa, I. Honjo and K. Honda, *Arch Otorhinolaryngol.* **215**, 317 (1977)
6. Y. Finkelstein, Y. P. Talmi, Y. Rubel and Y. Zohar, *Arch Otolaryngol Head Neck Surg.* **114**, 1134 (1988).
7. A. Mink, *Eur Arch Otorhinolaryngol.* **249**, 91 (1992).
8. H. Virtanen, *Acta Otolaryngol.* **86**, 93 (1978a).
9. H. Virtanen, *Acta Otolaryngol.* **86**, 401(1978b).
10. S. Kano, T. Kawase, Y. Baba, T. Sato and T. Kobayashi, *Acta Otolaryngol.* **124**, 43 (2004).

11. Y. Hori, T. Kawase, J. Hasegawa, T. Sato, N. Yoshida, T. Oshima, M. Suetake and T. Kobayashi, *Otol Neurotol.* **27**, 596 (2006).
12. Y. Suzuki, F. Asano, H. Y. Kim and T. Sone, *J Acoust Soc Am.* **97**, 1119 (1995).
13. J. Sade and M. Luntz, *Ann Otol Rhinol Laryngol.* **98**, 630 (1989).
14. M. Sudo, I. Sando, A. Ikui and C. Suzuki, *Ann Otol Rhinol Laryngol.* **106**, 583 (1997).
15. J. L. Pulec, *Laryngoscope* **77**, 1543 (1967).
16. C. D. Bluestone and E. I. Cantekin, *Laryngoscope* **91**, 149 (1981).
17. T. Kobayashi, *Practica Otologica.* **93**, 897 (2000).

MINIATURIZED MICROFLUIDIC BIOFUEL CELLS

MATSUHIKO NISHIZAWA [1,2)*]

1) Department of Bioengineering and Robotics, Graduate school of Engineering, Tohoku University, 6-6-01 Aramaki-Aoba, Sendai 980-8579, Japan

2) JST, CREST, Sanbancho, Chiyoda-ku,Tokyo,102-0075, Japan

We have been studying enzyme-based biofuel cells that utilize enzymes as electrocatalysts for the oxidation of fuels at anode and the reduction of oxygen at cathodes. Here we will describe our recent attempts to develop miniaturized microfluidic biofuel cells that have greater than 1 V this being achieved by automatic series-connection.

1. Introduction

The interface between biomolecules and device materials is one of the most important subjects for both *in-vitro* and *in-vivo* medical devices. We have been studying the potential use of microelectrode techniques as a tool for controlling bionic interfaces: special control of protein adsorption and cell adhesion / growth [1]. The technique is sufficiently simple so that it can be integrated into small and closed systems such as microfluidic devices [2]. In addition, the *in-situ* microcircuit formation with conducting polymers has been studied as a potential interface to cellular networks [3].

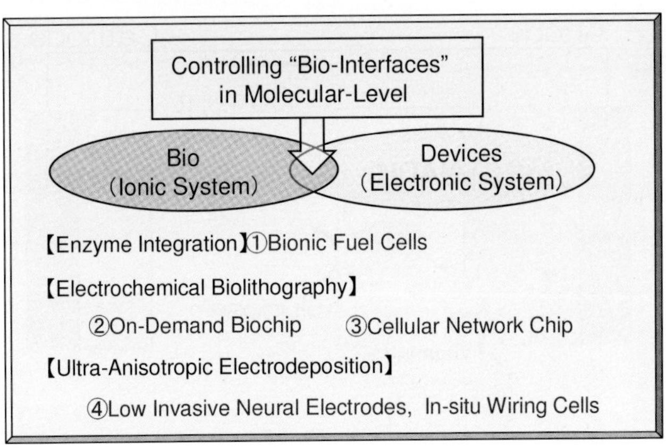

* Matsuhiko Nishizawa is a Tohoku University Global COE Member.

The biological fuel cell is another electrochemical bionic device we are studying. The energy conversion efficiency of this living system is very high. Reconstruction of the protein-based bioenergetic system *in vitro* is a challenge in developing bio-adaptive fuel cells [4-10]. Here we will describe our recent attempts to develop miniature microfluidic biofuel cells.

2. Bionic Fuel Cell

Electric power derived from dispersed ambient energy has attracted attention as ace of ubiquitous portable power. A potential option for a portable power source is a biological fuel cell that uses enzymes as an electrocatalyst to generate electricity from biological fuels such as alcohols and carbohydrates. The high reaction selectivity of enzymes would allow separator-free design and power generation from complex natural fuel solutions without purifications, that is, direct utilization of refreshments containing sugar, plant saps, and biological fluids such as blood, containing glucose (*ca.* 5 mM), lactate (*ca.* 1 mM) and oxygen (0.1 mM in arterial blood) [4,5].

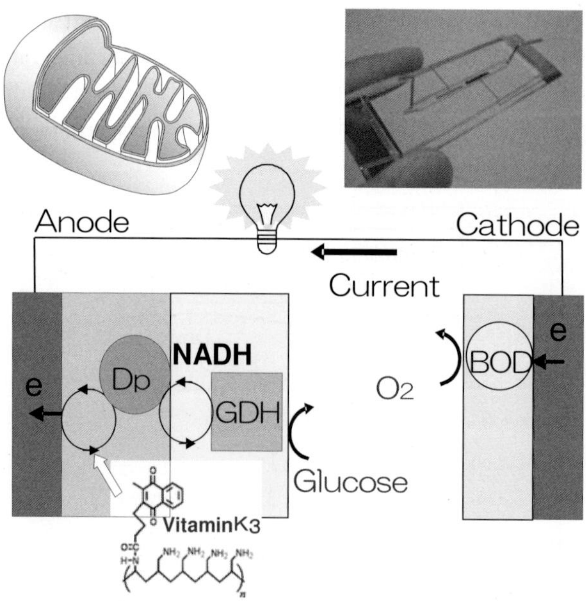

Figure 1. Enzymatic Glucose / Oxygen Fuel Cell.

We have studied the enzyme anode for glucose oxidation composed of a bi-layer polymer membrane with the inner layer containing diaphorase (Dp) and the outer, glucose dehydrogenase (GDH) [8-10]. The Dp membrane was formed from a synthesized Vitamin K3-based mediator polymer. The enzyme cathode for oxygen reduction was prepared with bilirubin oxidase (BOD).

The important advantage of the enzymatic fuel cells is ease in miniaturization, and the structural design of the cells in microscale should directly affect the total performance of the cells. We will present here our recent research on the microfluidic bionic cell along with the use of the semi-automatic air valve for series-connection.

3. Electrodes Preparation

The synthesis of the redox polymer VK_3-modified poly-L-lysine (PLL-VK_3) has been described previously [8-10]. Poly(1-vinylpyridine) complexed with Os(4,4'-dimethyl-2,2'-bipyridine)$_2$Cl (PVI-Os) and Poly (4-vinylpyridine) complexed with Os(2,2'-bipyridine)$_2$Cl and quaternized with bromoethyl amine (PVP-Os) were synthesized and donated by Daiichi pure chemicals Co., Ltd. The diaphorase (Dp) from *Bacillus stearothermophilus* (EC 1.6.99, 1090 U mg^{-1}), NAD^+-dependent glucose dehydrogenase (GDH; EC 1.1.1.47, 250 U mg^{-1}),

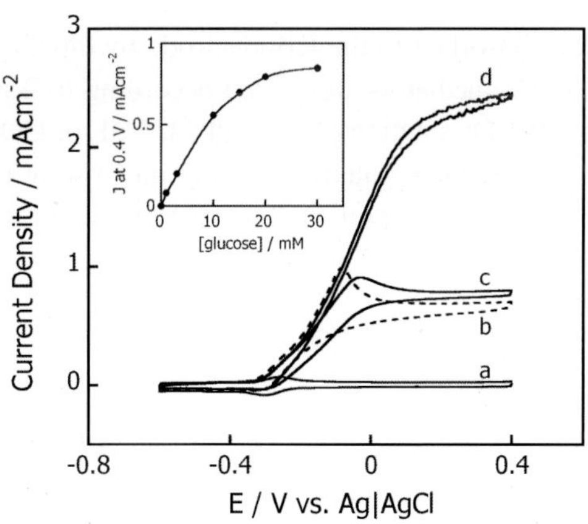

Figure 2. (a) A cyclic voltammogram for a KB/PLL-VK_3/Dp/GDH-modified GC electrode in a N_2-saturated pH 7.0 phosphate-buffered electrolyte solution at 37°C. (b-d) Cyclic voltammograms in the solution with 20 mM NADH (b), or 20 mM glucose, 1.0 mM NAD^+ (c), or 30 mM glucose, 1.0 mM NAD^+ (d). For (d), the electrolyte solution was stirred at 1000 rpm. In all cases the scan rate was 5 mV s^{-1}. The inset shows the current density at 0.4 V vs. glucose concentration.

Bilirubin oxidase (BOD) from *Myrothecium sp.* (EC 1.3.3.5, 2.45 U mg^{-1}) and Ketjenblack (KB; EC-600JD) were used as received.

A PLL-VK_3/Dp/GDH-coated KB electrode was prepared as follows. An 8 μL PLL-VK_3 solution (4.83 mM VK_3) was mixed with a 2 μL Dp solution (14 μg μL^{-1}) and 1 μL of KB dispersed water (*ca.* 13 mg mL^{-1}). A 5.3 μL portion of the resulting solution was put onto a gold film electrode (surface area, 0.028 cm^2) on a glass substrate, and was left to dry in air. To create the enzymatic bilayer, the surface of a PLL-VK_3/Dp-coated KB electrode was coated with 1.6 μL of a solution composed of equal volumes of a 16 μg μL^{-1} GDH solution and a 16 mg mL^{-1} PLL solution.

As shown in Fig. 2a catalytic oxidation current appeared on addition of 20 mM NADH (Fig. 2b, dashed curve). When both 20 mM glucose and 1.0 mM NAD^+ were present, glucose-oxidizing catalytic currents resulted (Fig. 2c), suggesting that the presumptive electron-relay system diagramed in Fig. 1 functioned in the enzyme bilayer. Although the shapes of the CVs suggest that an enzymatic reaction controlled the value of the catalytic current density, it could be further increased as high as 2 mA cm^{-2} when the electrolyte solution was stirred (Fig. 2d).

A BOD-adsorbed KB electrode was prepared as described below. A mixture of 3:1 weight ratio of KB and Poly(tetrafluoroethylene) (MW 5,000~20,000, Wako) was dispersed in isopropanol (2 mg mL^{-1}), and applied to a gold electrode (0.11 mL cm^{-2}) followed by overnight drying in a 70 C oven. And then, BOD was adsorbed to the KB electrode by dipping in the 5 mg mL^{-1} BOD solution. The Os mediators were used according to Tsujimura [6,7]. As for anode, 5 μL of PVI-Os solution (25 mg mL^{-1}), 2 μL of GOD solution (20 mg mL^{-1}) and 1.2 μL of PEGDGE solution (2.5 mg mL^{-1}) were mixed. Similarly, a 5 μL of PVP-Os solution (25 mg mL^{-1}), 2 μL of BOD solution (20 mg mL^{-1}) and 1.2 μL of PEGDGE solution (2.5 mg mL^{-1}) were mixed and used for cathode preparation.

4. Microfluidic Biofuel Cell

Enzyme-based biofuel cells were formatted into microfluidic systems as power sources for independent power-on-chip and implantable devices. We constructed microfluidic fuel cells of the type shown in Fig. 3. Both anode's (2.8 mm width, 1 mm long) and cathode's (2.8 mm width, 10 mm long) Au current collectors were modified by KB and enzymes. The gap between anode and cathode was 0.5 mm. The PDMS channel produced channels of 3 mm width and 0.1 mm or 1 mm height.

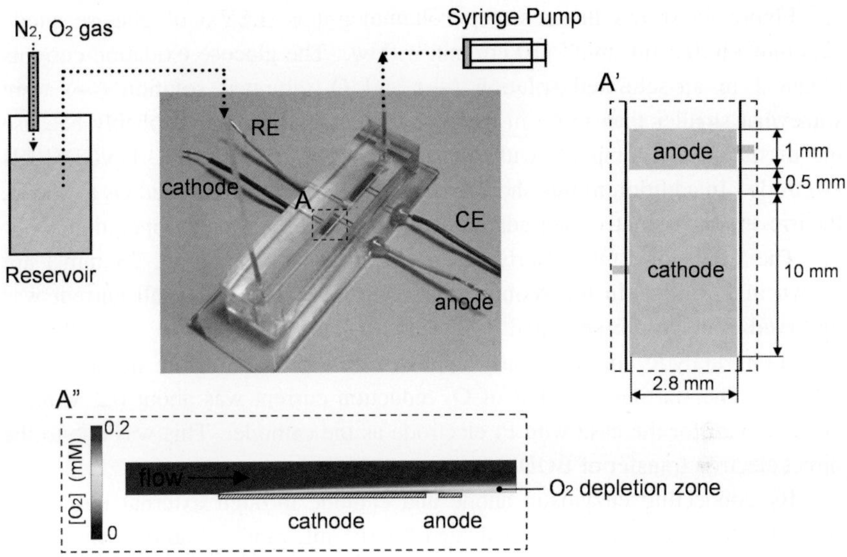

Figure 3. Schematic illustration and photograph of microfluidic biofuel cell with (A') the close up top view of the electrodes and (A") the cross-sectional view of a COMSOL simulation of an O_2 depletion layer forming along the channel.

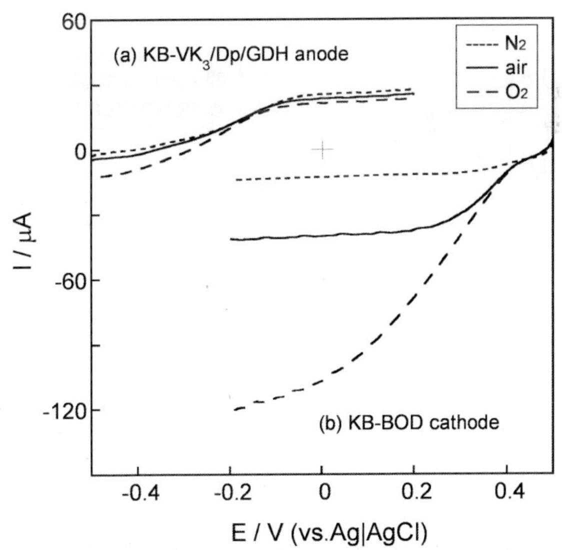

Figure 4. LSVs of (a) anode and (b) cathode in N_2-bubbled (···), air-saturated (—), and O_2-bubbled (---) 50 mM phosphate buffer (pH 7) containing 0.1 M NaCl, 1 mM NAD^+ and 10 mM glucose at room temperature, with a flow rate of 0.3 mL min^{-1}. Scan rate: 2 mV s^{-1}. Channel height: 1 mm.

Figure 4a shows linear sweep voltammograms (LSVs) of glucose anode (2.8 mm^2) in 0.3 mL min^{-1} (10 cm min^{-1}) flow. The glucose oxidation currents obtained in air-saturated solution (—) and O_2-saturated solution (---) were somewhat smaller than those in the N_2-bubbled solution (···), probably because the dissolved O_2 competes with electron relay at the PLL-VK_3/Dp/GDH/KB electrode. In addition to this short-term adverse effect of dissolved O_2 to anode, the irreversible oxidative degradation would occur during longer operation.

The LSVs of BOD-adsorbed KB electrode (10 mm long, 28 mm^2) are shown in Fig. 4b. In the N_2-bubbled solution (···), only a small current was observed, but in air-saturated (—) and O_2-bubbled solution (---), the O_2 reduction catalytic current clearly appeared at a potential more negative than 0.45 V. The starting potential of O_2 reduction current was about 0.2 V more positive than for the case with Pt electrode as the cathode. This was due to the direct electron transfer of BOD.

By connecting enzymatic anode and cathode through external resistance, biofuel cell performance was evaluated for 0.3 mL min^{-1}. Figure 5 shows the cell performance (V-I curve) in an air-saturated solution with an open circuit voltage (V_{oc}) of around 0.8 V and maximum current (I_{max}) of over 20 μA. This is in agreement with the prediction from performance of each anode and cathode.

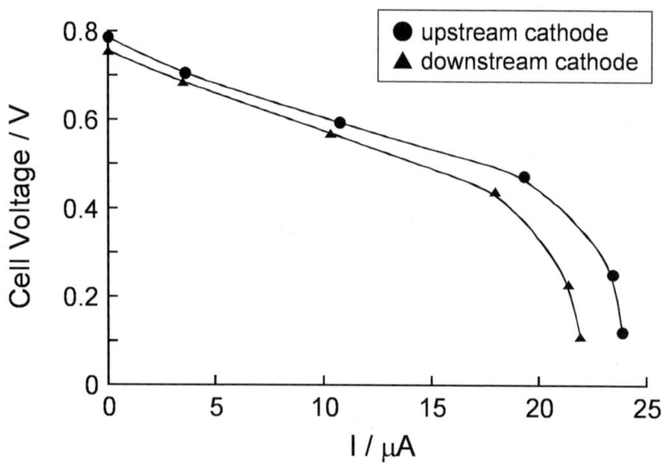

Figure 5. V-I curve of microfluidic biofuel cell operating under air-saturated phosphate buffer (pH 7) containing 0.1 M NaCl, 10 mM glucose and 1 mM NAD^+ at room temperature, with a flow rate of 0.3 mL min^{-1}. The cells were operated with upstream cathode (•) or downstream cathode (■).

The V-I curve obtained with the upstream-cathode cell (●) showed larger current than the downstream-cathode cell (▲), especially in the higher current region. The I_{max} increased 10% by placing the cathode upstream, mainly reflecting the improved anode as judged from the analogy between the shape of the V-I curve of the cell (—) and the E-I curve of the anode (···).

5. Automatic Air Valve

The output voltage of single biofuel cell is in practice lower than 1 V. Therefore, many applications require cell-stacking (series connection), which is however often troublesome due to short-circuiting of cells through an ion-conductive fuel solution. The series connection of biofuel cells requires a system for ionic isolation between each cell.

Our strategy is based on the air-trapping at a superhydrophobic area prepared in the fluidic channel so that each cell is ionically isolated. We prepared a lotus leaf-like micropillar array which was designed to be 20 μm-height, 15 μm-diameter and 15 μm-interspaced on the wall of a 40 μm-deep microchannel, as shown in Fig. 6. Three glucose/O_2 biofuel cell with open circuit voltage (OCV) of 0.35 V were connected with superhydrophobic gates. If this automatic air-valve system works as expected, users of this power device never have to introduce fuel solution to each separate chamber.

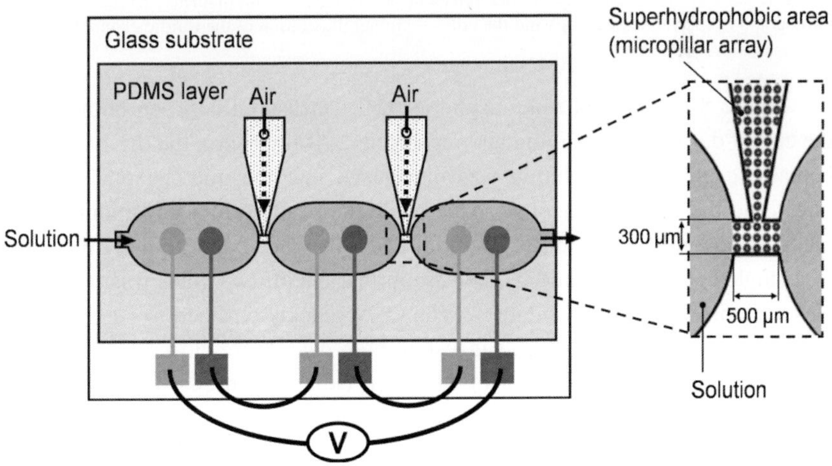

Figure 6. Illustration of the principle of series-connected biofuel cells on a fluidic chip. Inset shows close up top view of valve-area.

Figure 7 shows scanning electron microscope (SEM) images of the glass substrate etched by RIE. The squarely-arranged micropillar array (20 μm-long, 15 μm-diam., 15 μm-spaced) was successfully fabricated. The parylene-coated micropillar array showed superhydrophobicity that means θ_c exceeding 150°. The microscopic image of the superhydrophobic gate clearly shows the cells are divided by air introduced from the air-reservoir.

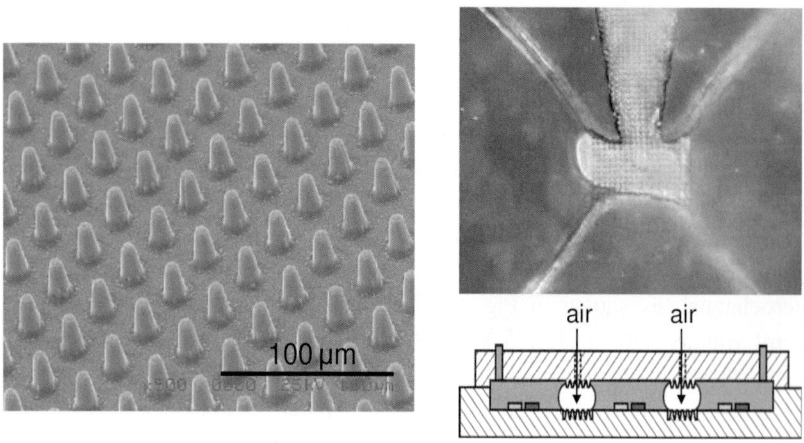

Figure 7. SEM image of the array of micropillars of 20 μm-long, 15 μm-diameter, 15 μm-interspaced. Photograph of the valve area (top) and the cross-sectional illustration of the fluidic channel.

Figure 8 is a typical time display of OCV measured between both ends of the circuit during fuel charging and recharging. At time zero, the 0.1 M glucose-containing electrolyte solution was introduced into the microchannel at 5 μL min^{-1}, and the whole channel was filled with the solution in a few minutes. The OCV was about 0.65 V during solution flow, this value being roughly twice that of a single cell. This should be the case in that one of two gates was practically closed (ionically almost insulated). The OCV quickly changed to *ca.* 1 V when the solution flow was stopped, suggesting both gates were closed by air valves, this time, for series-connection of the three cells. Such change in OCV was reversible for successive stop and flow operations.

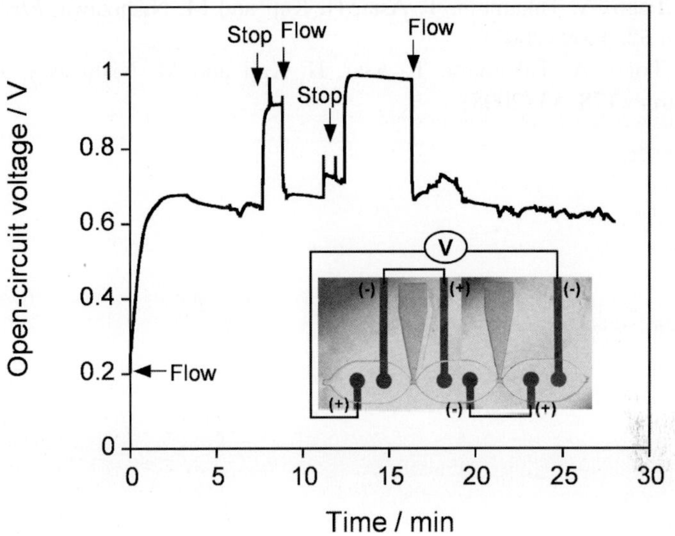

Figure 8. The time display of open-circuit voltage (OCV) measured between both ends of the electrode circuit during the charge and recharge of a 0.1 M NaCl and 0.1 M glucose-contained 50 mM phosphate buffer solution (pH 7).

Acknowledgements

This work is supported by Tohoku University GCOE Program "Grobal Nano-Biomedical Engineering Education and Research Network Centre" and Health and Labor Sciences Research Grant "Nano Medicine" from the Ministry of Health, Labor and Welfare of Japan.

References

1. H. Kaji, T. Kawashima and M. Nishizawa, *Langmuir* **22**, 10784 (2006).
2. H. Kaji, M. Hashimoto and M. Nishizawa, *Anal. Chem.* **78**, 5469 (2006).
3. M. Nishizawa, T. Kamiya, H. Nozaki and H. Kaji, *Langmuir* **23**, 8304 (2007).
4. A. Heller, *Phys. Chem. Chem. Phys.* **6**, 209 (2004).
5. S. C. Barton, J. Gallaway and P. Atanassov, *Chem. Rev.* **104**, 4867 (2004).
6. S. Tsujimura, K. Kano and T. Ikeda, *J. Electroanal. Chem.* **576**, 113 (2005).
7. S. Tsujimura, T. Nakagawa, K. Kano and T. Ikeda, *Electrochemistry* **72**, 437 (2004).
8. F. Sato, M. Togo, M. K. Islam, T. Matsue, J. Kosuge, N. Fukasaku, S. Kurosawa and M. Nishizawa, *Electrochem. Commun.* **7**, 643 (2005).

9. M. Togo, A. Takamura, T. Asai, H. Kaji and M. Nishizawa, *Electrochim. Acta* **52**, 4669 (2007).
10. M. Togo, A. Takamura, T. Asai, H. Kaji and M. Nishizawa, *J. Power Sources* **178**, 53 (2008).

DEVELOPMENT OF AN ENDOSCOPIC TACTILE SENSOR USING PVDF FILMS

MAMI TANAKA *, TAKESHI OKUYAMA, MIKIKO SONE
Tohoku University, Aobayama 6-6-04, Sendai, Japan 980-8579

YOSHIKATSU TANAHASHI
Office Urology Tanahashi

SEIJI CHONAN
Akita Prefectural University

In this work, a prototype polyvinylidene fluoride (PVDF) tactile sensor for endoscopic application was developed. The aim of the sensor is to measure hardness, which is one piece of information determined from tactile perceptions. This sensor is composed of two PVDF films, a silicone cylindrical column, and an aluminum cylinder. The classification of hardness is concerned with the ratio between the outputs of these PVDF sensors. In this study, two sensors were fabricated using two silicone cylindrical columns with different Young's moduli. The performance evaluation of each sensor was conducted using six silicone rubbers as measuring objects. The experimental results corresponded with the simplified theoretical analysis and the proposed sensor can distinguish differences in elastic property.

Keywords: Tactile sensor, Minimally invasive surgery, Endoscope, PVDF film.

1. Introduction

Recently, minimally invasive surgery (MIS) has been used in the treatment of various diseases because it has many advantages including relief of pain, early recovery and small operative wounds. A catheter and/or endoscope are used during MIS. The operation of these tools is difficult because human cavities have a complex shape and it is hard for the operator to know the position/force information between the tools and human cavities. This difficulty might lead to a critical accident. To avoid undesirable results in MIS, the progression of endoscope technique is necessary. To ensure safe insertion, many studies have been performed, assessing an active endoscope with a shape memory alloy [1,2],

---

* Mami Tanaka is a Tohoku University Global COE Member.

a force sensor endoscope [3], and a sensor for monitoring the contact condition of an endoscope [4].

One of the disadvantages of MIS is the lack of tactile sensation. Information on tactile perception is important for the safety MIS and for diagnosis during MIS, but is not sufficient. To overcome this disadvantage, an endoscopic grasper with a tactile sensor [5] has been developed to manipulate biological tissue safely. Tanaka et al have developed an active palpation sensor using a polyvinylidene fluoride (PVDF) film to detect prostatic cancer and hypertrophy [6]. The PVDF film, used as the sensory material in the above studies, is a piezoelectric polymer material. It has good features, such as high sensitivity, lightweight, thin film, flexibility, and low cost. It also has the interesting feature that the output signal induced by an applied pressure is similar to the response of the Pacinian corpuscle, a sensory receptor in the human dermis, and is more sensitive to oscillatory pressure input [7]. In other words, in the case of applying a pressure pulse on PVDF, a brief potential signal is obtained at the up edge and down edge of the pulse, but there is no response to static pressure. Therefore, PVDF film is used to measure various tactile properties.

In this work, a tactile sensor using PVDF films was developed for endoscopic application. This sensor measures the hardness of biological tissue to assist with diagnosis. There are some indicative values of hardness. In this work, the elastic property, Young's modulus, is used to evaluate hardness. This sensor is composed of two PVDF films, a silicone cylindrical column and an aluminum cylinder. The classification of Young's modulus is concerned with the ratio between these PVDF outputs. In this paper, two sensors are fabricated using two silicone cylindrical columns with different Young's moduli. The performance evaluation of each sensor was conducted using six silicone rubbers simulating the human prostate as measuring objects. The experimental results were compared with the theoretical analysis and the availability was confirmed.

2. Structure and Principle of the Tactile Sensor

A cross-section of the proposed tactile sensor is shown in Fig. 1. This sensor consists of two PVDF films, a cylindrical column of silicone rubber and a rigid aluminum cylinder attached to an aluminum block. The PVDF films are 28 μm in thickness. The silicone rubber is 3 mm in diameter and 3 mm in height. The aluminum cylinder is 6 mm in outer diameter, 4 mm in inner diameter and 3 mm in height. Aluminum blocks are 10 mm x 10 mm square and 3 mm in thickness. One of the PVDF films is laid under the silicone rubber; this film is called the upper PVDF. The other is sandwiched between the aluminum blocks and called the lower PVDF. Figure 2 shows a photograph of the prototype sensor. To reduce the overlap of noise, the connection between electrodes of PVDF and coaxial cables are shrouded by copper foil tape. In this study, two sensors, sensor H and sensor S, were fabricated using silicone rubbers with different Young's moduli. The Young's moduli of silicone rubbers used in sensors are shown in Table 1.

In measurements, PVDF output was obtained by vibrating the tactile sensor during contact with the object. When this sensor is pressed on the measuring object, forces F_a and F_s are applied on aluminum cylinder and silicone cylindrical column, respectively. The ratio between these forces depends on the Young's moduli of the aluminum, silicone rubber and measuring object. When a force is applied to a PVDF film, an electrical charge is induced by polarization in the PVDF film. Because the magnitude of its charge is proportional to the applied force, the applied force on a PVDF film can be obtained by measuring the output voltage through a charge amplifier connected to the PVDF film. The upper PVDF measures the force applied to the silicone rubber, F_s, and the lower PVDF measures the force applied to the whole sensor, namely, $F_a + F_s$. The ratio between these two PVDF outputs was calculated and the elastic property of the object being measured was determined.

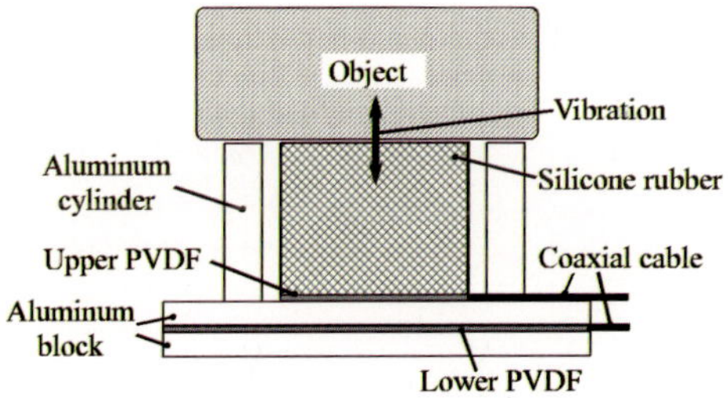

Figure 1. Across-section of tactile sensor.

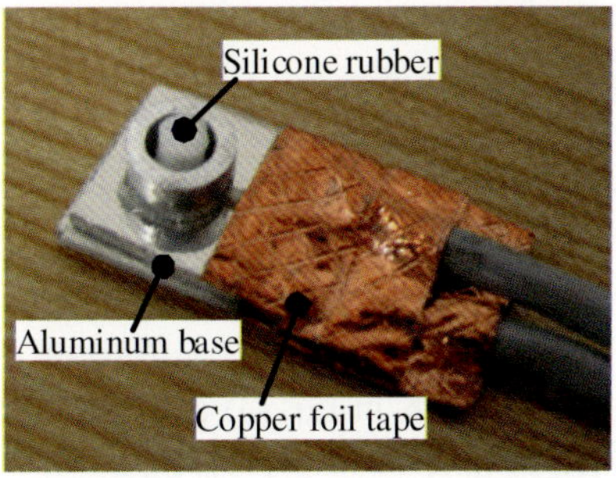

Figure 2. Photograph of the tactile sensor.

Table 1. Young's modulus of the silicone column, E_s.

Sensor symbol	H	S
Young's modulus (MPa)	0.24	0.16

3. Theoretical Analysis

A theoretical model of the developed sensor is shown in Fig. 3. This modeling is based on some assumptions. One is that there is no shear force in either the measuring object or the sensor. Another is that the aluminum parts are considered as rigid bodies. It is also assumed that the surface of the measured object is flat. In the analytical model, A_a and A_s are the cross-sectional areas of the aluminum cylinder and the silicone rubber, respectively. T_o is the thickness of the measuring object. T_s is the thickness of the silicone cylindrical column and the aluminum cylinder. X_1 is the deformation of the measuring object on the rigid aluminum cylinder and X_2 is the deformation of the silicone cylindrical column. E_o and E_s are the Young's moduli of the measuring object and the silicone rubber, respectively. F_a and F_s are mentioned above, and force F is applied to the whole sensor. Using the theory of the mechanics of materials, these forces can be calculated from following equations.

$$F_a = E_o \frac{X_1}{T_o} A_a \tag{1}$$

and

$$F_s = E_s \frac{X_2}{T_s} A_s = E_o \frac{X_1 - X_2}{T_o} A_s \tag{2}$$

Combining the above expressions, we can obtain the ratio between the applied forces as

$$\frac{F_s}{F} = \frac{F_s}{F_a + F_s} = \frac{1}{\frac{A_a}{A_s}\left(1 + \frac{T_s E_o}{T_o E_s}\right) + 1} \tag{3}$$

As mentioned in Sec. 2, F_s and F in Eq. (3) are measured by the upper PVDF film and the lower PVDF film, respectively. T_o, T_s, E_s, A_a and A_s are given. Then, we can calculate the Young's modulus of the measuring object, E_o. Figure 4 shows the relationship between the force ratio, F_s/F, and the ratio between Young's moduli, E_o/E_s, using Eq. (3), designed to satisfy the conditions that A_s is equal to A_a and T_s is equal to T_o. When the measuring object is infinitely stiffer or softer than the silicone column, that is $E_o >> E_s$ or $E_o << E_s$, the ratio of forces asymptotically approaches zero or $A_s/(A_a+A_s)=0.5$. The force ratio changes significantly near $E_o/E_s=1$. In this case, a silicone rubber with stiffness near the measuring object is chosen as the sensor material for accuracy.

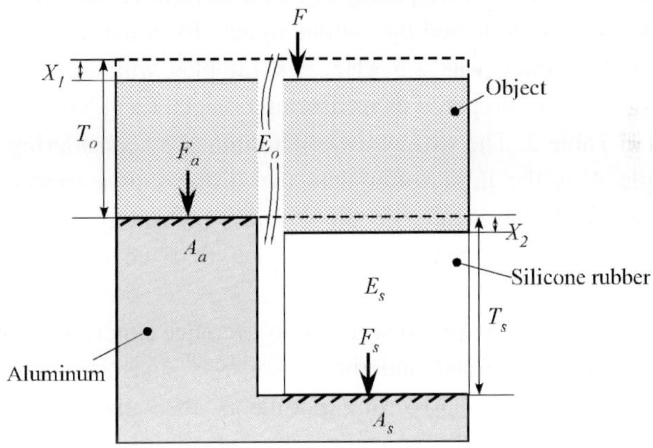

Figure 3. An analytical model of the sensor and the measuring object.

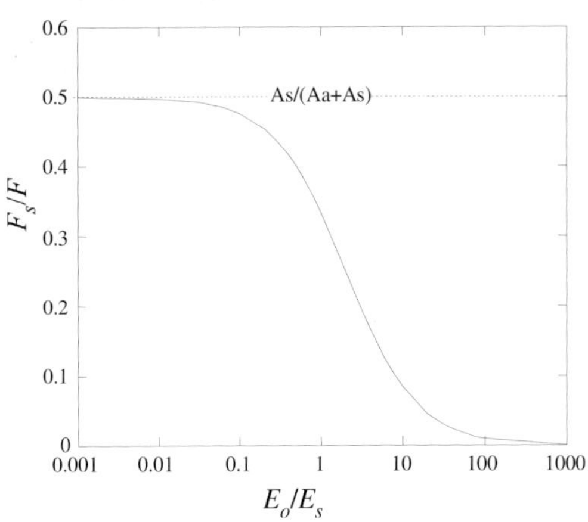

Figure 4. A theoretical relationship between the force ratio and the ratio of Young's modulus E_o/E_s, in the case that $A_s=A_a$ and $T_s=T_o$.

4. Experiments

Figure 5(a) shows the experimental setup to confirm the performance of the developed tactile sensor. For simplicity, the sensor was fixed on the rigid table and the measuring object was attached to a vibro-machine. In this experiment, the measuring object was sinusoidally vibrated in the vertical direction by the vibro-machine. The amplitude and frequency of vibration were 2 mm and approximately 40 Hz, respectively. Maximum indentation depth was 0.5 mm. The upper and lower PVDF films were connected to each charge amplifier and a low-pass filter with a cut-off frequency of 1 kHz to remove the overlap of noises. Then, an oscilloscope collected the output signals for a measuring time 0.5 ms. The sampling frequency was 2.5 kHz. Six samples with different stiffnesses of silicone rubber were prepared as measuring objects. Each Young's modulus is summarized in Table 2. The stiffness was determined by considering the prostate gland. Sample A is the little softer than the stiffness of a cancerous prostate gland, and sample F was similar to the stiffness of healthy prostate gland [8]. Samples were 10 mm in thickness, width and length. Each sample was measured five times.

In addition, to take the practical use into account, experiments in which the measuring objects were fixed and the sensor was sinusoidally vibrated were conducted as shown in Fig. 5(b). In experiments, the same experimental setup was used except for cut-off frequency. The cut-off frequency of 45 Hz was chosen to reduce the noise due to vibration of the sensor.

Table 2. Young's modulus of sample objects, E_o.

Sample symbol	A	B	C	D	E	F
Young's modulus (MPa)	0.76	0.56	0.28	0.14	0.08	0.04

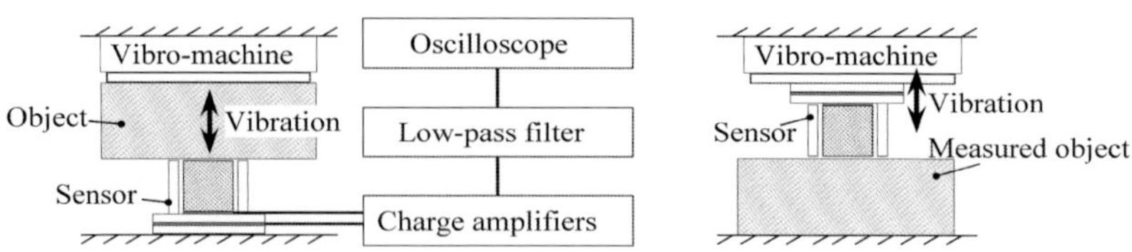

(a) Sensor is fixed. (b) Sensor is vibrated.

Figure 5. Experimental setups in which (a) the sensor is fixed and the measuring object is vibrated, and (b) the sensor is vibrated and the measuring object is fixed. The same measurement system was used in all experiments.

5. Results

Representative output waveforms from sensor S and sensor H are shown in Fig. 6 and 7, respectively. Panels (a) and (b) of each figure show the results for sample A. Panels (c) and (d) show the results for sample F. By comparison between samples, the sensor outputs for sample A were larger than those for sample F with both sensor S and sensor H. With regard to the comparison between the two sensors, although the output of the lower PVDF is unchanged, the outputs of the upper PVDF using sensor H are larger than those using sensor S.

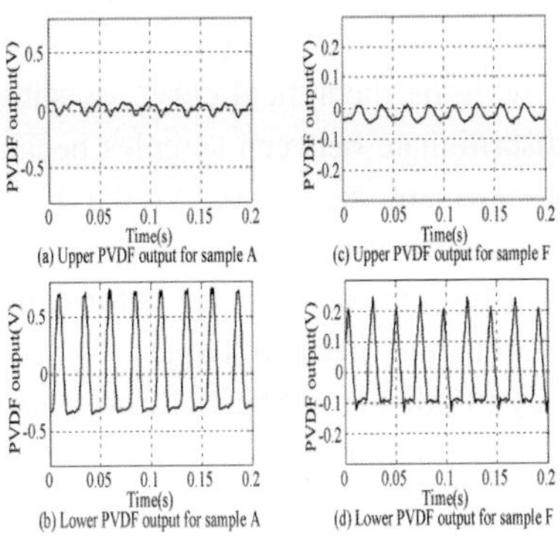

Figure 6. The output waveforms of sensor S for sample A and F.

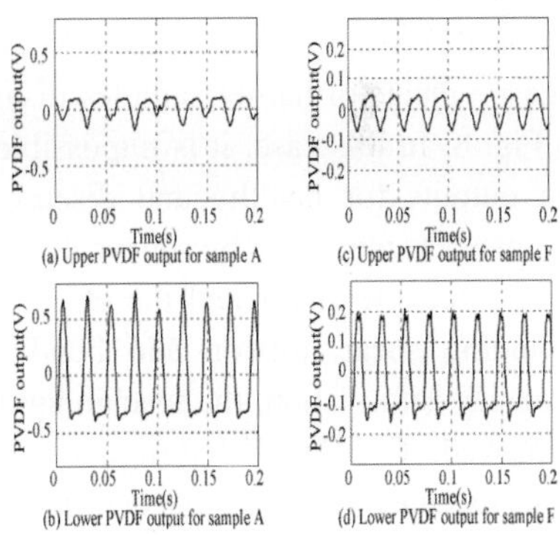

Figure 7. The output waveforms of sensor H for sample A and F.

The average amplitude $V_{p\text{-}p}$ was determined from each waveform. As $V_{p\text{-}p}$ of the upper PVDF and lower PVDF are described as V_{upper} and V_{lower}, respectively, V_{upper} is proportional to the force applied on the silicone cylindrical column, F_s, and V_{lower} is proportional to the force applied on the whole sensor, F. Therefore, the left side of Eq. (3) is represented as follows.

$$\frac{F_s}{F} = \frac{F_s}{F_a + F_s} = \frac{V_{upper}}{V_{lower}} \tag{4}$$

Using the above equation, the relationship between the experimental force ratio, F_s/F, and Young's modulus of the measuring object is shown in Fig. 8. The theoretical behaviors using Eq. (3) are also shown in Fig. 8. It was found that the experimental results with sensor H obviously corresponded with the simplified theoretical analysis. Also, in the case of sensor S, the experimental results roughly corresponded with the simplified theoretical ones, but the force ratio for harder samples grew larger than the theoretical one. In addition, by comparing the results with sensor H and sensor S, in both theoretical and experimental results, it was confirmed that the classification performance depends on the Young's modulus of the cylindrical columnar material. The Young's modulus of the silicone cylinder can be determined by considering the threshold of classification. From these results, we confirmed that the proposed tactile sensor has enough availability to distinguish the hardness of the measurement object, including the biomedical tissue. Next, Fig. 9 shows the results from a case in which the sensors are vibrated. With regard to sensor H, the experimental results corresponded with the theoretical results, and it is possible to discriminate a diseased prostate from a healthy one. In the results using sensor S, the measurements of harder samples disagreed with the theoretical curve, as shown in Fig. 8. In this case, it is impossible to discriminate between samples because the outputs for healthy and diseased prostates are the same. As the harder samples are measured by sensor S, the displacement of the silicone rubber of the sensor became larger and the effect of viscoelasticity or irregular deformation, including shear, torsion and bending deformation, became larger. To avoid irregular deformation, the machining accuracy of silicone rubber will need to be improved. For detailed design of sensors, we should conduct finite element analysis considering the viscoelasticity and shear force.

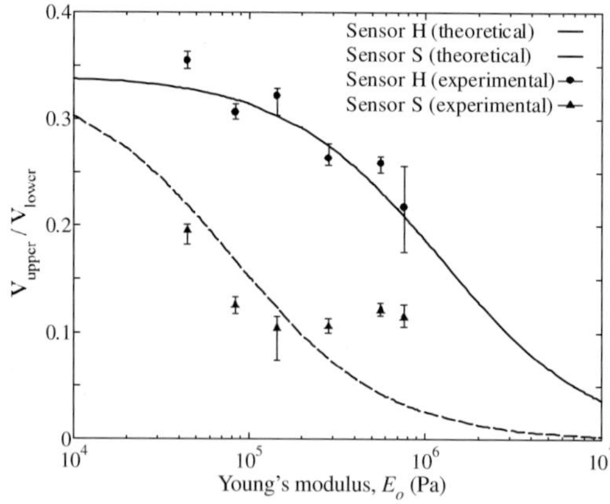

Figure 8. Comparison between theoretical values and experimental results where the measured objects are vibrated.

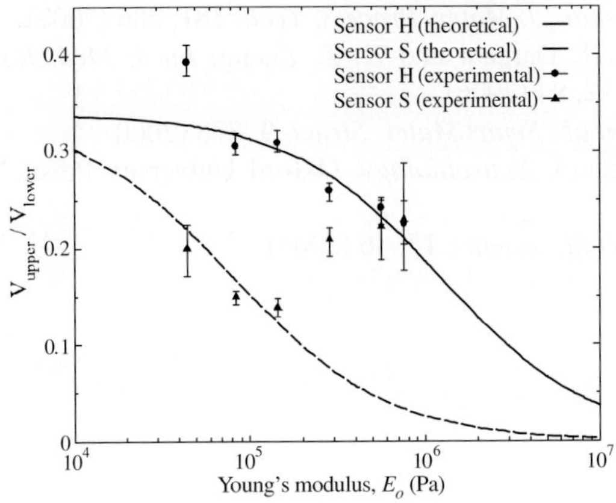

Figure 9. Comparison between theoretical values and experimental results where the sensors are vibrated.

6. Conclusions

In this paper, a tactile sensor to measure hardness under an endoscope was developed using PVDF films. Two sensors with different cylindrical columns were fabricated and their performance evaluation was conducted using silicone samples simulating the Young's modulus of the human prostate. The experimental relationships between sensor output and the Young's modulus of the object roughly corresponded with the simplified theoretical ones. From the experimental results it was found that the proposed tactile sensor can distinguish between samples on the basis of the elastic properties. However our simplified analysis is not sufficient to predict whole sensor behaviors, so we will need to conduct a calculation based on the finite element method to design and predict sensor output.

Acknowledgments

The authors acknowledge the support of Tohoku University Global COE Program "Global Nano-Biomedical Engineering Education and Research Network Centre".

References

1. S. Hirose, K. Ikuta and M. Tsukamoto, *Advanced Robotics* **4**, 3 (1990).
2. M. C. Montesi *et al.*, *J. Micromech. Microeng.* **5**, 180 (1995).
3. C. A. Mosse *et al.*, *Med. Biol. Eng. Comput.* **36**, 186 (1998).

4. M. Tanaka *et al.*, *J. Mater. Process. Tech.* **181**, 286 (2007).
5. S. Najarian, J. Dargahi and X. Z. Zheng, *Int J. Med. Robotics Comput. Assist. Surg.* **2**, 84 (2006).
6. M. Tanaka *et al.*, *Smart Mater. Struct.* **9**, 878 (2000).
7. G. M. Shepherd, *Neurobiology*, Oxford University Press, New York, 272 (1994).
8. M. Tanaka *et al.*, *Science* **17**, 46 (2005).

DEVELOPMENT OF A FULLY IMPLANTABLE RETINAL PROSTHESIS WITH A THREE-DIMENSIONALLY STACKED LARGE-SCALE INTEGRATED CIRCUIT

TETSU TANAKA [1)*], TAKAFUMI FUKUSHIMA [2)], MITSUMASA KOYANAGI [2)]

1) Department of Biomedical Engineering, Graduate School of Biomedical Engineering, Tohoku University

6-6-01 Aza-Aoba, Aramaki, Aoba-ku, Sendai 980-8579, Japan

2) Department of Bioengineering and Robotics, Graduate School of Engineering, Tohoku University

6-6-01 Aza-Aoba, Aramaki, Aoba-ku, Sendai 980-8579, Japan

To restore the vision of blind patients, we developed a fully implantable retinal prosthesis using a three-dimensionally stacked chip consisting of several large-scale integrated circuit (LSI) chips that are stacked vertically and connected electrically using three-dimensional integration technology. Our retinal prosthesis chip is small and lightweight, with a high resolution, which increases the patients' quality of life (QOL). In this study, we fabricated a retinal prosthesis module consisting of the retinal prosthesis chip, a flexible cable, and a stimulus electrode array. Our retinal prosthesis chip was not degraded in the module fabrication process. For the first time, we successfully implanted a retinal prosthesis module into a rabbit eyeball. We also fabricated and evaluated coils for power transmission between the extra- and intraocular units in the fully implantable retinal prosthesis.

Keywords: Retinal prosthesis, Three-dimensionally stacked LSI, Flexible cable, Stimulus electrode array.

1. Introduction

Worldwide, several million patients are blind because of diseases such as retinitis pigmentosa (RP) and age-related macular degeneration (AMD). RP and AMD result from impairment of the photoreceptor cells that convert optical signals into electrical signals in the retina. There is no effective medical treatment for RP or AMD. While the photoreceptor cells degenerate in RP and AMD, many other retinal cells, including the bipolar, horizontal, amacrine, and ganglion cells, remain normal [1]. Accordingly, it should be possible to restore

---

* Tetsu Tanaka is a Tohoku University Global COE Member.

one's vision by electrically stimulating the remaining retinal cells. Many studies of retinal prostheses are in progress [2-5].

Figure 1 shows the configuration of a conventional retinal prosthesis. It has three key components: photodetector, signal processing circuit, and stimulus current generator and stimulus electrode array. The photodetector receives optical signals and converts these into electrical signals. The processing circuits perform image processing, such as edge extraction and motion detection, using electrical signals. The stimulus current generator placed on the surface of the retina generates appropriate patterns of electrical current, and the stimulus electrode array stimulates the remaining retinal cells. When the remaining retinal cells are activated in this manner, blind patients will perceive a dot of light at each point stimulated.

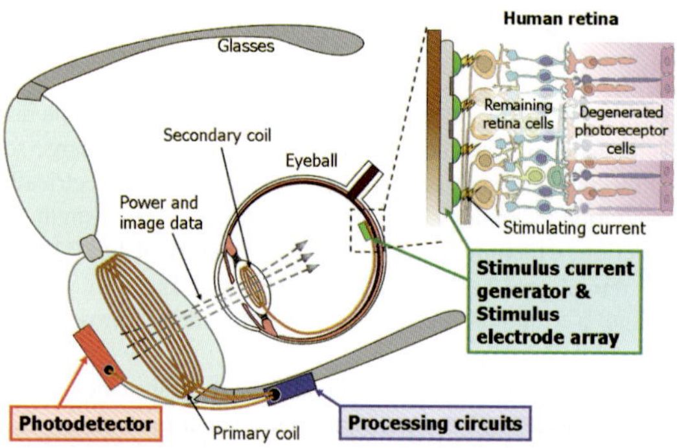

Figure 1. Configuration of a conventional retinal prosthesis.

In the conventional retinal prosthesis, a stimulus current generator with tens of pixels is implanted in the eyeball. Its size is limited by the small retinal area (~3 mm2) suitable for retinal chip implantation. As the photodetector and signal processing circuits are placed outside the eyeball, the conventional retinal prosthesis is large, heavy, and complicated. Moreover, the patient cannot use saccadic effects based on high-speed eyeball movement. These disadvantages result in a low quality of life (QOL) for the patients.

We have been developing a three-dimensional (3D) stacked retinal prosthesis chip using 3D integration technology [6-7]. Figure 2 is a conceptual drawing of our 3D stacked retinal prosthesis, in which all of the key components are stacked vertically into one chip that is implanted on the retina surface, unlike other retinal prostheses. By implanting the 3D stacked retinal prosthesis chip in the eyeball, patients can use their own lens and cornea, and can shift the point of gaze by moving the eyeball, which leads to high-speed visual information processing using saccadic effects. The 3D stacked retinal prosthesis chip has a layered structure similar to that of the human retina, as shown in Fig. 3. Therefore, photodetectors with more than 1000 pixels can be fabricated in the top layer of the 3D stacked chip, resulting in a small, lightweight, large fill-factor, high-resolution prosthesis, which allows a high QOL.

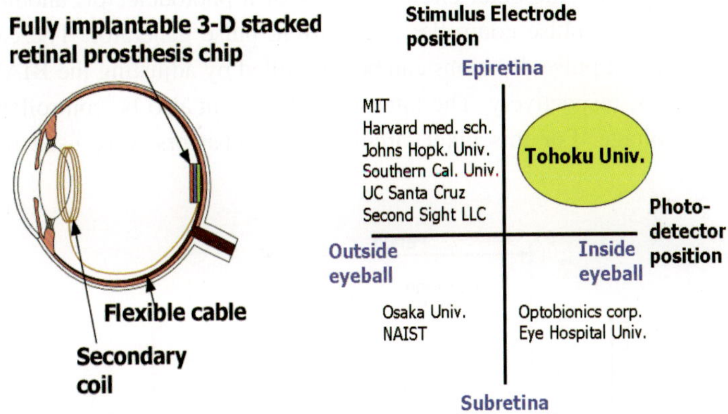

Figure 2. Conceptual drawing of a fully implantable retinal prosthesis with a 3D stacked large-scale integrated circuit (LSI) and the classification of retinal prostheses.

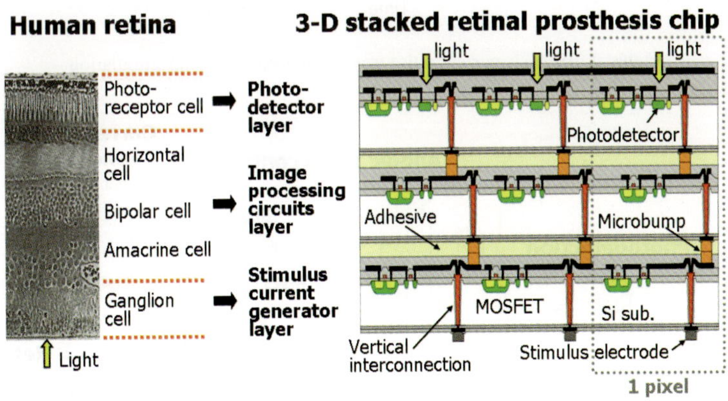

Figure 3. Structural similarity between the human retina and our 3D stacked retinal prosthesis chip.

In this paper, we studied a retinal prosthesis module consisting of a retinal prosthesis chip and a flexible cable with a stimulus electrode array. We also investigated the power transmission characteristics of coils.

2. Design of the Retinal Prosthesis Chip

Figure 4 shows the biphasic current pulses that are usually used to stimulate retinal cells. Cathodic current pulses activate retinal cells, and anodic current pulses keep the charge balance. It is very important that the parameters of the stimulus current are optimized for each patient. Therefore, we designed pixel circuits that enable us to adjust the current pulse waveforms with BIAS voltages, as shown in Fig. 5. This pixel circuit consists of a photodetector, anodic pulse controller, cathodic pulse controller, and current pulse generator. The cathodic and anodic current pulse durations can be controlled by adjusting the BIAS2 and BIAS5 voltages, respectively. The interphase delay can also be controlled using the BIAS4 voltage. For this pixel circuit, phototransistors were used for light detection.

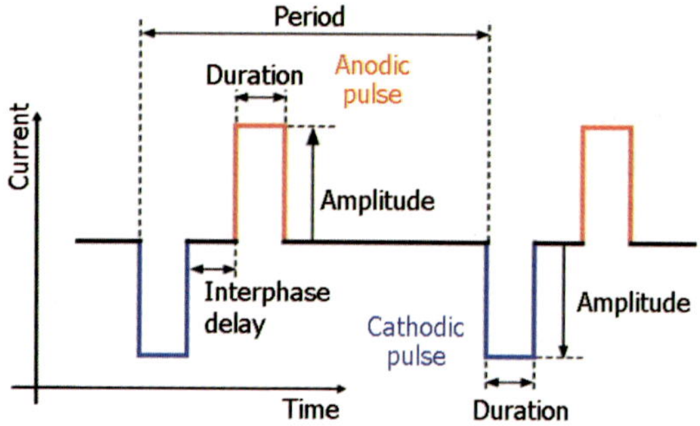

Figure 4. Biphasic current pulse waveform used for retinal stimulation.

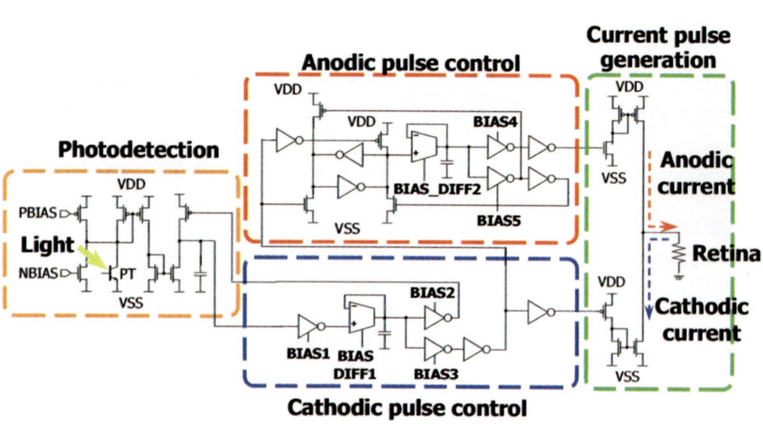

Figure 5. The pixel circuit diagram of the retinal prosthesis chip.

In this study, we fabricated and used a two-dimensional (2D) retinal prosthesis chip as a prototype for a 3D retinal prosthesis chip. There is no functional difference between the 2D and 3D designs, except for the area of the pixel circuit. Figure 6 shows a photograph of the prototype 2D retinal prosthesis chip with 16 pixels and the pixel circuit layout. We used 0.35-μm double poly-Si and triple metal complementary metal oxide semiconductor (CMOS) process technology for this retinal prosthesis chip. Each pixel has photodetector, biphasic pulse control, and stimulus current generator circuits. In the photodetector, the photodetection area is 20 × 20 μm^2.

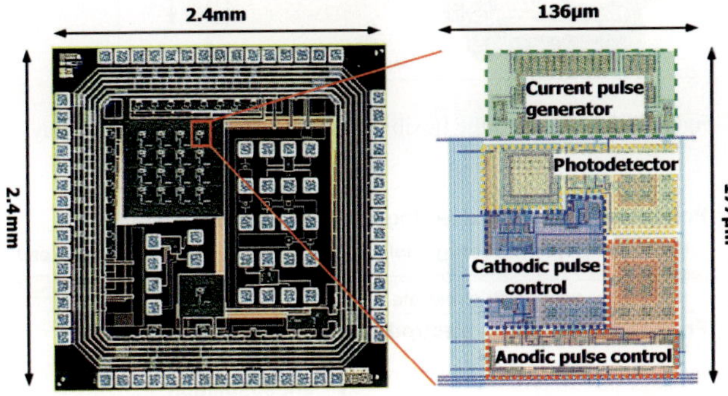

Figure 6. Photograph of the retinal prosthesis chip and pixel circuit layout.

3. Fabrication and Evaluation of the Retinal Prosthesis Module

In order to implant the retinal prosthesis chip in the eyeball, the chip needs to be attached to a flexible cable with a stimulus electrode array. Figure 7 illustrates both the flexible cable and stimulus electrode array used for chip implantation experiments. An array of 4 × 4 Pt stimulus electrodes was formed at the end of the flexible cable. The stimulus electrodes were placed at a pitch of 200 μm. Three rings were formed and used for module fixation on the retina. One ring was inserted inside the eyeball using a retinal tack, and the other two rings were sewn on the sclera outside the eyeball. The other end of the flexible cable had pads for connecting to a power supply. Figure 8 shows the fabrication process of a retinal prosthesis module consisting of a stimulus electrode array, flexible cable, and retinal prosthesis chip.

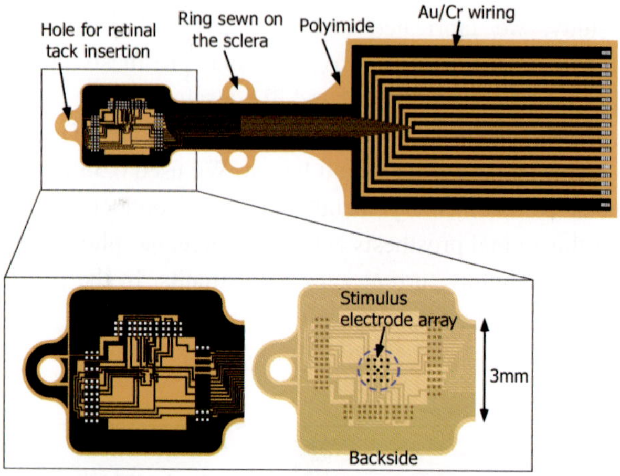

Figure 7. Structures of the flexible cable and stimulus electrode array.

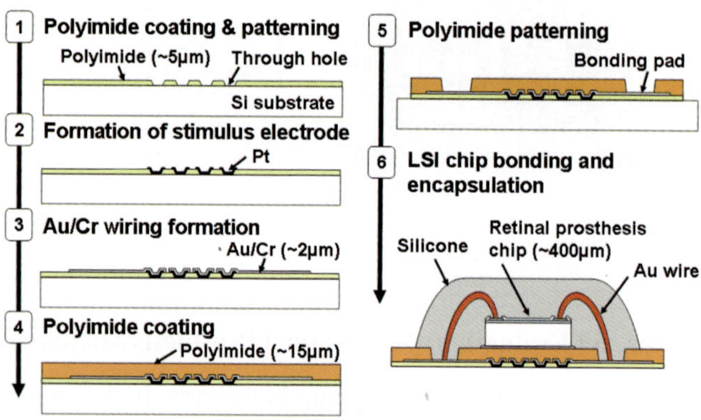

Figure 8. Fabrication of the retinal prosthesis module.

A biocompatible, photosensitive polyimide was used as the flexible substrate of the cable. Pt was used for the stimulus electrodes, and Au/Cr was used for the wiring connected to the power supply. After patterning the Au/Cr wiring, the cable was spin-coated with 15-μm-thick polyimide. Then, the bonding pads were formed by oxygen ashing. The retinal prosthesis chip was bonded on the cable with epoxy resin and connected electrically with Au/Cr wiring using the conventional wire bonding technique. The bonded chip was encapsulated with silicone for protection from corrosive biological fluids. Photographs of the fabricated retinal prosthesis module are shown in Fig. 9. The silicone cap has been removed in the enlargement to show the retinal prosthesis chip.

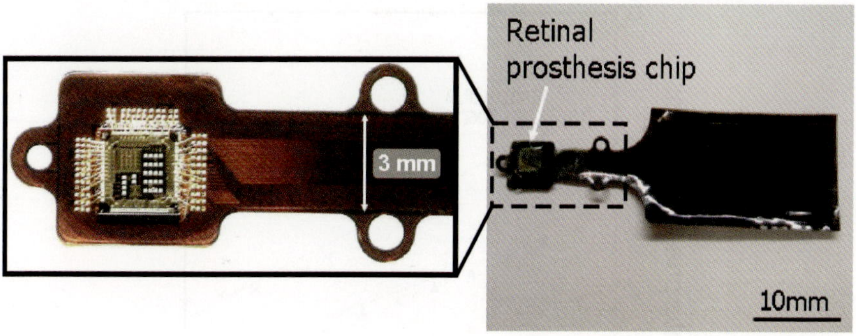

Figure 9. Photographs of the retinal prosthesis module.

To confirm chip functions after the module fabrication process, its electrical characteristics were evaluated. Figures 10 and 11 show that the anodic pulse duration and interphase delay of the stimulus current pulses can be adjusted sufficiently using the respective BIAS voltages.

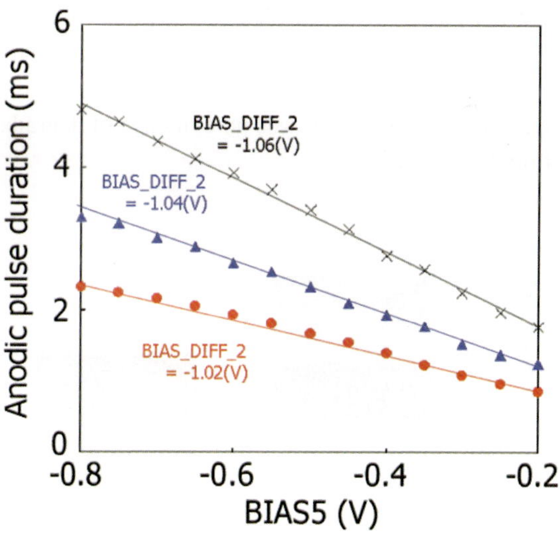

Figure 10. Adjusting the anodic pulse duration.
[Plots: measured values, Lines: SPICE simulation]

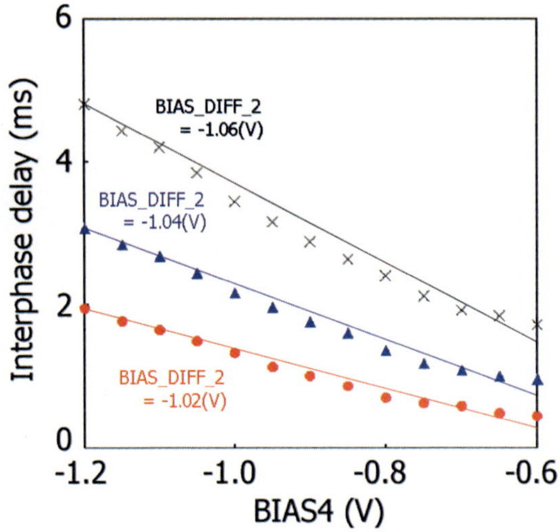

Figure 11. Adjusting the interphase delay.
[Plots: measured values, Lines: SPICE simulation]

Moreover, the incident light was successfully converted into electrical pulse trains, and the current pulse frequency changed in proportion to the intensity of the incident light, as shown in Fig. 12. The diamonds indicate measurements before the module fabrication process. The triangles indicate measurements after the module fabrication process. There was no evidence of degradation during the module fabrication process.

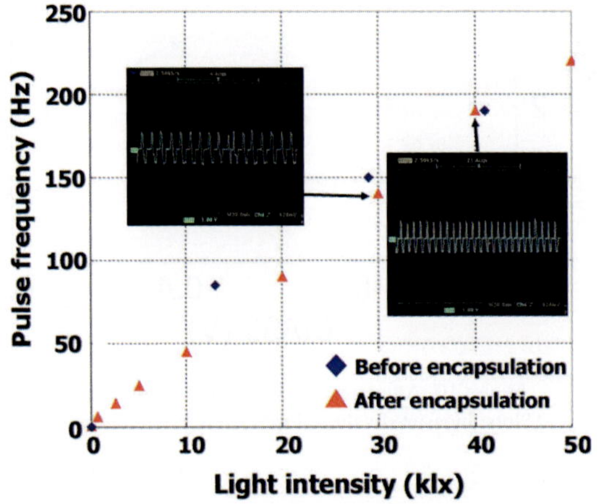

Figure 12. Relationship between the incident light intensity and generated current pulse frequency.

To verify the usability of the retinal prosthesis module and to optimize the parameters of the stimulus current pulse, it is very important to perform fundamental animal experiments. All of the procedures used in the animal experiments in this study adhered to the Association for Research in Vision and Ophthalmology (ARVO) Resolution on the Use of Animals in Research and the guidelines of the University of California at San Francisco Committee on Animal Research. Japanese white rabbits weighing 2~3 kg were anesthetized with ketamine hydrochloride (66 mg/kg) and xylazine hydrochloride (33 mg/kg) and were kept under surgical anesthesia with additional injections of the mixture. In this study, the retinal prosthesis chip was implanted completely into the rabbit eyeball and fixed on the surface of the retina. It is obvious that our module has sufficient durability for implantation in the eyeball. Figure 13 shows the apparatus used to record the electrically evoked potential (EEP). The EEP is the electrical potential elicited from the brain cortex when the retina is stimulated by an electrical current. A Pt electrode with a 4 × 4 matrix was implanted in the rabbit eyeball and used to stimulate the rabbit retina. The EEP was recorded with ground, reference, and measurement electrodes.

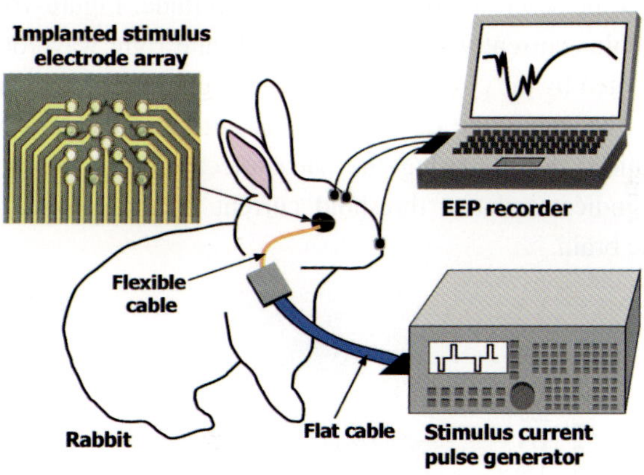

Figure 13. Apparatus for recording the EEP in a rabbit.

Figure 14 compares the waveforms of the EEPs recorded in different situations. The inset in Fig. 14 is the stimulus current pulse waveform used for the EEP recording experiments. Stimulus current pulses were applied using an implanted stimulus electrode array, to elicit EEPs of the rabbit brain. There was no EEP response without electrical stimulation or with electrical stimulation after axotomy. Therefore, the EEP response with electrical stimulation before axotomy indicates that some visual information was transferred to the primary visual cortex through the visual pathway. Consequently, the rabbit would perceive light via electrical stimulation of the retina.

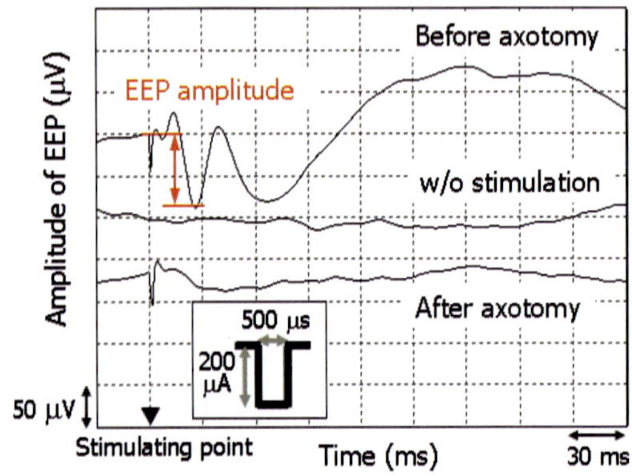

Figure 14. Comparison of the EEP waveforms before and after axotomy.

Figure 15 investigates the effects of both the duration and amplitude of the stimulus current pulse on the recorded EEP amplitude. Figure 16 schematically shows the stimulus current waveforms. The left and right waveforms are for the conditions denoted by the circles to the left and right in Fig. 15, respectively, and are identical. The amplitude of the stimulus current pulse was more effective than the duration, despite using the same electrical charge for the stimulus current. This indicates that a threshold current is necessary to elicit an EEP signal from the brain.

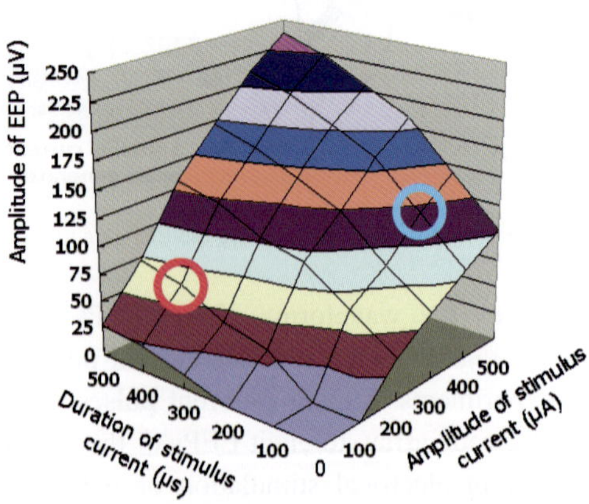

Figure 15. Effects of both the duration and amplitude of the stimulus current pulse on the recorded EEP amplitude.

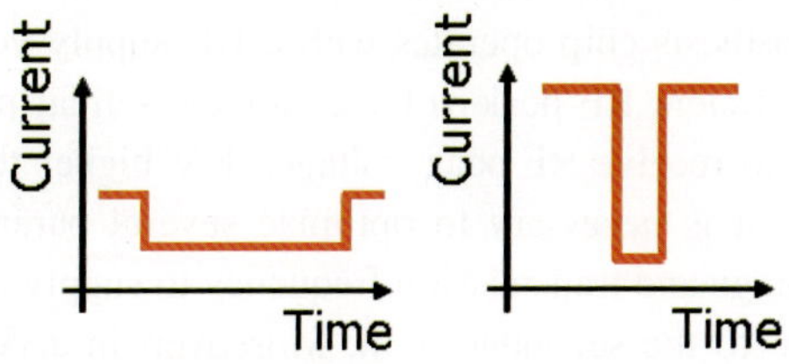

Figure 16. Schematic of the stimulus current waveforms used for the animal experiments.

4. Fabrication and Evaluation of the Coil

Electromagnetic induction was used as the power supply method to prevent the eyeball from being infected. Our power supply units consist of a primary coil, an extraocular power supply circuit, a secondary coil placed at the eyeball, and an RF/DC voltage conversion chip for converting AC voltage into DC voltage, as shown in Fig. 17.

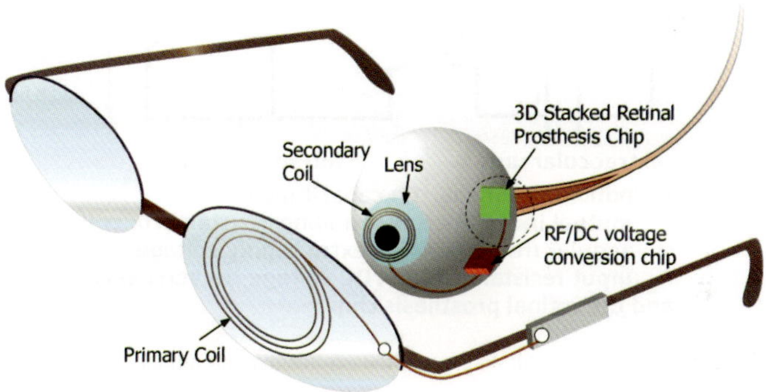

Figure 17. Conceptual drawing of the power supply unit used for a fully implantable retinal prosthesis chip.

The RF/DC voltage conversion chip has three functions: rectification, smoothing, and stabilization. The rectification and smoothing functions convert the induced AC voltages of the secondary coil into a DC voltage with ripple voltages, and the stabilization function converts the DC voltage into the appropriate DC voltage required to operate the retinal prosthesis chip.

Few papers have reported power supply systems suitable for a fully implantable retinal prosthesis, because the human eyeball is very small and there are many restrictions. Regarding the power supply system for a fully implantable retinal prosthesis, this section describes the design and fabrication of the secondary coil for power transmission.

Our retinal prosthesis chip operates with a DC supply voltage of 3.3 V. In order to receive sufficient RF peak voltages for the retinal prosthesis chip, the secondary coil has to receive RF peak voltages 1 V higher than the DC supply voltage. Therefore, it is necessary to optimize several parameters such as the external supply voltage and transmission frequency to supply an RF peak voltage of more than 4.3 V to the secondary coil. Moreover, in accordance with these parameters, structural optimization is also required for the primary and secondary coils.

Figure 18 shows the equivalent circuit diagram of the extraocular and intraocular units, including the inductance of the secondary coil and the input resistance of the retinal chip.

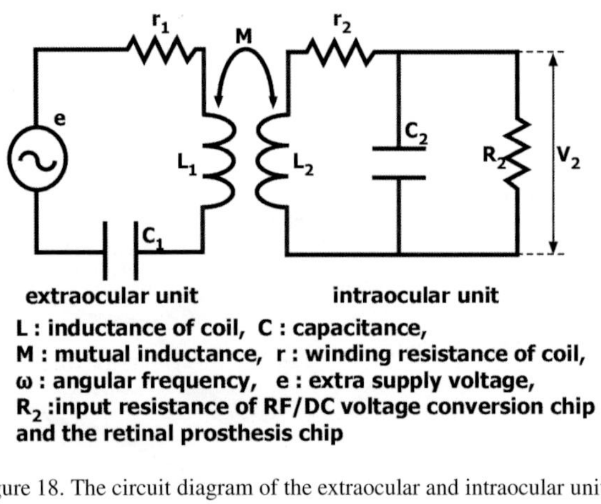

Figure 18. The circuit diagram of the extraocular and intraocular units.

To improve the transmission efficiency, capacitors C_1 and C_2 are designed to resonate with coils L_1 and L_2, respectively. The RF peak voltage V_2 can be calculated using the following formula:

$$|V_2| = \frac{\varpi M|e|R_2}{\sqrt{(r_2 r_1 + \varpi^2 M^2)^2 + (\varpi L_2 r_1 + \varpi^3 C_2 M^2 R_2 + \varpi C_2 r_2 R_2 r_1)^2}}$$

where L is the inductance of the coil, C is the capacitance, M is the mutual inductance, r is the winding resistance of the coil, ω is the angular frequency, e is the external supply voltage, and R_2 is the input resistance of the RF/DC voltage-conversion chip and the retinal prostheses chip.

Mutual inductance and the resulting transmitted voltages increase with an increasing number of turns in the coils. However, the number of turns in the secondary coil is limited to 20, because the secondary coil needs to be thinner so that it can be implanted in the crystalline lens. Figure 19 shows the calculation results using the above equation. The peak RF voltage of 4.5 V can be obtained by using a primary coil with 50 turns, a secondary coil with 20 turns, an external supply voltage of 4.1 V, and frequencies ranging from 3 to 300 MHz. Our power supply units using electromagnetic induction can supply sufficient peak RF voltage for the retinal prosthesis chip.

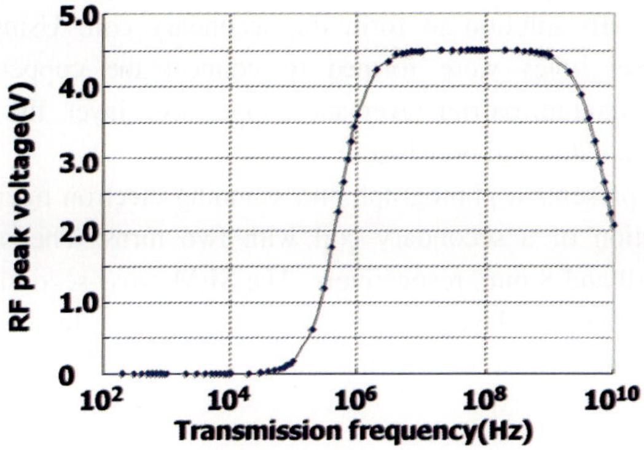

Figure 19. Relationship between the transmission frequency and RF peak voltage of the secondary coil (calculated).

In this study, we fabricated secondary coils with one and two turns. The secondary coils were fabricated on a 2-inch Si substrate using copper electroplating and damascene techniques, as shown in Fig. 20.

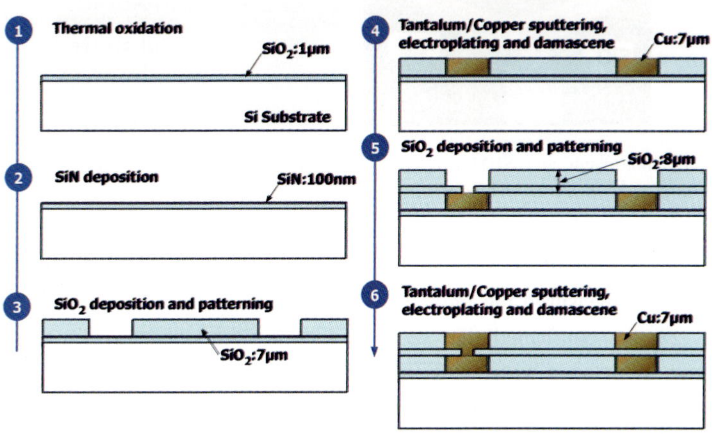

Figure 20. Fabrication of the secondary coil.

First, 1-μm-thick silicon dioxide (SiO_2) was formed by thermal oxidation. Then, 100-nm-thick silicon nitride (SiN) was deposited as an etching stop layer, using low-pressure chemical vapor deposition (LPCVD). Next, 7-μm-thick SiO_2 was deposited using chemical vapor deposition (CVD) and was shaped into the secondary coil using buffered HF solution. After this, a 100-nm-thick tantalum barrier layer and 100-nm-thick copper seed layer were formed using conventional sputtering. The copper was electroplated and polished. Then, 8-μm-thick SiO_2 was deposited using CVD, and a 7-μm layer of this was removed using buffered HF solution to form the secondary coil. Using buffered HF solution, contact holes were formed to connect the copper layers. After sputtering the tantalum barrier layer and copper seed layer, the second copper layer was electroplated and polished.

Figure 21 presents a photograph and scanning electron microscopy (SEM) of a cross-section of a secondary coil with two turns. The outer and inner diameters are 10 and 8 mm, respectively. The SEM cross-section shows that the first and second copper layers are isolated successfully. In this case, the first copper layer is thinner than the target value.

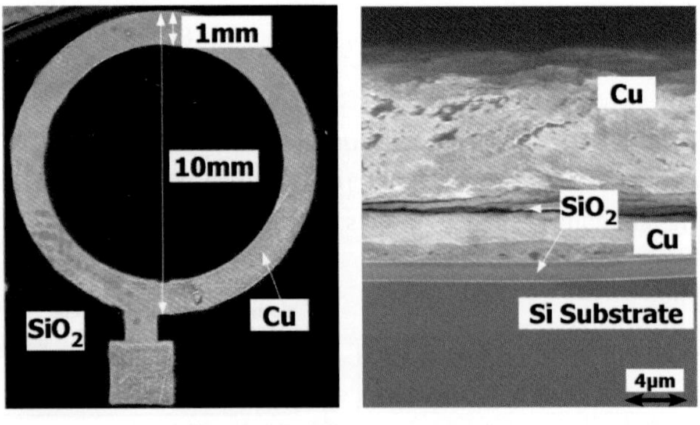

Figure 21. Photograph and SEM cross-section of a secondary coil fabricated using copper electroplating and damascene techniques.

To verify the inductance values of secondary coils fabricated with one and two turns, we measured the inductance using an impedance analyzer. As shown in Fig. 22, we confirmed that secondary coils with one and two turns have measured inductance values of 7 and 19 nH, respectively, at 3 MHz. The calculated inductance values for secondary coils with one and two turns are 7 and 30 nH, respectively. For the coil with two turns, the inductance value is off by approximately 33%. This is partly attributable to the parasitic capacitance between the first and second copper layers. From our calculations, the secondary coil needs inductance values exceeding 1 μH. Therefore, we are now trying to fabricate a secondary coil with optimized parameters, including the number of turns.

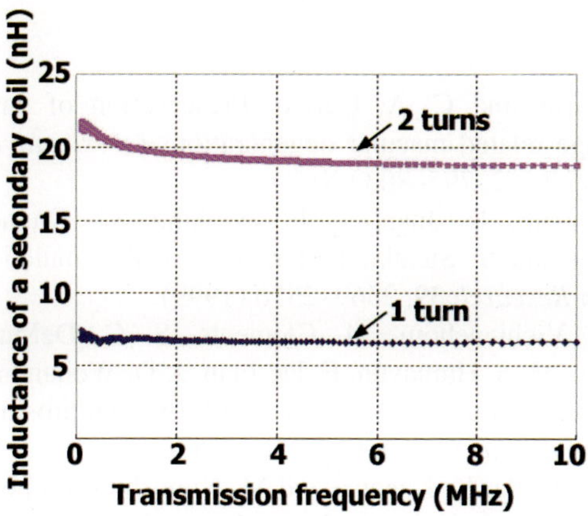

Figure 22. Relationship between the transmission frequency and inductance of the secondary coil.

5. Conclusion

We fabricated a fully implantable retinal prosthesis chip with a photodetector and stimulus current generator bonded on a flexible cable. Our chip had excellent electrical characteristics, even after the module fabrication process. For the first time, we completely implanted a retinal prosthesis chip bonded on a flexible cable with a stimulus electrode array into a rabbit eyeball. In animal experiments, the recorded EEP indicated that the rabbit would perceive light via electrical stimulation of the retina and that a threshold current is necessary to elicit an EEP from the brain.

The design and fabrication of the secondary coil in the power supply unit was studied in detail. From calculation results, we concluded that the power supply units can transmit a peak RF voltage of 4.5 V to the secondary coil when we use a primary coil with 50 turns, a secondary coil with 20 turns, an external supply voltage of 4.1 V, and frequencies ranging from 3 to 300 MHz. We fabricated secondary coils using copper electroplating and damascene techniques. Our power supply units can supply sufficient electric power for the retinal prosthesis chip.

Acknowledgments

The authors acknowledge the support of the Tohoku University Global COE Program, Global Nano-Biomedical Engineering, Education, and Research Network Centre. This work was performed in the Micro/Nanomachining Research and Education Center, Tohoku University.

References

1. N. E. Medeiros and C. A. Curcio, Preservation of ganglion cell layer neurons in age-related macular degeneration. *Investigative Ophthalmology & Visual Science* **42**, 795–803 (2001).
2. M. S. Humayun, E. Juan, J. E. Weiland, G. Dagnelie, S. Katona, R. Greenberg and S. Suzuki, Pattern electrical stimulation of the human retina. *Vision Research* **39**, 2569–2576 (1999).
3. W. Liu, K. Vichienchom, M. Clements, S. C. DeMarco, C. Hughes, E. McGucken, M. S. Humayun, E. De Juan, J. D. Weiland and R. Greenberg, A neuro-stimulus chip with telemetry unit for retinal prosthetic device. *IEEE Journal of Solid-State Circuits* **35**, 2569–2576 (2000).
4. J. Ohta, N. Yoshida, K. Kagawa and M. Nunoshita, Proposal of application of pulsed vision chip for retinal prosthesis. *Japanese Journal of Applied Physics* **41**, 2322–2325 (2002).
5. M. S. Humayun, E. Juan, J. E. Weiland, G. Dagnelie, S. Katona, R. Greenberg and S. Suzuki, Pattern electrical stimulation of the human retina. *Vision Research* **39**, 2569–2576 (1999).
6. H. Kurino, K. W. Lee, T. Nakamura, K. Sakuma, K. T. Park, N. Miyakawa, H. Shimazutsu, K. Y. Kim, K. Inamura and M. Koyanagi, Intelligent image sensor chip with a three-dimensional structure. *International Electron Devices Meeting Technical Digest*, 879–882 (1999).
7. M. Koyanagi, Y. Nakagawa, K. W. Lee, T. Nakamura, Y. Yamada, K. Inamura, K. T. Park and H. Kurino, Neuromorphic vision chip fabricated using three-dimensional integration technology. *International Solid-State Circuit Conference Digest of Technical Papers*, 270–271 (2001).

VISUALIZATION OF ION DISTRIBUTION BY A CHEMICAL IMAGING SENSOR

TATSUO YOSHINOBU *

Department of Biomedical Engineering, Tohoku University
6-6-05 Aza Aoba, Aramaki, Aoba-ku,
Sendai 980-8579, Japan

KO-ICHIRO MIYAMOTO, SHIN'ICHIRO KANOH

Department of Electronic Engineering, Tohoku University
6-6-05 Aza Aoba, Aramaki, Aoba-ku,
Sendai 980-8579, Japan

TORSTEN WAGNER, MICHAEL J. SCHÖNING

Institute of Nano- and Biotechnologies, Aachen University of Applied Sciences,
Ginsterweg 1, 52425 Jülich, Germany

Institute of Bio- and Nanosystems (IBN-2), Research Centre Jülich,
52425 Jülich, Germany

Based on the principle of the light-addressable potentiometric sensor (LAPS), a chemical imaging sensor was developed which can visualize the two-dimensional distribution of ions. To realize rapid scanning of a large area, a prototype chemical image scanner was developed, in which the measurement is simultaneously carried out at a plurality of positions on the sensing surface using a linear array of LEDs. The total measurement time required to obtain a chemical image at a resolution of 16 x 128 was reduced to 6.4 sec. The resolution can be enhanced by increasing the number of light sources on the scanner.

1. Introduction

For the visualization of chemical species in a solution, markers or labels such as fluorescent dyes are often utilized. For biological applications, however, the toxicity of such dyes may be a problem, especially in the case of cultured cells and tissues. In this study, a label-free method of ion imaging based on a semiconductor device was developed.

* Tatsuo Yoshinobu is a Tohoku University Global COE Member.

Chemical sensors based on semiconductor devices are advantageous for miniaturization, integration with peripheral circuits and fabrication of various structures on their surfaces with the help of microfabrication techniques such as photolithography. Figure 1 schematically compares the structures of the ion-sensitive field-effect transistor (ISFET) [1], the EIS capacitive sensor [2] and the light-addressable potentiometric sensor (LAPS) [3], all of which are based on the electrolyte-insulator-semiconductor (EIS) structure. They detect the change of the carrier concentration in the semiconductor layer, which responds to the change of the ion concentration of the solution in contact with the sensing surface. The ISFET detects the change of the conductance of the channel between the source and drain electrodes, whereas the EIS capacitive sensor and the LAPS detect the change of the capacitance of the depletion layer at the semiconductor-insulator interface. In the case of a p-type semiconductor, the depletion layer grows thicker and its capacitance becomes smaller when the sensing surface is more positively charged.

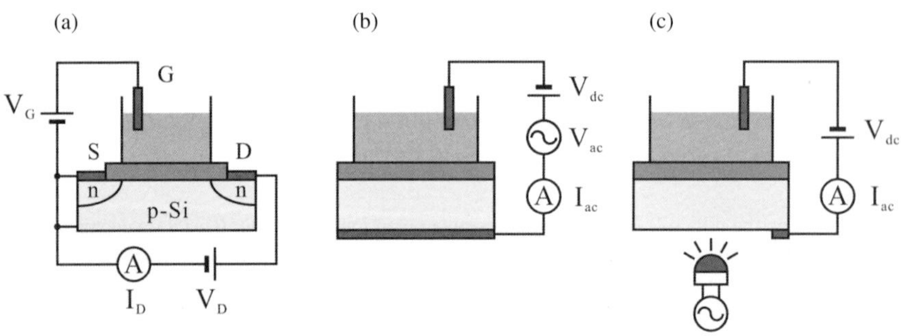

Figure 1. Structures of (a) ISFET, (b) EIS capacitive sensor and (c) LAPS.

In the LAPS measurement, a photocurrent is generated to detect the change of the capacitance of the depletion layer. The bottom surface of the sensor plate is illuminated with a light beam modulated at a frequency of several kHz, and the amplitude of the ac photocurrent is measured as a function of the bias voltage applied to the EIS system.

Figure 2(a) shows typical current-voltage characteristics of a pH-sensitive LAPS with a Ta_2O_5 film as an insulating layer. In Fig. 2(a), a positive bias (applied to the solution with respect to the semiconductor substrate) corresponds to the depletion and inversion states, and a negative bias corresponds to the accumulation state. In Fig. 2(b), the bias voltage at the inflection point of the current-voltage curve is plotted as a function of the pH value, showing a nearly-Nernstian value of pH sensitivity, 57.9 mV/pH.

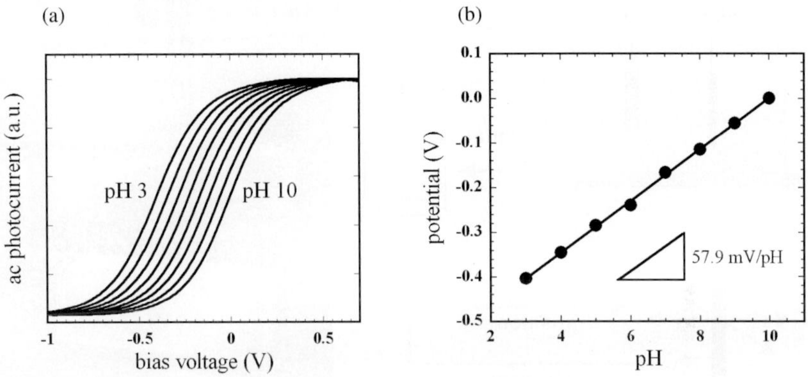

Figure 2. (a) Current-voltage characteristics and (b) pH sensitivity of LAPS.

In addition to the measurement of pH, LAPS can be also applied to selective measurement of various ions and molecules by modifying the sensing surface with respective sensing materials such as ionophores or enzymes [4].

2. Chemical Imaging Sensor

2.1. *Principle*

The chemical imaging sensor [4-6] is based on the measurement principle of the LAPS. At the semiconductor-insulator interface of the chemical imaging sensor, the width of the depletion layer is spatially distributed in response to the spatial distribution of the ion concentration on the sensing surface. When the sensor plate is illuminated with a modulated laser beam as shown in Fig. 3(a), an AC photocurrent is generated, the amplitude being dependent on the local value of the capacitance at the illuminated position. Therefore, the photocurrent image obtained by scanning the sensor plate with a focused laser beam yields a map of the ion concentration on the sensing surface.

One of the advantages of the LAPS-based chemical imaging sensor is the simple structure of the sensor plate. A semiconductor substrate with an insulating layer on the front surface and an ohmic contact on the back surface functions as a sensor plate without the necessity of fabricating device structures. In addition to the low fabrication cost of the sensor plate, the flatness of the sensing surface may be an advantage for certain applications where uniformity is required.

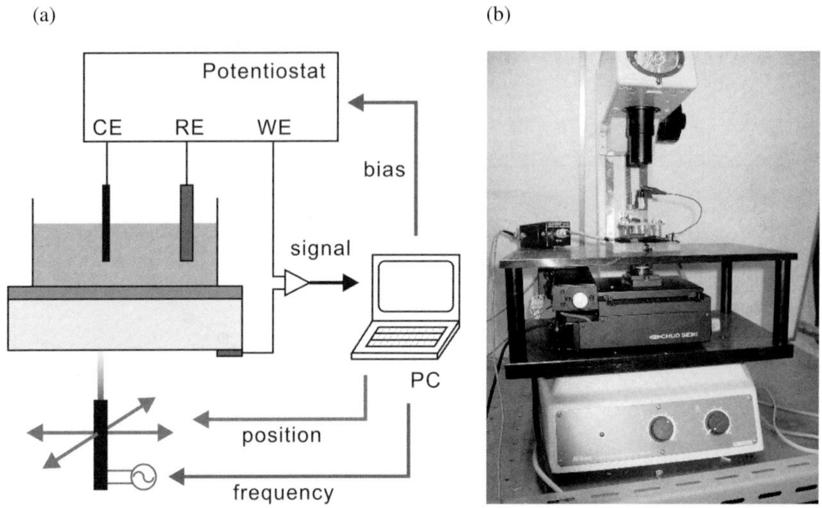

Figure 3. (a) Schematic diagram of the chemical imaging sensor system. (b) Measurement system integrated with an optical microscope.

More importantly, in contrast to other devices such as MEA (microelectrode array) or ISFET array, in which the number of pixels and their positions are predetermined by the layout of electrodes or transistors, these parameters can be arbitrarily chosen in the case of a LAPS-based chemical imaging sensor, simply by changing the scan pattern of the light beam. The measured area can be zoomed in, for example, from the size of the silicon wafer down to less than a millimeter.

The spatial resolution of the chemical imaging sensor is determined by various parameters such as the diameter of the focused laser beam, the thickness of the sensor plate and the diffusion length of minority carriers in the semiconductor layer. When a combination of silicon and visible light is used, the absorption coefficient in the silicon layer is large and most of the photocarriers are generated in the proximity of the back surface. In this case, the photocarriers diffuse both vertically and horizontally before they arrive at the depletion layer, and therefore, the spatial resolution in this case approximately equals the thickness of the silicon substrate. Thus, one of the strategies to achieve a higher spatial resolution is to make the silicon substrate thinner, which, however, makes the sensor plate fragile. Another approach is the use of infrared (IR) light instead of visible light. Since the absorption coefficient for IR light in silicon is smaller, the light penetrates deeper into the silicon substrate before generating photocarriers. The photocarriers generated near the depletion layer contribute to the photocurrent signal after travelling a shorter distance, which has the same effect as the reduced thickness of the silicon substrate. By combining these two approaches, *i.e.*, a thinner silicon plate and an IR light beam, imaging of a 5-μm line-and-space pattern has already been demonstrated [7].

Yet another approach to realize a higher spatial resolution is the use of a thin-film semiconductor layer on a transparent glass substrate. A submicron resolution can be achieved by using amorphous Si thin film [8].

2.2. *Measurement System*

A schematic diagram of the chemical imaging sensor system is shown in Fig. 3(a). The computer controls the bias voltage, the modulation frequency and the position of the light beam. The bias voltage is applied to the EIS system through a potentiostat, which defines the potential of the semiconductor substrate with respect to the reference electrode. The light source is typically a laser diode mounted on a scan stage. A focusing optics is used for high-resolution measurements. The photocurrent signal is amplified and converted into a voltage signal, the amplitude of which is recorded.

Figure 3(b) shows the external view of the chemical imaging sensor system integrated with an optical microscope. The sensor plate and the specimen are mounted on the stage, and the back surface of the sensor plate is illuminated with a focused laser beam from the objective lens mounted on the scanning unit under the stage. An IR camera on top is used for focusing and positioning the laser beam.

2.3. *Examples of Chemical Images*

Possible applications of the chemical imaging sensor include visualization of electrochemical reactions and observation of biological specimens [9-13].

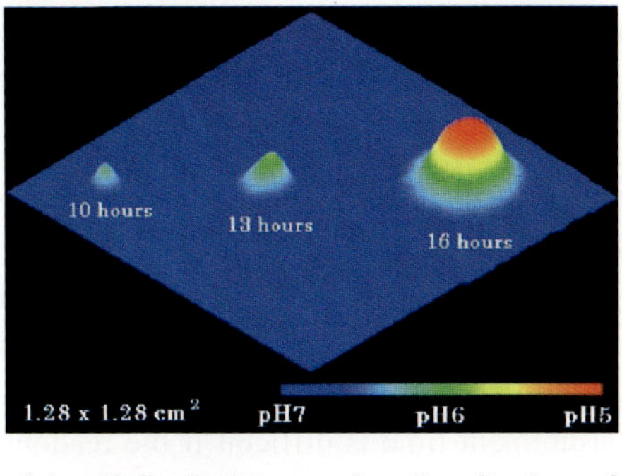

Figure 4. Visualization of the pH distribution around an *E. coli* colony after incubation of 10, 13 and 16 hours.

Figure 4 is an example of the visualization of biological specimens. In this example, the area surrounding an *E. coli* colony cultured on agar is acidified by metabolic activity. Figure 5 shows an example of visualization of an electrochemical system. In Fig. 5(a), the temporal change of pH distribution due to diffusion of ions after electrolysis of 0.01 M NaCl solution is visualized. In this experiment, a pair of Pt electrodes, 6 mm apart, was used for electrolysis, and the subsequent change of pH distribution was recorded by the chemical imaging sensor. In Fig. 5(b), the pH profiles on a line perpendicular to the electrodes at 50, 100, 150 and 200 sec after electrolysis are shown. By fitting the pH profiles with the diffusion equation, the diffusion constants of ions could be determined.

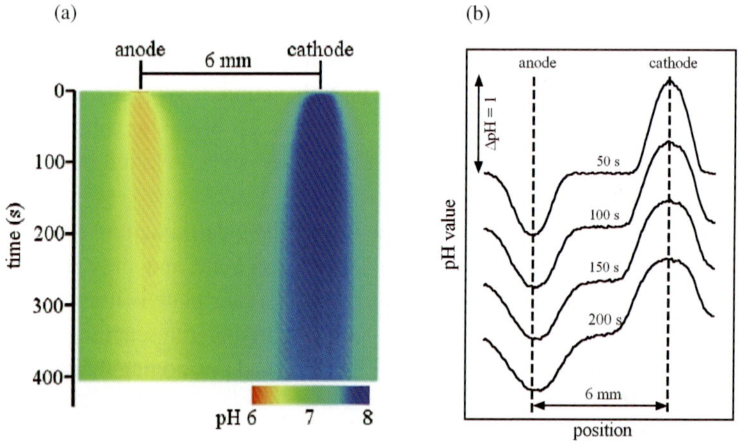

Figure 5. (a) Observation of pH redistribution between two electrodes due to diffusion of ions after electrolysis. (b) pH profiles at 50, 100, 150 and 200 sec after electrolysis.

3. Rapid Scanning of Chemical Images

3.1. *Principle*

In the chemical imaging sensor, the total measurement time is a major issue. For example, acquisition of an image at a resolution of 128 x 128 = 16,384 pixels requires nearly 3 minutes, assuming that the measurement time per pixel is 10 ms. Due to this rather long measurement time, the chemical imaging sensor has not been applicable to observation of rapidly changing specimens. A drastic reduction of the measurement time is difficult if the readout is done in a pixel-by-pixel manner.

A method for simultaneous measurement at multiple positions on the LAPS sensing surface has been proposed on the basis of frequency division multiplexing [14,15]. In this method, a plurality of positions on the sensing surface are illuminated with light beams modulated at different frequencies. The measured photocurrent in this case is a superposition of AC signals with different frequencies at respective positions, which can be separated by Fourier analysis as shown in Fig. 6.

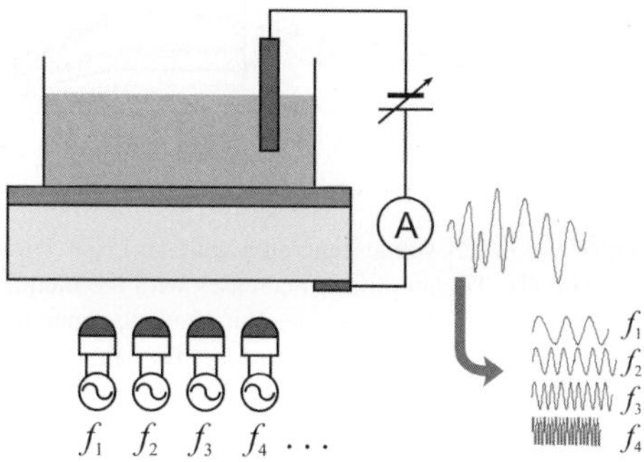

Figure 6. Simultaneous measurement at different positions on the sensing surface based on frequency division multiplexing.

In the present study, this principle was applied to rapid scanning of chemical images. With the developed system, a linear array of light sources is used to simultaneously illuminate multiple positions with different frequencies, and this array scans the sensing area to obtain a two-dimensional image. In this method, pixels on one line are simultaneously measured, and therefore, the total time of measurement can be reduced depending on the multiplicity. A prototype system with a linear array of 16 LEDs was developed. The current of each LED is sinusoidally modulated at an independently specified frequency generated with a programmable waveform generator, the frequency of which can be programmed by the control PC.

Figure 7 shows the FFT spectrum of the photocurrent signal from a LAPS sensor illuminated at 16 different positions with 16 LEDs. The frequency of modulation at each position was set at 2000 Hz, 2100 Hz, ..., and 3500 Hz with intervals of 100 Hz, and the sampling frequency was 100 kHz. In the FFT spectrum, signals corresponding to these modulation frequencies are observed. The peak height decreases with the modulation frequency due to the low-pass filtering effect of carrier diffusion in the semiconductor layer [16].

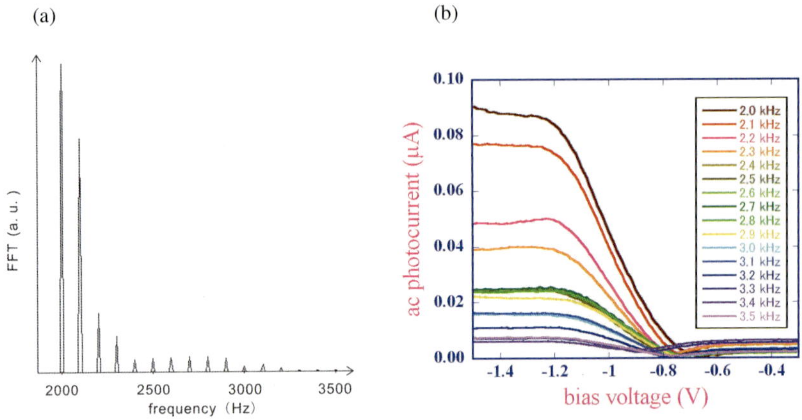

Figure 7. (a) FFT spectrum of the LAPS signal generated with 16 LEDs with different frequencies, 2000 Hz, 2100 Hz, ..., and 3500 Hz. The amplitude decreases with the modulation frequency due to the low-pass filtering effect of carrier diffusion. (b) Simultaneous measurement of the current-voltage characteristics at 16 positions on the sensing surface. By calculating the horizontal shift of each curve, the ion concentration can be determined at 16 different positions.

In Fig. 7(b), the amplitude of the photocurrent at each modulation frequency is plotted as a function of the bias voltage applied to the EIS system. In this way, the current-voltage curves at 16 different positions on the sensing surface are simultaneously obtained, which means that ion concentrations at these positions can be simultaneously determined.

For imaging applications, it is necessary to equalize the signals at different frequencies. Additional circuits and software were developed, which automatically adjust the intensities of light beams from 16 LEDs.

3.2. *Chemical Image Scanner*

A prototype of the chemical image scanner system was developed (Fig. 8(a)), which has a flatbed structure and scans the sensor plate with a linear array of LEDs. The scanner has 16 LEDs with intervals of 3.6 mm, mounted on a one-dimensional scan stage, which moves perpendicular to the array. The 16 points on the line are measured simultaneously, and the measurement is repeated 128 times during the one-dimensional scan. The width of the scanned area is 54 mm. A 4-inch sensor plate is mounted on the stage with a measurement cell and the reference electrode.

(a) (b)

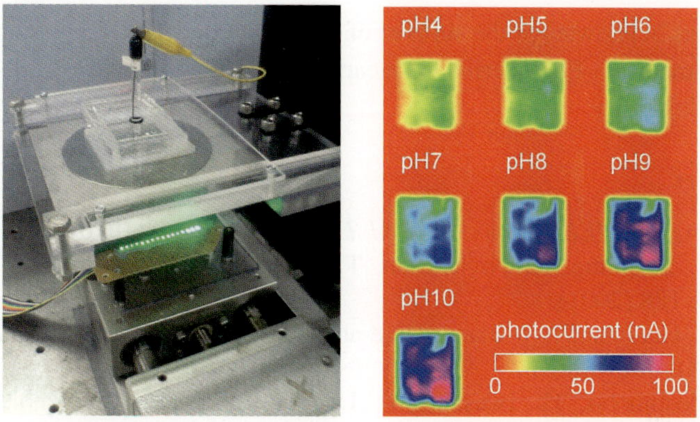

Figure 8. (a) A prototype of the chemical image scanner system with a linear array of 16 LEDs. The sensor is 4 inches in diameter. A chemical image with a resolution of 16 x 128 pixels can be obtained in 6.4 sec. (b) Photocurrent images of pH buffer solutions measured with the new scan system.

With this system, the total measurement time needed to obtain a chemical image at a resolution of 16 x 128 is 6.4 sec when the measurement time per line is set to be 50 ms. Figure 8(b) shows an example of chemical images of buffer solutions with pH 4 to 10 obtained with this system. The photocurrent increases with the pH value in accordance with the current-voltage characteristics of a LAPS sensor with an n-type Si substrate. By increasing the number of LEDs on the array, the resolution can be enhanced without increasing the total time of measurement.

4. Summary

Based on the principle of LAPS, a chemical imaging sensor was developed which can visualize the two-dimensional distribution of ions. To realize rapid scanning of a large area, a prototype system of the chemical image scanner was developed, in which the measurement is simultaneously carried out at a plurality of positions on the sensing surface using a linear array of LEDs. The total measurement time needed to obtain a chemical image at a resolution of 16 x 128 was reduced to 6.4 sec. The resolution can be enhanced by increasing the number of light sources on the scanner.

Acknowledgments

The authors acknowledge the support of the 2007 Global COE Program "Global Nano-Biomedical Engineering Education and Research Network Centre," Tohoku University.

References

1. P. Bergveld, *IEEE Trans. Biomed. Eng.* **BME-17**, 70 (1970).
2. M. J. Schöning, A. Poghossian, T. Yoshinobu and H. Lüth, *SPIE Proc.* **4205**, 188 (2001).
3. D. G. Hafeman, J. Wallace Parce and H. M. McConnel, *Science* **240**, 1182 (1988).
4. T. Yoshinobu, H. Iwasaki, Y. Ui, K. Furuichi, Yu. Ermolenko, Yu. Mourzina, T. Wagner, N. Näther and M. J. Schöning, *Methods* **37**, 94 (2005).
5. M. Nakao, T. Yoshinobu and H. Iwasaki, *Sensors and Actuators B* **20**, 119 (1994).
6. M. Nakao, T. Yoshinobu and H. Iwasaki, *Jpn. J. Appl. Phys.* **33**, L394 (1994).
7. M. Nakao, S. Nomura, S. Takamatsu, K. Tomita, T. Yoshinobu and H. Iwasaki, *Trans. IEE of Japan* **118-E**, 584 (1998) (in Japanese).
8. T. Yoshinobu, W. Moritz, F. Finger and M. J. Schöning, *Mat. Res. Soc. Symp. Proc.* **910**, 0910-A20-01 (2006).
9. M. Nakao, S. Inoue, R. Oishi, T. Yoshinobu and H. Iwasaki, *J. Fermentation and Bioengineering* **79**, 163 (1995).
10. T. Yoshinobu, N. Oba and H. Iwasaki, *J. Electrochem. Soc.* **144**, 3919 (1997).
11. T. Yoshinobu, H. Iwasaki, M. Nakao, S. Nomura, T. Nakanishi, S. Takamatsu and K. Tomita, *Jpn. J. Appl. Phys.* **37**, L353 (1998).
12. T. Yoshinobu, T. Harada and H. Iwasaki, *Jpn. J. Appl. Phys.* **39**, L318 (2000).
13. T. Yoshinobu, H. Ecken, A. B. M. Ismail, H. Iwasaki, H. Lüth and M. J. Schöning, *Electrochimica Acta* **47**, 259 (2001).
14. Q.-T. Zhang, P. Wang, W. J. Parak, M. George and G.-Y. Zhang, *Sensors and Actuators B* **73**, 152 (2001).
15. T. Wagner, R. Molina, T. Yoshinobu, J. P. Kloock, M. Biselli, M. Canzoneri, T. Schnitzler and M. J. Schöning, *Electrochimica Acta* **53**, 305 (2007).
16. M. Satore, M. Adami, C. Nicolini, L. Bousse, S. Mostarshed and D. G. Hafeman, *Sensors and Actuators A* **32**, 431 (1992).

A CAR TRANSPORTATION SYSTEM WITH MULTIPLE ROBOTS: ICART

MITSURU ENDO [*], KENJI HIROSE, YASUHISA HIRATA, KAZUHIRO KOSUGE [†]

Department of Bioengineering and Robotics, Tohoku University
Aoba 6-6-01, Aramaki, Sendai 980-0857, Japan

KOKI SUZUKI, TAKASHI KANBAYASHI

Ishikawajima Transport Machinery
Akashi 6-4, Chuoh-ku, Tokyo 104-0044, Japan

We propose a novel car transportation system termed as iCART (intelligent cooperative autonomous robot transporters). This system transports cars using two robots that move in coordination. This car transportation system can transport cars of any size and offers various functions such as parking, valet parking, towing service, and the transportation of cars inside a factory, onto a ferry, and in complicated parking areas. The robots in this car transportation system comprise three modules: the mobile base module, the lifter module, and the connecting module. This paper describes the details of the operation mechanism of each module. A leader-follower-type distributed motion control algorithm is applied to the proposed system and is used in an experiment performed to verify the validity of its function.

1. Introduction

Cars are one of the most popular transportation devices and are being used by a large population; the ease of control offered by a car to both novice and skilled drivers is a major reason for its popularity. As a result, the production and sale of cars has been constantly increasing over the last two years. However, a certain amount of skill is required for driving the vehicle, particularly in narrow spaces, because of its non-holonomic motion constraint.

In this study, we propose the use of a robot system to assist the drivers to overcome this non-holonomic constraint. There are two methods to develop robot systems that assist drivers: installation of an automatic driving system in an intelligent car [1], and development of a robot system that will transport the car. Additionally, the construction of an intelligent car requires sensors, actuators, and controllers to be installed in the car. Such a system has been put

[*] Mitsuru Endo is a Tohoku University Global COE Research Assistant.

[†] Kazuhiro Kosuge is a Tohoku University Global COE Member.

to practical use; however, the number of cars equipped with this system is very limited, and they are being used by only a small fraction of the population. Moreover, in this system, the trajectory of a source-destination path occasionally becomes complex owing to the non-holonomic motion constraint on the car. In contrast to the above system, the system that employs an omni-directional mobile robot to transport the car does not have non-holonomic motion constraints. In addition, because the robot system is not directly applied to the car but its surrounding, physical modifications to the car are not required. Thus, this system is easy to apply and can be used by everyone.

In this study, a novel car transportation system termed as iCART (intelligent cooperative autonomous robot transporters) is proposed. In narrow spaces, to assist the drivers, the robot approaches the car, grasps it, and transports it to the required location autonomously. This car transportation system can transport a car of any size and can be used for various functions such as parking, valet parking, towing, and the transportation of cars to a factory, ferry, parking areas, *etc.*

There are a several systems that achieve the transportation of stationary object or objects [2] *etc.* However, certain problems are encountered during the transportation of cars, such as the vehicle's weight, size, and the variation in the bodies of different cars, which are designed depending on the car's specific purpose. Moreover, most cars are painted and coated with certain materials. In order to prevent damage to the painted or coated surface, it should be ensured that the robot does not touch, grasp or hold the body of the car directly.

In the case of the system employing a single robot, the robot must be sufficiently large to transport the largest of cars and have good rigidity for grasping and holding the heaviest car. However, the operation of this system, which employs such a large robot, requires vast spaces; therefore, a system using a single robot cannot be used for providing assistance to drivers in narrow spaces.

Moreover, the system must be such that it can transport a variety of cars because it is aimed at being useful for a large population. To achieve this aim, the system must have a provision to accommodate itself depending on the size and shape of the car it will transport. As mentioned above, the robot should not come in direct contact with the body of the car, which is more vulnerable to damage.

Keeping the abovementioned objectives in mind, we propose a system with two robots that are equipped with end effectors to grasp and hold the wheels of a car. The two robots transport the car in coordination. Figure 1 shows the concept of the proposed system iCART. The robots in this system can be designed to be relatively small because the weight of the car is now being shared by the two robots. Additionally, the robot becomes smaller due to the end-effecter consistently connects to ground and the car is putted on it.

The system transports the car using two robots, as shown in Fig. 1. The two robots approach the car from opposite sides: right and left. By adjusting their relative positions, the system can suitably accommodate the tread width of different cars. In addition, by adjusting the grasping position according to the position of the wheel, the system can also fit different wheelbase lengths. Thus, the system can be employed for transporting various cars with different sizes and shapes.

In the proposed system, the car transportation is possible using small robots. Therefore, in contrast to the system that requires a large robot, this system can be used in narrow spaces. Further, the use of the end effector prevents damages to or scratches on the body of the car during its transportation. Finally, the above features, along with the fact that this system can adjust itself to accommodate various sizes and shapes of cars, make it preferable over others.

Figure 1. Concept of the car transportation system iCART.

2. Robot in Car Transportation System: iCART

In this section, the robot used in the car transportation system is described. Figure 2 shows the developed robot, and Tab. 1 lists its specifications. Two such robots in coordination with each other are used to transport the car. To achieve the coordinated motion of these two robots, each robot is designed such that it comprises three modules: the mobile base module, the lifter module, and the connecting module.

2.1. *Mobile Base Module*

The fuction of the mobile base module is to drive the robot, which follows an omnidirectional mechanism. The nonholonomic motion constraint of the car does not restrict its transportation due to the omnidirectional mechanism of the robot. Figure 2 shows the developed mobile base module.

Many omnidirectional mechanisms that require a particular kind of wheel have been proposed to acheive omnidirectional motion. The wheel used in this module is equipped with rollers, which permit a free rotation along the circumference of the wheel, in the axial direction. By using an appropriate number of such wheels, the omnidirectional movement of the robot is possible. However, this kind of wheel offers low bump-climbing performance and low load capacity. In addition, this kind of wheel gets easily damaged because of exposure to dust and gravel from the ground. This disadvantage of the wheel poses a challenge in our study, in that it is difficult to develop an omnidirectional mechanism using this wheel because this system is meant for outdoor operations.

In order to achieve omnidirectional motion using simple wheels, the mobile base module involves two caster mechanisms that control the steering angle and the angular velocity of the wheel using servomotors [3]. The rotational axis of steering locates on the contact point of the wheel with ground.

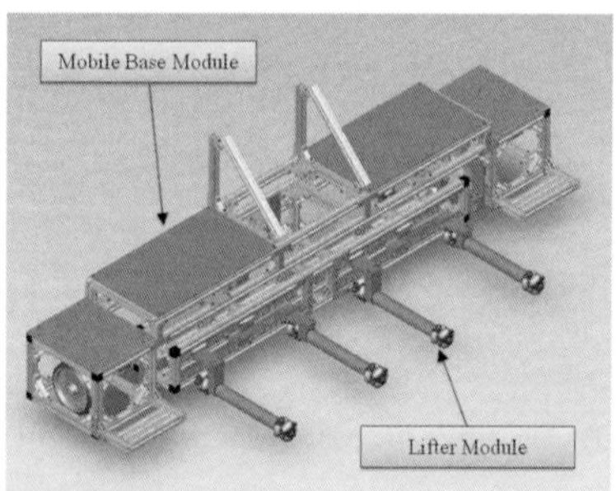

Figure 2. Robot of the car trasnportation system: iCART.

2.2. *Lifter Module*

If the robot were to grasp and lift the car directly, the body of the car could easily get damaged. On the other hand, if the car were to be lifted from the jack-up point, it would be prevented from being damaged. However, it is difficult to locate the position of the jack-up point because it is situated underneath the car. Moreover, its position is different in different cars. Thus, it is difficult for robots to lift the car from the jack-up point. In this study, we consider grasping and holding the wheels of a car, which the system can locate easily using a laser range finder or a camera, so that damages to the body of the car are avoided.

Figure 2 shows the lifter module, *i.e.*, the end effector, proposed in this study. The robot grasps the wheel by fitting it between two bars: one in front and the other behind, as shown in Fig. 3. These bars of the lifter module are equipped with casters on both their ends to lift up and hold a heavy object such as a car. Thus, the lifter module can independently support almost the entire weight of the car.

In addition, the mechanism of holding the wheel by two bars causes the car to float along the front-rear direction. Because of this floating mechanism, the robot can detect the center of the wheel automatically by controlling the relative position of the two bars, and not their individual positions. Figure 4 shows the sequence of movements involved in lifting of the wheel by the lifter module. In this mechanism, the presence of only one actuator is sufficient.

Because of the above property of this mechanism, of detecting the center of the wheel automatically, the lifter module can adapt to a change in the position of the wheel. Thus, the lifter module can adjust itself to accommodate various wheelbase lengths. Moreover, as mentioned above, the system can also accommodate various tread widths by adjusting the relative positions of the two robots.

Figure 3. Lifting of the wheel using two bars.

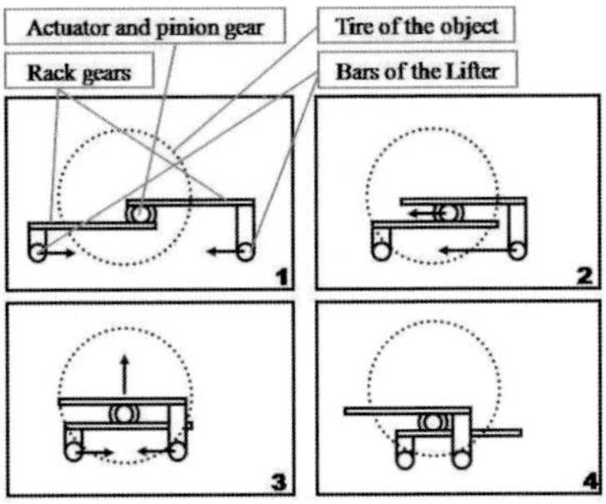

Figure 4. Automatic detection of the center of the wheel.

2.3. *Connecting Module*

The connecting module connects the mobile base module to the lifter module; it consists of three links, as shown in Fig. 5. Each link has a ball-joint assembly at both its ends and creates a parallel link mechanism with the mobile base module and the lifter module. The length of each link is fixed.

Each link has a force sensor that can measure the pushing or pulling force along one axis. Thus, the entire connecting module can function as a force sensor that measures the planar 3-axis force/moment. The planar 3-axis force/moment is calculated geometrically from the 1-axis force measured by each link [4].

The parallel link mechanism using three links, each having a ball-joint assembly on both ends, constrains 3 DoF of spatial 3 DoF. By varying the position of the three links, the connecting module restricts the planar 3 DoF and allows free movement of other 3 DoF. Because of this freedom in movement, the connecting module acts as buffer that mitigates the difference between the movements of the mobile base module and the lifter module, which occurs during undulation of the path.

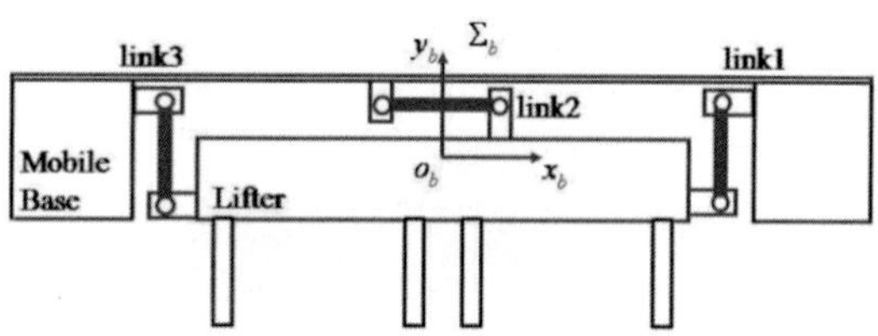

Figure 5. Connecting module.

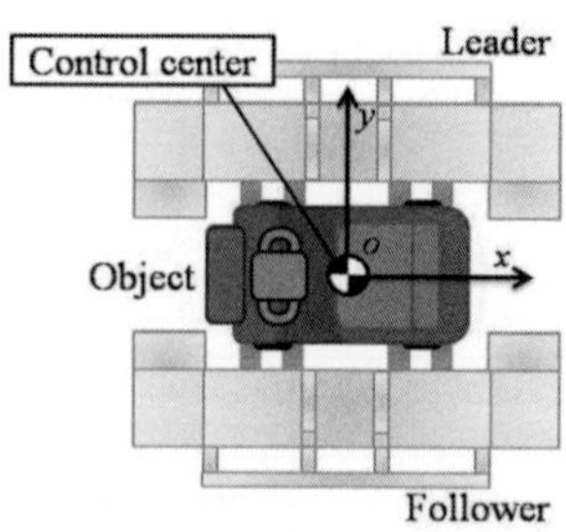

Figure 6. Coordinate system for the experiment.

3. Experiment

An experiment to illustrate the validity of the proposed car transportation system was performed on it. An algorithm for leader-follower-type distributed motion coordination proposed in [5] was applied to this system for achieving coordination between the two robots.

A model car (1100×600×1090 [mm], 86 [kg]) was used as the object to be transported. The system travels a distance of 5 [m] in 60 [sec] along the y-axis of the coordinate system shown in Fig. 6. Further, the system can rotate by 90 [deg] during transportation.

The experimental results are shown in Fig. 7. The pictures show the sequence of the process of transportation using two robots.

4. Conclusion

In this study, a novel car transportation system termed as iCART was proposed. By applying the distributed motion coordination algorithm proposed in [6], an experiment on the car transportation system was performed, and its validity was verified.

(a) 0.00[sec] (b) 8.57[sec] (c) 17.14[sec] (d) 25.71[sec]

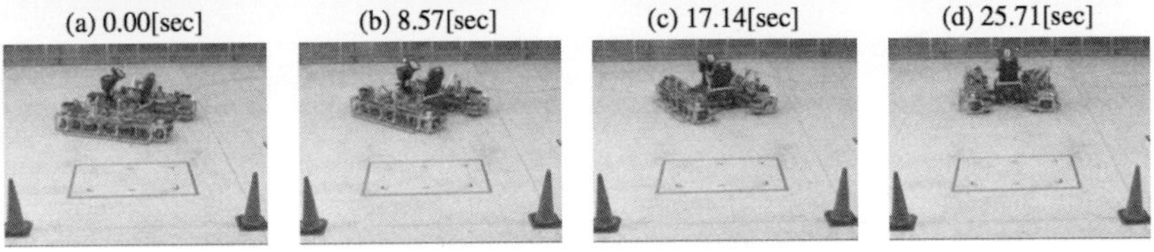

(e) 34.28[sec] (f) 42.85[sec] (g) 51.72[sec] (h) 60.00[sec]

Figure 7. Experimental result.

References

1. R. M. Murray *et al.*, Nonholonomic Motion Planning: Steering Using Sinusoids, *IEEE Trans. on Automatic Control*, **38**, 700–716 (1993).
2. K. Kosuge *et al.*, Decentralized Control of Multiple Robots Handling an Object, *Proc. of the 1996 IEEE/RSJ Int. Conf. on Intelligent Robots and Systems*, 318–323 (1996).

3. M. Wada *et al.*, Caster Drive Mechanisms for Holonomic and Omnidirectional Mobile Platforms with no Over Constraint, *Proc. of the 2000 IEEE Int. Conf. on Robotics and Automation*, 1531–1538 (2000).
4. K. Kosuge *et al.*, Handling of a Single Object by Multiple Autonomous Mobile Robots in Coordination with Body Force Sensor, *Proc. of the 1998 IEEE/RSJ Int. Conf. on Intelligent Robots and Systems*, 1419–1424 (1998).
5. Y. Hirata *et al.*, Coordinated Transportation of a Single Object by Mutiple Mobile Robots without Position Information of Each Robot, *Proc. of the 2000 IEEE/RSJ Int. Conf. on Intelligent Robots and Systems*, 2024–2029 (2000).

A PHOTORECEPTIVE STIMULATOR FOR A RETINAL PROSTHESIS WITH 3D STACKED LSI

KOUJI KIYOYAMA [1)], KEIGO SATO [2)], YOSHIYUKI KAIHO [2)], HIROSHI TOMITA [3)], ERIKO SUGANO [3)], TAKAFUMI FUKUSHIMA [2)], HIROYUKI KURINO [2)], TETSU TANAKA [2)*], MITSUMASA KOYANAGI [2)]

1) Department of Micro-Nano Research and Education Center, Tohoku University, 6-6-01 Aza-Aoba, Aramaki, Aoba-ku, Sendai 980-8579, Japan

2) Department of Bioengineering and Robotics, Graduate School of Engineering, Tohoku University, 6-6-01 Aza-Aoba, Aramaki, Aoba-ku, Sendai 980-8579, Japan

3) Institute of International Advance Research and Education Organization, Tohoku University, 6-3 Aoba, Aramaki, Aoba-ku, Sendai 980-8578, Japan

In this paper, we describe a high-sensitivity, high-resolution, low-power circuit for a fully implantable three-dimensionally stacked retinal prosthesis. This device is composed of a photo-sensor with a current mode readout circuit, an image processing circuit, stimulus current generator circuit, and an electrode array on an intraocular chip. The prototype intraocular chip was fabricated with standard 0.35-μm CMOS technology. Our experimental results confirmed that the frequency of the stimulus current pulse has a linear dependence on the incident light intensity.

1. Introduction

The demand for biomedical implants, such as vital-signal sensors and neuromuscular stimulators that depend on the development of small low-power circuits, is on the rise. Moreover, interest has been growing in visual prostheses as the number of blind patients with intact optic nerves and damaged photoreceptor cells increases. Retinal prostheses have been developed to restore the vision of patients via electrical stimulation of their remaining retinal cells. In the case of degenerative retinal conditions, mostly caused by retinitis pigmentosa and age-related macular degeneration, the retinal bipolar, horizontal, and ganglion cells of patients have been demonstrated to still work even after several years of blindness and electrical stimulation of these cells yields visual sensations [1-4]. These retinal prostheses can be classified by their general approach, either epiretinal or subretinal stimulation of the retina. In epiretinal stimulation, the prosthesis device is placed on the surface of the retina while a

---

* Tetsu Tanaka is a Tohoku University Global COE Member.

subretinal stimulation prosthesis is implanted in the space between the choroid and inner nuclear layer. Typically, image data and power are transmitted wirelessly from an extraocular unit to the intraocular unit in such systems.

We propose an epiretinal stimulation retinal prosthesis with a three-dimensional (3D) stacked chip, and photo-sensor and stimulation electrodes located on the intraocular unit [5]. Figure 1 shows the configuration of the proposed retinal prosthesis system.

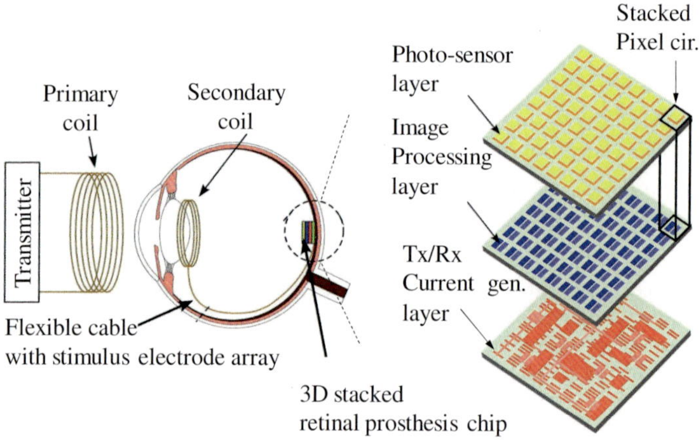

Figure 1. Overview of the proposed retinal prosthesis with 3D stacked chips.

This system consists of an extraocular and an intraocular unit. On a two-dimensional (2D) retinal prosthesis, the photo-sensor and image processing circuits are fabricated on the same layer. Therefore, the fill-factor of the photo-sensor, the resolution, and the sensitivity are low. In contrast, in the 3D stacked prosthesis chip, the layers, each with a different function, are stacked and electrically interconnected vertically using through-Si vias (TSVs) [6,7]. The image data in every photo-sensor are simultaneously transferred in the vertical direction and processed in parallel on each layer. Using 3D technology results in a photo-sensor with a high fill-factor, high resolution, and high sensitivity because several successive processing circuits with large areas are formed on the layers beneath the photo-sensor layer. In this system, the retinal prosthesis is implanted on the surface of the retina, so the patients can employ their own lens and cornea, and shift the point of their gaze by moving the eyeball.

2. Retinal Prosthesis Chip Design

2.1. *Stimulating Current Waveform Parameters*

Figure 2 shows a current pulse that is usually used to stimulate retinal cells. In general, a cathodic pulse followed by an anodic pulse (*i.e.*, biphasic pulse) is used as stimulus current for the retinal prosthesis. The cathodic pulse is used as the stimulus current while the anodic current maintains the charge balance of the retinal cells. Seven parameters are used to define the stimulus current, including amplitude, duration, interphase delay between the cathodic and the anodic pulse, and period. These parameters must be tunable for proper stimulation of the remaining retina cells because the optimum values of the parameters vary from patient to patient depending on the degree of damage in the retina.

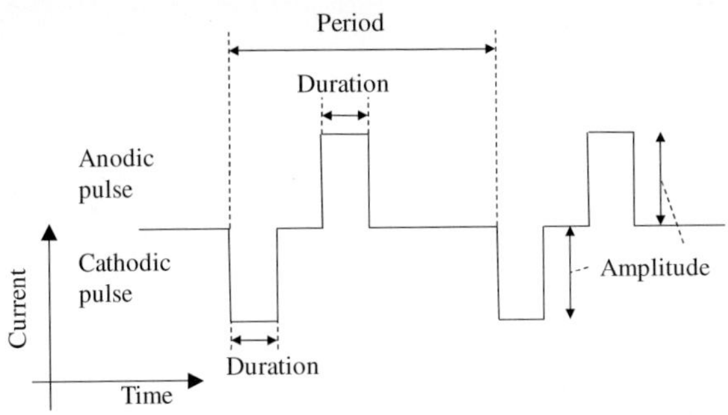

Figure 2. Diagram of the stimulating current waveform.

2.2. *Pixel Circuit*

The pixel circuit diagram for the 3D stacked retinal prosthesis is shown in Fig. 3. It consists of the photo-sensor, cathodic pulse control, anodic pulse control, and stimulus current output block (Current-Gen). The first layer contains the photo-sensor, and the second layer consists of the circuit for controlling the stimulus current waveform. The circuit for generating the stimulating current is on the third layer. The stimulus electrode is formed on the backside of that third layer, and the pulse is controlled by the incident light. The pixel circuit acts as a variable oscillator. The photodiode (PD) acts as a variable current source controlled by the incident light intensity on the photo-sensor circuit, which is designed as a current mode readout circuit. The voltage *net1* gradually decreases to "Low" voltage, depending on the photo current. When the voltage reaches the threshold voltage of INV1, the voltage of *net2* increases to “High” voltage immediately, whereas the voltage of *net3* increases to “High” voltage gradually. The transistor M3 charges the capacitor Cp, and the voltage of *net1* and *net2* decrease to “Low” voltage immediately after the voltage of *net4* reaches the threshold voltage of INV2. Thus, the photo-sensor circuit detects incident light and then the cathodic pulse control circuit converts the light signal into output pulses. The amplitude level of the stimulus current for each pixel is set by the 3-bit digital-to-analog converter (DAC) using the current amplitude memory. To reduce the power consumption of Current-Gen, we added a shutoff circuit that cuts off the DC current to the current generator during the stimulation off-state. This circuit consists of a pMOS transistor, an inverter, and a NOR circuit as the “Shutoff switch” section in Fig. 3. Therefore, the Current-Gen consumes power only when the stimulus current is required.

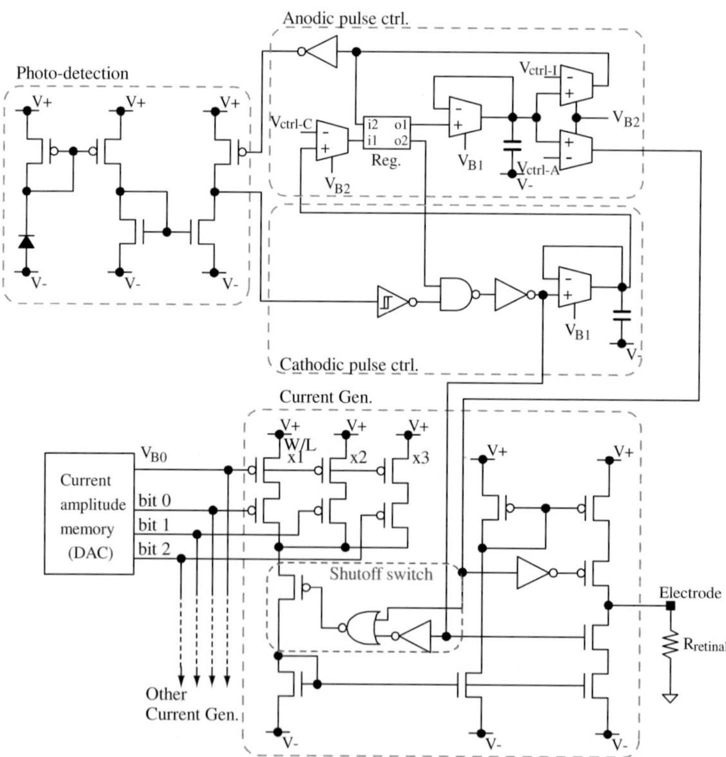

Figure 3. Pixel circuit diagram.

3. Experimental Results

Figure 4 shows a photograph of the fabricated retinal prosthesis chip with 100 pixels and the layout of the pixel circuit with a vertical interconnection TSV. We fabricated a prototype chip using standard 0.35-μm double poly-Si, triple-metal CMOS technology. The chip size of the retinal prosthesis was 2.4 × 2.4 mm and the active die area of the pixel circuit was 155 × 144 μm. As previously mentioned, the circuit of the retinal prosthesis was designed for the 3D structure, but it is shown here as a single layer (2D) implementation for the functional test. With the 3D structure, the photodiodes and photo-circuits are placed on the first layer, and the signal processing circuits (cathodic and anodic pulse control, and current pulse generator) are placed on the second and third layers connected through the TSVs.

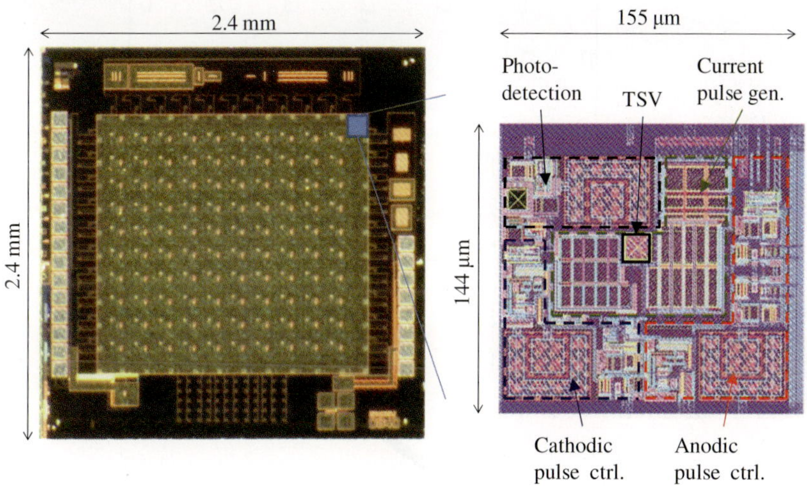

(a) (b)

Figure 4. (a) Microphotograph of the fabricated retinal prosthesis chip and (b) a layout of the pixel circuit.

The interface between the electrode and retina is made of resistive components. Therefore, we measured the current output with a resistive load of 10 kΩ. Figure 5 shows the stimulation waveforms of the 100-output stimulator. The waveforms are the voltages measured at the 10-kΩ loads. Figure 5(b) shows the incident light intensity and the output frequency of the retinal prosthesis. A halogen lamp was used for the input light. The output frequency is proportional to the incident light intensity over a dynamic range of 40 dB, and the power consumption of the pixel circuit can be reduced 15.5% using the shutoff circuit. The pixel circuit consumes about 115 μW biphasic pulse for a 0.5-ms pulse duration and a 100-μA current amplitude at 100 Hz.

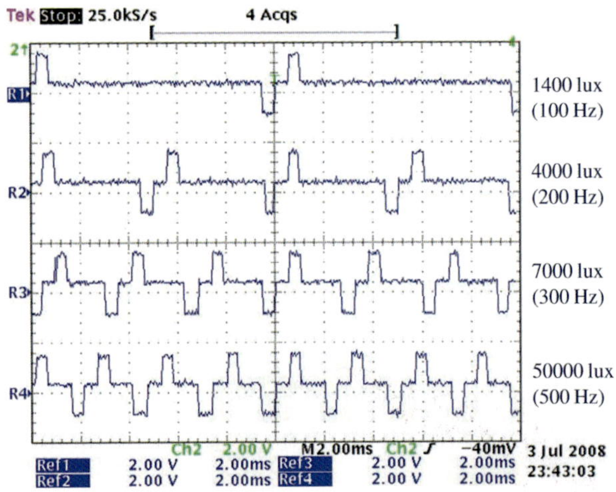

(a)

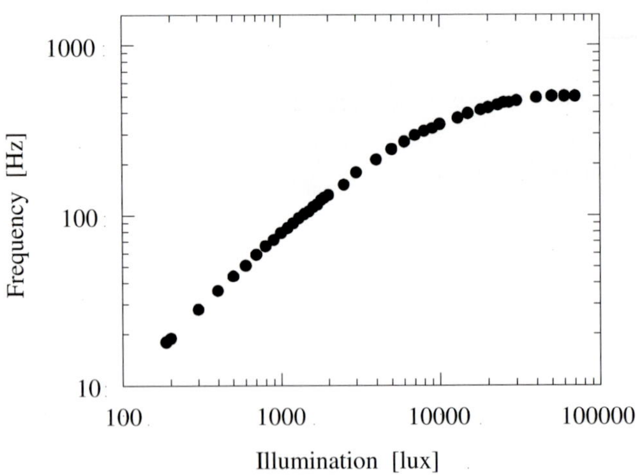

(b)

Figure 5. Experimental results: (a) captured waveform of stimulus current and (b) illumination vs frequency of output pulse.

4. Conclusions

To create a high-sensitivity, high-resolution, low-power retinal prosthesis, we designed and tested a novel 3D stacked LSI system. Our experiments confirmed that the stimulus current pulse frequency showed a linear dependence on the input light intensity, and the power consumption of the pixel circuit could be reduced 15.5% with the shutoff circuit.

References

1. M. S. Humayun, E. de Juan Jr., J. D. Weiland, G. Dagnelie, S. Katona, R. Greenberg and S. Suzuki, Pattern electrical stimulation of the human retina, *Vision Research* **39**, 2569–2576 (1999).
2. N. E. Medeiros and C. A. Curcio, Preservation of ganglion cell layer neurons in age-related macular degeneration, *Investigative Ophthalmology & Visual Science* **42**, 795–803 (2001).
3. W. Liu, K. Vichienchom, M. Clements, S. C. DeMarco, C. Hughes, E. McGucken, M. S. Humayun, E. de Juan, J. D. Weiland and R. Greenderg, A neuro-stimulus chip with telemetry unit for retinal prosthetic device, *IEEE Journal of Solid-State Circuits* **35**, 1487–1497 (2000).
4. J. Ohta, N. Yoshida, K. Kagawa and M. Nunoshita, Proposal of application of pulsed vision chip for retinal prosthesis, *Japanese Journal of Applied Physics* **41**, 2322–2325 (2002).
5. T. Tanaka, K. Sato, K. Komiya, K. Kobayashi, T. Watanabe, T. Fukushima, H. Tomita, H. Kurino, M. Tamai and M. Koyanagi, Fully implantable retinal prosthesis chip with photodetector and stimulus current generator, *International Electron Devices Meeting (IEDM) Technical Digest*, 1015–1018 (2007).
6. H. Kurino, K. W. Lee, T. Nakagawa, K. Sakuma, K. T. Park, N. Miyagawa, H. Shimazutsu, K. Y. Kim, K. Inamura and M. Koyanagi, Intelligent image sensor chip with three-dimensional structure, *International Electron Devices Meeting (IEDM) Technical Digest*, 879–882 (1999).
7. M. Koyanagi, Y. Nakagawa, K. T. Nakamura, Y. Yamada, K. Inamura, K. T. Park and H. Kurino, Neuromorphic vision chip fabricated using three-dimensional integration technology, *International Solid-State Circuit Conference (ISSCC) Digest of Technical Papers*, 270–271 (2001).

DEVELOPMENT OF A SILICON NEURAL PROBE FOR AN INTELLIGENT SILICON NEURAL PROBE SYSTEM

RISATO KOBAYASHI [1)], LEE SANGHOON [2)*], SOICHIRO KANNO [2)],
BEA JICHEOL [1)], TAKAFUMI FUKUSHIMA [1)], KAZUHIRO SAKAMOTO [3)],
NORIHIRO KATAYAMA [4)], HAJIME MUSHIAKE [5)], TETSU TANAKA [1,2)†],
MITSUMASA KOYANAGI [1)]

1)Department of Bioengineering and Robotics, Graduate School of Engineering, Tohoku University, 6-6-01 Aza-Aoba, Aramaki, Aoba-ku, Sendai 980-8579, Japan

2)Department of Biomedical Engineering, Graduate School of Biomedical Engineering, Tohoku University, 6-6-01 Aza-Aoba, Aramaki, Aoba-ku, Sendai 980-8579, Japan

3)Research Institute of Electrical Communication, Tohoku University, 2-1-1 Katahira, Aoba-ku, Sendai 980-8577, Japan

4)Department of Applied Information Science, Graduate School of Information Sciences, Tohoku University, 6-3-09 Aza-Aoba, Aramaki, Aoba-ku, Sendai 980-8579, Japan

5)Department of Physiology, Tohoku University School of Medicine, 2-1 Seiryo-machi, Aoba-Ku, Sendai 980-8575, Japan

We have proposed a new approach to implantable neural recording that we call the intelligent Si neural probe system. The key part of this system is a novel Si neural probe, 40 mm in length. By developing an equivalent circuit model of the Si neural probe, we estimated the optimum structural parameters for recording neuronal action potentials with a high S/N ratio. Si neural probes were successfully fabricated and showed an impedance value of 2.5 MΩ at 1 kHz. Using the fabricated Si neural probe, neuronal action potentials were successfully recorded from the CA1 area in hippocampal slices of guinea pig brain. Additionally, to realize high-density recording, we developed a novel Si double-sided neural probe, which has recording sites on both the front and back, and evaluated this Si double-sided neural probe.

Keywords: intelligent silicon (Si) neural probe system, Si neural probe, neural action potential, simultaneous multi-site recording, Si double-sided neural probe.

1. Introduction

With the progressive ageing of society, the number of patients with brain diseases, such as paralysis, severe epilepsy, and Parkinson's disease, has been increasing. However, medical cures for these diseases have not yet been

* Lee Sanghoon is a Tohoku University Global COE Research Assistant.
† Tetsu Tanaka is a Tohoku University Global COE Member.

established. Medical care and rehabilitation of patients with brain diseases cause stress not only for patients but also for their families. To improve this situation, much neurophysiological research is being conducted around the world. In addition to brain research having great potential for epochal discoveries, progress in neurophysiology may lead to medical cures and to also innovative computers and/or robots, based on lessons about information processing learned from studying the brain. In neurophysiology, recording of neuronal signals from the brain plays an important role in advanced studies of neuronal circuit dynamics and of the relationship between body actions and activity in parts of the brain.

In recent decades, various types of neural probes made of silicon (Si) have been developed to record neuronal signals from the brain of animals [1-5]. We have proposed an implantable neural probe system, which we call the intelligent Si neural probe system, and have been developing the components of this system [6-8]. Figure 1 shows a configuration of the intelligent Si neural probe system.

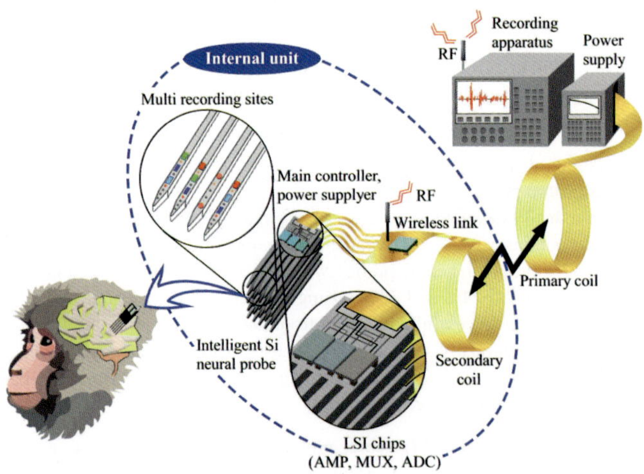

Figure 1. Configuration of intelligent Si neural probe system.

In this system, a large number of recording sites are located on the Si neural probe, for the simultaneous multi-site recording of neuronal responses. Additionally, various sensors, such as those for detecting blood flow and pH, can be mounted on the Si neural probe, enabling neurophysiologists to observe chemical reactions in the brain while neuronal action potentials are recorded. Because neuronal signals are extremely weak, electrical circuits such as amplifiers (AMPs), A/D converters (ADCs), and multiplexers (MUXs), are also integrated on the Si neural probe. Thus, weak neuronal responses recorded from the brain of experimental animals are amplified while maintaining a low noise level. Neuronal signals from the brain can be transmitted to the external recording apparatus through biocompatible polyimide-based flexible printed circuits (FPC) that provide two closely coupled coils. Electrical power is supplied by electromagnetic induction. Once the intelligent Si neural probe is embedded in the head of an experimental animal, neuronal responses from the brain of the unrestrained animal can be precisely recorded. Monolithic fabrication processes have previously been used in the fabrication of a conventional Si probe having electrical circuits [9-13]. In our intelligent Si neural probe, the Si neural probe and some circuits are fabricated by discrete processes. Fabricated LSI chips and some circuits are integrated on the Si neural probe using a multi-chip bonding technique we developed [14]. Thus, simple processes that are suitable for each circuit or device can be employed, resulting in high performance, high yield, and high degrees of freedom of the Si neural probe.

In this paper, we propose a novel Si neural probe 40 mm in length that can be used with the intelligent Si neural probe system. In general, a long neural probe has a large wiring area, and output signal intensity may be decreased by the capacitance of a large insulating film under and over the wiring. To solve this problem, we optimized structural parameters of our Si neural probe, using an equivalent circuit model, and used the resultant design to fabricate the Si neural probe. Using the probe, we performed animal experiments, recording neuronal action potentials from the CA1 area in hippocampal slices. Further, we developed a novel Si double-sided neural probe, which has recording sites on both the front and the back for high-density recordings. The electrical characteristics of the Si double-sided neural probe were evaluated.

2. Si Neural Probe

2.1. *Design of Si Neural Probe Using an Equivalent Circuit Model*

Si probes have been widely used because they can be fabricated using advanced CMOS processes; this approach has several important advantages, including mass production, high repeatability, and accuracy. Additionally, Si is a highly biocompatible material [15]. However, conventional Si probes are used only for acute *in vitro* animal experiments or analysis of the cortical surface because their length is very short, typically 10 mm or less. We are interested in analyzing many parts of the brain, including both shallow parts such as the cortex and deep parts such as the basal ganglia. In particular, the *substantia nigra* in the basal ganglia is associated with Parkinson's disease, and is a major focus of neurophysiology research. To accomplish this, we developed a novel Si neural probe 40 mm in length. Our Si neural probe can be used for analysis of brain areas where conventional Si probes are inadequate.

The overall structure of the Si neural probe is shown in Fig. 2. Our Si neural probe has a length of 40 mm for analysis of deep parts of the brain. The width and thickness are 120 μm and 140 μm, respectively. The recording sites, located 40 μm behind the tip of the probe, are made of tungsten/aluminum (W/Al), with a circular pattern. The diameter of the circular pattern is 20 μm and the separation between the two patterns is 100 μm, center to center. At the other end of the probe are bonding pads for connecting the probe to an external recording apparatus. Insulating films under and over the W/Al wiring are of SiO_2.

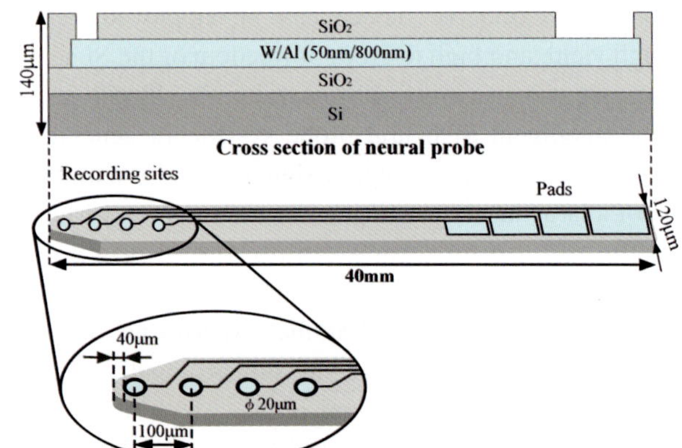

Figure 2. Structure of Si neural probe.

In a recording experiment, using the long Si neural probe, capacitance due to the large SiO_2 layer under and over the wiring leads to decreased neuronal signals recorded from the brain of animals. To overcome this and obtain high-intensity neuronal signals, we optimized the thicknesses of the SiO_2 layers using an equivalent circuit model (Fig. 3). This equivalent circuit model is composed of a resistance cell fluid (R_{sp}), a capacitance between the recording site and electrolyte (C_e), the resistance of the W/Al wiring (R_w), and the capacitance of SiO_2 layers under and over the wiring (C_s). We calculated the output signal amplitude based on C_s. Here, the input signal from the neuron is a sine signal, with an amplitude of 250 μV and frequency of 1 kHz.

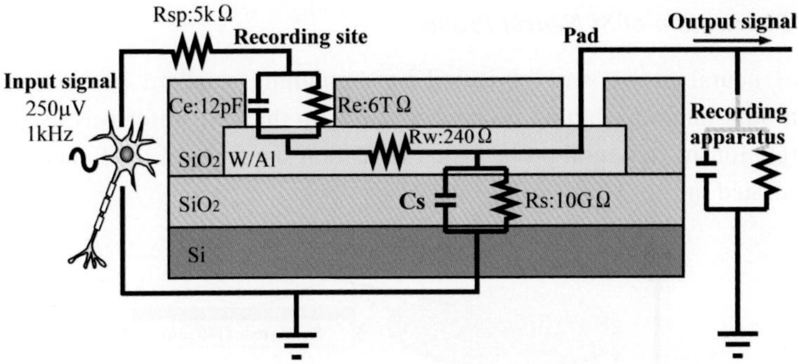

Rw : Resistance of wiring
Cs : Capacitance of SiO_2 layers
Rs : Resistance of SiO_2 layers
Rsp : Resistance of cell fluid
Ce : Capacitance of recording site
Re : Resistance of recording site

Figure 3. Equivalent circuit model of Si neural probe.

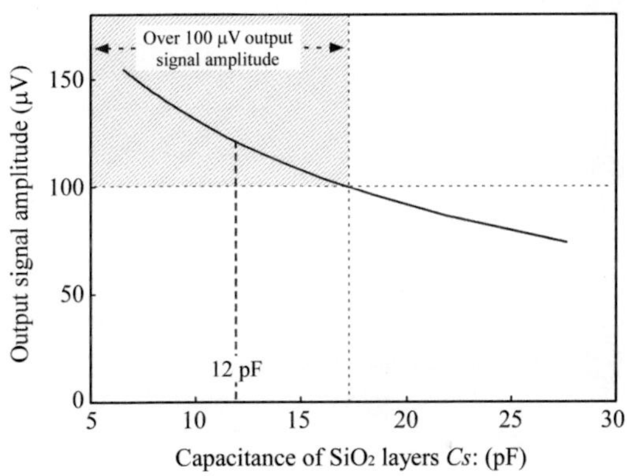

Figure 4. Relationship between capacitance of SiO_2 layers and amplitude of output signal.

Figure 4 shows the relationship between C_s and the amplitude of the output signal. The amplitude of the noise, which was observed in neuronal action potential recording, was on the order of about 12 μV. To discriminate neuronal action potentials from noise, a signal amplitude several times larger than the noise is needed, and we estimated the output signal amplitude to be approximately 100 μV. Thus, as shown in Fig. 4, C_s must be less than 17 pF. In this study, considering process compatibility, we employed SiO_2 layers with a thickness of 1.5 μm over the wiring and 1.0 μm under the wiring. As a result, C_s was estimated to be 12 pF, and the output signal amplitude was estimated at 120 μV.

2.2. *Fabrication of Si Neural Probe*

Our Si neural probe was fabricated by combining standard photolithography with a bulk micromachining process. A 140-μm thick Si wafer was used as the substrate for the Si neural probe. The fabrication sequence of the Si neural probe is illustrated in Fig. 3.

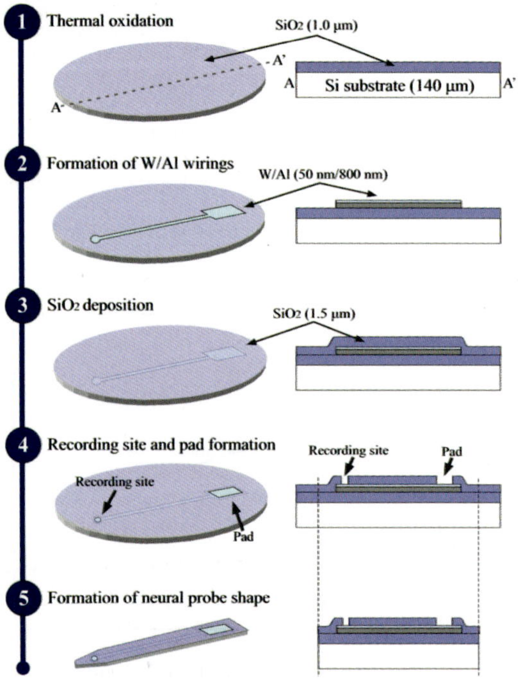

Figure 5. Fabrication sequence of Si neural probe.

First, SiO_2 was formed by thermal oxidation, with a thickness of 1.0 μm. Next, W/Al wirings were formed by sputtering and reactive ion etching (RIE) with BCl_3 and Cl_2 gases. The thicknesses of the W and Al were 50 nm and 800 nm, respectively. Then, SiO_2 with a thickness of 1.5 μm was deposited on the front surface of the wafer, using plasma enhanced chemical vapor deposition (PECVD) to insulate the wiring. Next, recording sites and bonding pads were formed by wet etching with buffered HF solution (HF/NH_4F 1/10), followed by the formation of the shape of the neural probe using deep RIE with SF_6 and C_4F_8 gases. The final product was a Si neural probe with four recording sites. Figure 6 shows a photograph of the fabricated Si neural probe with four recording sites, mounted on a printed circuit board to connect to the recording apparatus.

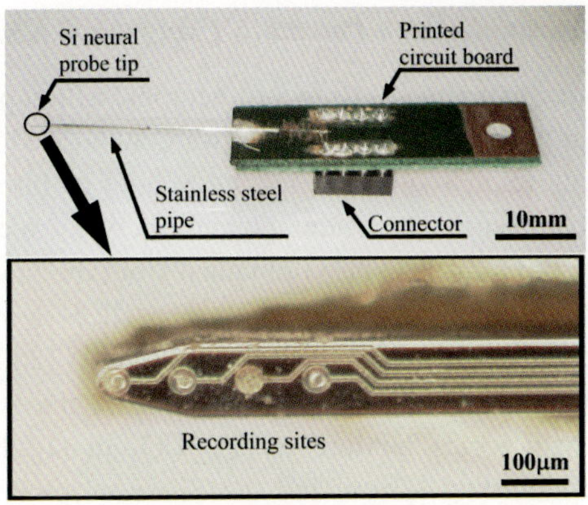

Figure 6. Photograph of fabricated Si neural probe.

3. Experimental Methods

3.1. *Impedance Measurement of Fabricated Si Neural Probe*

Before animal experiments were begun, we used a HP-4194A Precision Impedance Analyzer to perform an impedance measurement of the fabricated Si neural probe to characterize the interface between the recording site and electrolytes. Measurements were performed with the frequency ranging from 100 Hz to 10 MHz, using a 10-mV AC sine signal in 0.9% saline solution. A Ag/AgCl electrode and a Pt electrode were used as the reference and counter electrodes, respectively. Figure 7 shows the impedance characteristics of the fabricated Si neural probe. Our Si neural probe had an appropriate impedance value of about 2.5 MΩ at 1 kHz; neural probes usually require impedance values of several MΩ at 1 kHz to record neuronal signals [16].

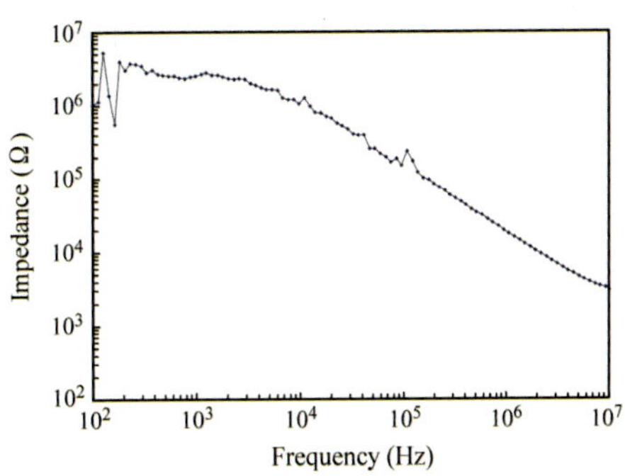

Figure 7. Impedance characteristics of the fabricated Si neural probe.

3.2. *Recording Neuronal Action Potentials Using the Si Neural Probe*

We performed animal experiments to record neuronal action potentials using the fabricated Si neural probe. The experiments were authorized by the Committee for Experiments on Animals of the Graduate School of Information Sciences, Tohoku University. Neuronal action potentials were recorded from thin hippocampal slices (400 μm thickness) obtained from the brains of guinea pigs (200-250 g) by conventional methods [17]. The slices were kept in normal artificial cerebrospinal fluid (ACSF), containing (in mM) 124 NaCl, 5 KCl, 1.25 NaH_2PO_4, 1 $MgSO_4$, 1.25 $CaCl_2$, 22 $NaHCO_3$, and 10 glucose (pH 7.4), which was equilibrated with a gas mixture of 5% CO_2 in O_2. Figure 8 shows a photograph of the recording system and the Si neural probe inserted into the CA1 area of the hippocampal slice.

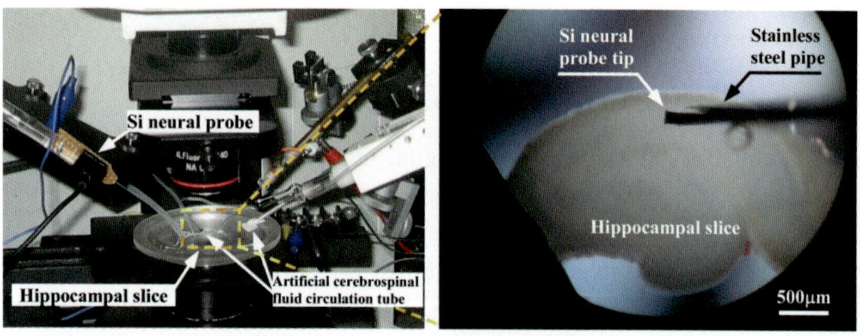

Figure 8. Photograph of the recording system and Si neural probe inserted into a hippocampal slice.

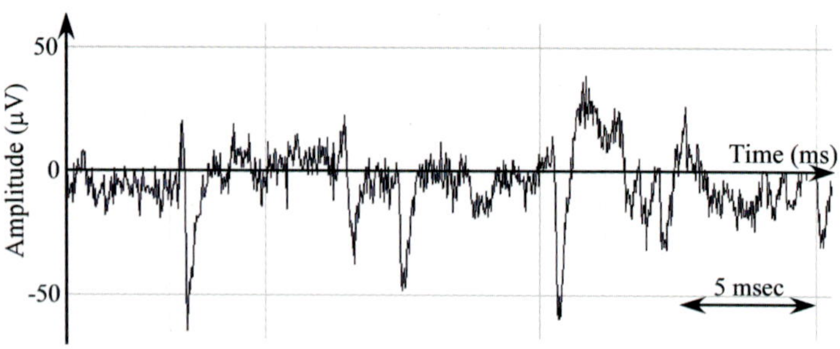

Figure 9. Recorded neuronal spike potentials from the hippocampal slice.

The hippocampal slice was put in a perfusing fluid chamber kept at 32±2°C. The signals were amplified and band-pass filtered (0.15-10 kHz) with Biosignal amplifiers (AB-610J, Nihon Kohden, Japan) and sampled at 30 kHz with an A/D converter box (Digidata 1320A, Axon Instruments, USA). Neuronal action potentials were recorded extracellularly in the pyramidal cell layer of the CA1 area of the hippocampus. As shown in Fig. 9, we successfully recorded neuronal action potentials from individual neurons using the fabricated Si neural probe.

4. Si Double-sided Neural Probe

Following the successful use of our Si neural probe in recording neuronal action potential, we developed a Si double-sided neural probe for further advanced recordings. The overall structure of the Si double-sided neural probe is shown in Fig. 10. Our Si double-sided neural probe has recording sites on the front and back of the Si, for high-density 3-D recording of neuronal action potentials. The back-side recording sites were connected to the recording apparatus by wire bonding through the Si. As shown in Fig. 10, our Si probe had four shanks with a pitch of 300 μm. Each shank has four recording sites on each side. This Si probe has a length of 40 mm. The width and thickness were each 100 μm. At the ends of the Si probe, bonding pads for connecting to an external recording apparatus were formed.

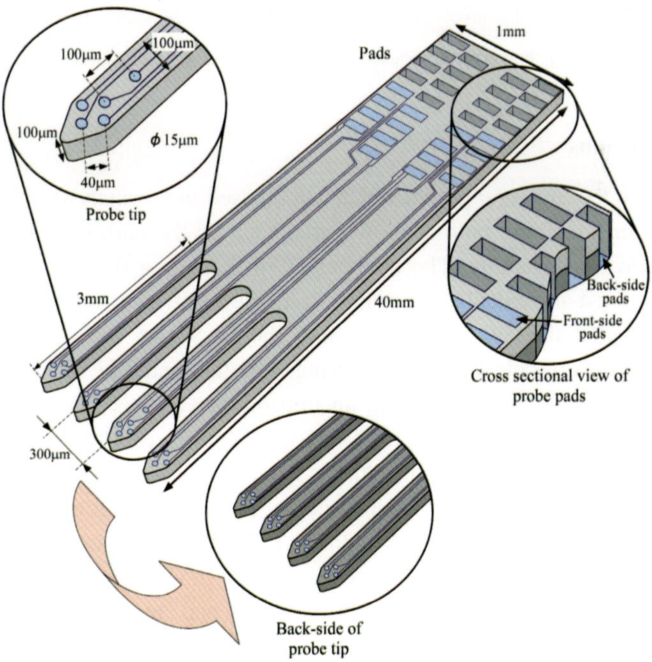

Figure 10. Si double-sided neural probe.

To characterize the interface between recording site and electrolyte, we measured the electrical impedance of the fabricated Si double-sided neural probe. Measurements were performed with the frequency ranging from 100 Hz to 10 MHz, using a 10-mV AC sine signal in 0.9% saline solution. Figure 11 shows the impedance characteristics of the fabricated Si probe.

The front-side recording site had an impedance value of 2.5 MΩ at 1 kHz, and the back-side recording site had an impedance value of 2.7 MΩ at 1 kHz. Thus, the two sides have almost the same electrical characteristics.

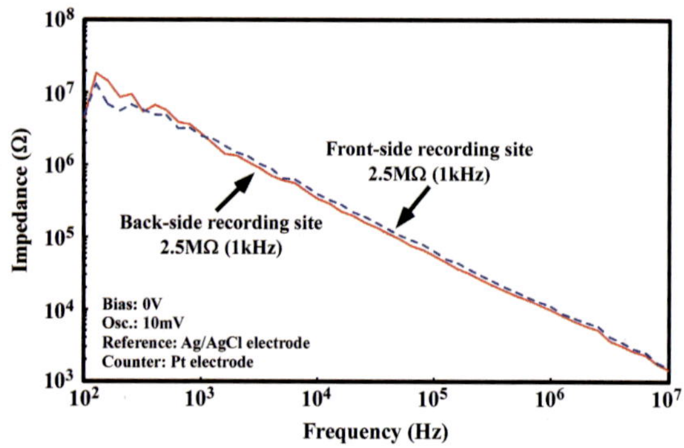

Figure 11. Impedance characteristics of the Si double-sided neural probe.

5. Conclusions

We propose the Si neural probe, which can be used in the analysis of many parts of the brain, including both shallow parts, such as the cortex, and deep parts, such as the basal ganglia. The thicknesses of insulating films of the Si probe were estimated by calculats deriving from the equivalent circuit model. We determined that the SiO_2 layer thickness should be 1.5 μm above the W/Al wiring and 1.0 μm below the W/Al wiring. The fabricated Si neural probe with SiO_2 layers of this thicknesses had an impedance value of 2.5 MΩ at 1 kHz. Using the fabricated Si neural probe, we successfully recorded neuronal action potentials, with an amplitude of 100 μV from hippocampal slices. In addition, we developed a novel Si double-sided neural probe, enabling high-density, 3-D recordings of neuronal action potentials. The front and back recording sites of the fabricated Si probe had impedances of 2.5 and 2.7 MΩ at 1 kHz, respectively, indicating that the recording sites had approximately equivalent characteristics.

Acknowledgments

The authors acknowledge the support of Tohoku University Global COE Program "Global Nano-Biomedical Engineering Education and Research Network Centre." This work was partially supported by a Grant-in-Aid for JSPS Fellows. The device was fabricated in the Micro/Nano-Machining Research and Education Center, Tohoku University, Japan.

References

1. P. K. Campbell, K. E. Jones, R. J. Huber, K. W. Horch and R. A. Normann, A silicon-based, three-dimensional neural interface: manufacturing processes for an intracortical electrode array. *IEEE Trans Biomed Eng* **38**, 758–768 (1991).
2. K. L. Drake, K. D. Wise, J. Farraye, D. J. Anderson and S. L. BeMent, Performance of planar multisite microprobes in recording extracellular single-unit intracortical activity. *IEEE Trans Biomed Eng* **35**, 719–732 (1988).
3. D. T. Kewley, M. D. Hills, D. A. Borkholder, I. E. Opris, N. I. Maluf, C. W. Storment, J. M. Bower and G. T. Kovacs, Plasma-etched neural probes. *Sens and Actuators A* **58**, 27–35 (1997).
4. G. T. Kovacs, C. W. Storment, M. Halks-Miller, C. R. Belczynski, C. Della Santina, E. R. Lewis and N. I. Maluf, Silicon-substrate microelectrode arrays for parallel recording of neural activity in peripheral and cranial nerves. *IEEE Trans Biomed Eng* **41**, 567–577 (1994).
5. K. D. Wise and K. Najafi, Microfabrication techniques for integrated sensors and Microsystems. *Science* **254**, 1335–1342 (1991).
6. T. Watanabe, K. Motonami, K. Sakamoto, J. Deguchi, T. Fukushima, J. C. Shim, H. Mushiake, H. Kurino and M. Koyanagi, Ultimate Functional Multi-Electrode System (UFMES) Based on Multi-Chip Bonding Technique. *Extended Abstracts of International Conference on Solid State Devices and Materials*, Tokyo, 380–381 (2004).
7. T. Watanabe, K. Motonami, K. Sakamoto, J. Deguchi, R. Kobayashi, K. Komiya, K. Okumura, T. Fukushima, H. Kurino, H. Mushiake and M. Koyanagi, Intelligent Neural Implant Microsystem Fabricated Using Multi-Chip Bonding Technique. *Extended Abstracts of International Conference on Solid State Devices and Materials,* Kobe, 462–463 (2005).
8. R. Kobayashi, T. Watanabe, K. Komiya, T. Fukushima, K. Sakamoto, H. Kurino, T. Tanaka, N. Katayama, H. Mushiake and M. Koyanagi, Development of Si Long Microprobe (SiLM) for Platform of Intelligent Neural Implant Microsystem. *Extended Abstracts of International Conference on Solid State Devices and Materials,* Tokyo, 898–899 (2006).

9. Q. Bai and K. D. Wise, Single-unit neural recording with active microelectrode arrays. *IEEE Trans Biomed Eng* **48**, 911–920 (2001).
10. A. C. Hoogerwerf and K. D. Wise, A three-dimensional microelectrode array for chronic neural recording. *IEEE Trans Biomed Eng* **41**, 1136–1146 (1994).
11. K. Najafi and K. D. Wise, An implantable multielectrode array with on-chip signal processing. *IEEE J Solid-State Circuits* **21**, 1035–1044 (1986).
12. K. Takahashi and T. Matsuo, Integration of multi-microelectrode and interface circuits by silicon planar and three-dimensional fabrication technology. *Sens Actuators* **5**, 89–99 (1984).
13. K. D. Wise, D. J. Anderson, J. F. Hetke, D. R. Kipke and K. Najafi, Wireless Implantable Microsystems: High-Density Electronic Interfaces to the Nervous System. *Proc IEEE* **92**, 76–97 (2004).
14. K. Hiroyuki, W. L. Kang, S. Katsuyuki, N. Tomonori and K. Mitsumasa, A New Wafer Scale Chip-on-Chip (W-COC) Packaging Technology Using Adhesive Injection Method. *Jpn J Appl Phys* **38**, 2406–2410 (1999).
15. J. K. Niparko, R. A. Altschuler, X. L. Xue, J. A. Wiler and D. J. Anderson, Surgical implantation and biocompatibility of central nervous system auditory prostheses. *Ann Otol Rhinol Laryngol* **98**, 965–970 (1989).
16. N. A. Blum, B. G. Carkhuff, H. K. Charles, R. L. Edwards and R. A. Meyer, Multisite microprobes for neural recordings. *IEEE Trans Biomed Eng* **38**, 68–74 (1991).
17. K. H. Ito, H. Miyakawa, Y. Izumi and H. Kato, Longterm potentiation induced by repeated paired stimulation of the stratum radiatum and alveus in CA1 pyramidal neurons of hippocampal slices. *Biomed Res* **10**, 111–124 (1989).

EVOKED POTENTIALS IN RESPONSE TO ELECTRICAL STIMULATION OF THE COCHLEAR NUCLEUS BY MEANS OF MULTI-CHANNEL SURFACE MICROELECTRODES

KIYOSHI ODA

Department of Otolaryngology-Head and Neck Surgery, Tohoku University Graduate School of Medicine, Sendai, Miyagi 980-0872, Japan

TETSUAKI KAWASE *

Laboratory of Rehabilitative Auditory Science, Tohoku University Graduate School of Biomedical Engineering, Sendai, Miyagi 980-8574, Japan

Department of Audiology, Tohoku University Graduate School of Medicine, Sendai, Miyagi 980-0872, Japan

Department of Otolaryngology-Head and Neck Surgery, Tohoku University Graduate School of Medicine, Sendai, Miyagi 980-0872, Japan

DAISUKE YAMAUCHI, HIROSHI HIDAKA, TOSHIMITSU KOBAYASHI

Department of Otolaryngology-Head and Neck Surgery, Tohoku University Graduate School of Medicine, Sendai, Miyagi 980-0872, Japan

A auditory potentials in response to electrical stimulation of the cochlear nucleus using two types of multi-channel surface microelectrodes, in which the inter-electrode distance is 100 μm and 200 μm respectively, were examined in guinea pigs. Even when surface bipolar stimulations with this very short separation of two electrodes were employed, the unequivocal waves of electrically evoked auditory brainstem responses (EABR), which increased in amplitude with increasing stimulation current, were constantly derived. Electrophysiological mapping using the multi-channel surface microelectrodes indicated that more precise electro-physiological mapping may be possible based on the EABR threshold distribution using bipolar stimulation with very short separation. It is expected that the multi-channel surface microelectrodes can be usefully applied in clinical use in electrophysiological mapping to determine the optimal location for the positioning of auditory brainstem implants.

1. Introduction

Auditory brainstem implants (ABI) are a clinical application of artificial organ technology in which electrodes are implanted in the brainstem in order to restore

---

* Tetsuaki Kawase is a Tohoku University Global COE Member.

the hearing of patients who have lost their hearing as a result of auditory nerve dysfunction [1]. ABI have reached the clinical trial stage in Europe and the United States and satisfactory outcomes have been demonstrated in successful cases, bringing great hope to these patients. However, the success rate has been less than satisfactory [2,3].

Thus, it will be necessary to map the cochlear nucleus with greater accuracy. To achieve this, multi-channel surface microelectrodes were employed in the present study to measure electrical evoked auditory responses, and these responses were utilized to map the cochlear nucleus in an animal model.

2. Methods

Guinea pigs weighing 400-700g were used as experimental animals. After intramuscular anesthetization of the animals with ketamine (60 mg/kg) and xylazine (5 mg/kg), a click stimulus was used to measure the auditory brainstem response (AABR) (Neuropack μ, Nihon Kohden) in order to confirm normal auditory function.

The animals were then placed in a supine position, and an incision was made in the skin behind the ear to expose and remove the skull to allow cerebellar retraction and subsequent visualization of the cochlear nucleus. A custom-made 64-channel microelectrode was placed on the surface of the cochlear nucleus. This specialized device has 64 electrodes in an area of only 0.7 mm x 0.7 mm, with the inter-electrode distance being set at only 100 μm. A 260-channel microelectrode (inter-electrode distance, 200 μm; electrode size, 3.8 mm x 2.4 mm), which sufficiently cover the border of the cochlear nucleus, was also used.

Of the multi-channel electrode channels of this microelectrode, two points were stimulated arbitrarily and the resulting electrical evoked brainstem responses (EABR) were recorded. In order to record the EABR, active electrodes were placed on the surface of the auditory region, and reference electrodes, consisting of stainless steel needles, were placed behind the ear. EABR were recorded using the same type of system as for AABR (Neuropack μ, Nihon-Kohden) under the following measurement conditions: average number, 500 times; low-cut filter, 200-1000 Hz; high-cut filter, 3000 Hz. As regards the electrical stimuli, a SEN-3041 pulse generator (Nihon Kohden) and an SS-203J Isolator were used to generate bipolar pulses of 80 μsec in duration.

3. Results

3.1. *EABR Waveform Recording System*

Using the 64 and 260-channel systems, we succeeded in recording reproducible EABR when the cochlear nucleus was stimulated electrically. Typical EABR waveforms after electrical stimulation of the cochlear nucleus are shown in Fig. 1.

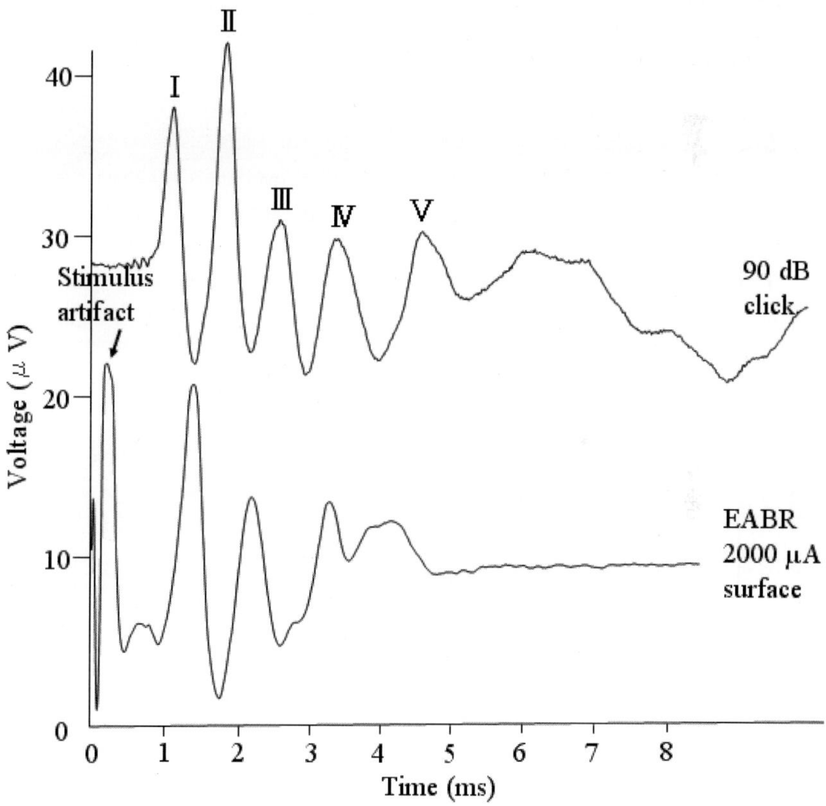

Figure 1. EABR vs. ABR auditory brainstem response (above), auditory brainstem response to electrical stimulation of the cochlear nucleus (below).

In the case of acoustically evoked responses (AABR), 5-6 voltage deflections (wave peaks) were recorded, but in AABR generated by electrical stimulation of the cochlear nucleus, the peaks show individual variability. In humans, 1-4 peaks have been reported [2].

There are various theories regarding the origin of these waveforms; however, it is considered that wave I is related to the ipsilateral cochlear nerve, wave II is related to the ipsilateral cochlear nucleus, and wave III is related to the contralateral superior olive nuclei [3]. In EABR, wave II can't be observed as a result of an artifact generated by the electrical stimulation. Therefore, the auditory brainstem response to electrical stimulation of the cochlear nucleus is thought to appear after wave III.

3.2. *EABR Input/output Characteristics*

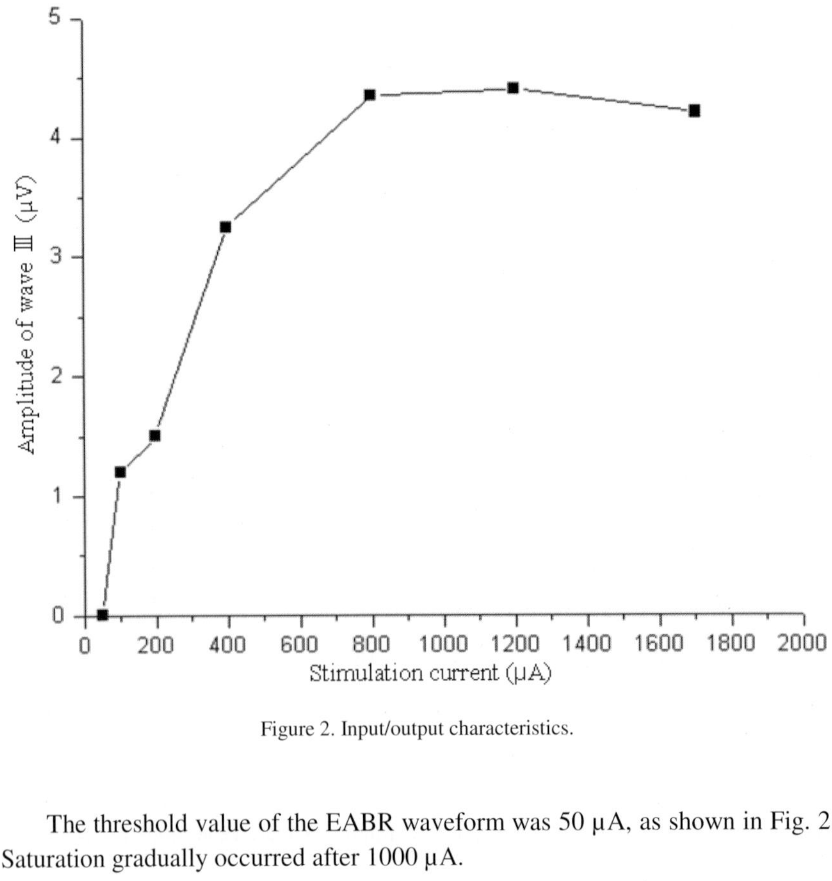

Figure 2. Input/output characteristics.

The threshold value of the EABR waveform was 50 μA, as shown in Fig. 2. Saturation gradually occurred after 1000 μA.

3.3. *Electrophysiological Mapping of the Cochlear Nucleus Using the 260 Channel Microelectrode System*

Mapping is considered to be most effective when the stimulation used is capable of recording reproducible waveforms with high sensitivity and an excellent signal-to-noise ratio. Therefore, after taking the input/output characteristics of the EABR shown in Fig. 2 into consideration, electrical stimulation from the dynamic portion of the signal (600 μA -1000 μA) was, in most of the experiments, used to carry out the mapping. Figure 3 shows the results of mapping of the cochlear nucleus using representative EABR. In this case, the 260-channel microelectrode was placed on the dorsal portion of the cochlear nucleus. We could distinguish positive EABRs from undetectable EABRs in a 3.8 by 2.4 mm area.

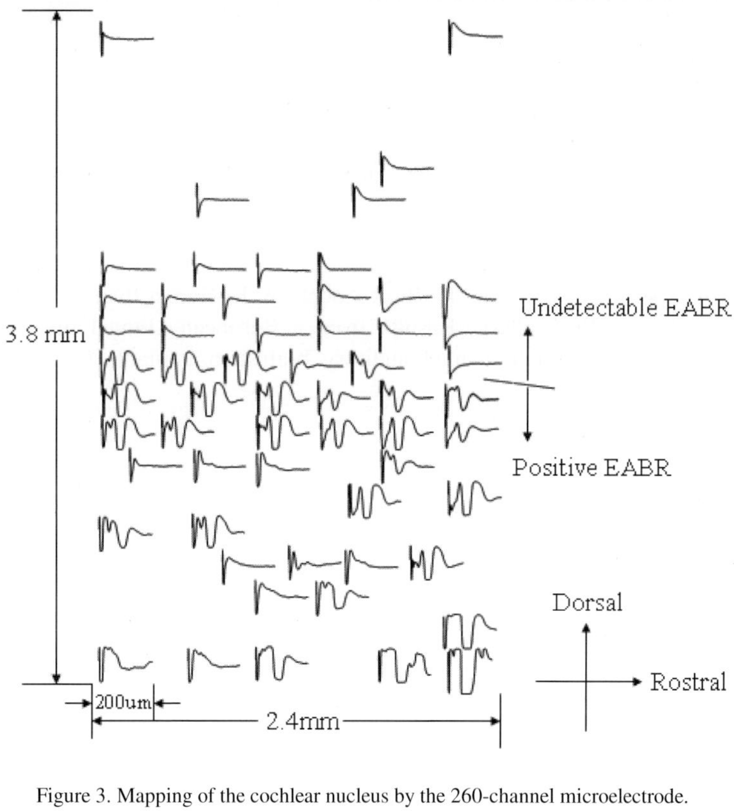

Figure 3. Mapping of the cochlear nucleus by the 260-channel microelectrode.

4. Conclusion

In the present study, EABR were recorded using surface microelectrode with inter-electrode distance of only 100 μm and 200 μm, respectively, less than one-fifth the electrode separations of the electrode devices presently in clinical use. These same microelectrodes were used to conduct mapping of the cochlear nucleus, which revealed that the electrophysiological boundaries could be identified quite clearly with this method.

Acknowledgments

This study was supported by a grant-in-aid from the Japanese Ministry of Education, Culture, Sports, Science and Technology for fiscal 2006-2007 and Tohoku University Global COE Program "Global Nano-Biomedical Engineering Education and Research Network Centre".

References

1. K. B. Jackson, G. Mark, J. Helms, J. Mueller and R. Behr, An auditory brainstem implant system. *Am J Audiol.* **11**, 128–133 (2002).
2. S. R. Otto, D. E. Brackmann, W. E. Hitselberger, R. V. Shannon and J. Kuchta, Multichannel auditory brainstem implant: update on performance in 61 patients. *J Neurosurg.* **96**, 1063–1071 (2002).
3. B. Nevison, R. Laszig, W. P. Sollmann, T. Lenarz, O. Sterkers, R. Ramsden, B. Fraysse, M. Manrique, H. Rask-Andersen, E. Garcia-Ibanez, V. Colletti and E. von Wallenberg, Results from a European clinical investigation of the Nucleus multichannel auditory brainstem implant. *Ear Hear* **23**, 170–183 (2002).

DEVELOPMENT OF A TACTILE SENSOR FOR EVALUATION OF DETERGENTS

DAISUKE TSUCHIMI [*], MAMI TANAKA [†]

Department of Biomedical Engineering, Tohoku University
Aobayama 6-6-04, Sendai, 980-8579, Japan

This paper is concerned with the development of a tactile sensor using PVDF (polyvinylidene fluoride) film as a receptor for the sensor to evaluate a detergent. Along with eyesight, tactile sensation is one of the most important senses in humans. When a dish that has been washed clean is rubbed with a finger, good tactile sense and sound (vibration signal) like 'Squeak-squeak' are obtained. From this tactile sense and sound, we judge that a dish becomes squeaky-clean. This tactile sense and sound are evaluation parameters when a consumer selects a detergent. In this study, a tactile sensor using PVDF film as the receptor was fabricated. Sensory tests of detergents were conducted. The experimental results show that sensor outputs have a good correlation with the times a consumer judges that a washed object becomes squeaky-clean using detergents.

1. Introduction

When a dish that has been washed clean is rubbed with a finger, good tactile sense and a sound (vibration signal) like 'Squeak-squeak' (QQ) are obtained. A good detergent removes dirt cleanly and quickly, and can be washed off easily with tap water. The tactile sense is a slipping and stopping force when a finger moves across the surface of a dish or other object. The QQ feeling is defined as a combination of tactile sense and sound. Detergents that quickly give this QQ feel are considered to be good detergents. Therefore, the QQ feeling is one of the evaluation parameters a consumer considers when selecting a detergent. However, the QQ feeling depends on the tactile perception of the human finger, which is said to be ambiguous and subjective. This situation has drawn interest in the development of objective instrumentation that measures and evaluates the QQ feeling of dishes washed with different detergents.

PVDF film is flexible and very sensitive to sudden variations in stress/strain. An electrical current generated by mechanical stress in piezoelectric materials decays as a result of the movement of electric charge. The voltage signal takes

[*] Daisuke Tsuchimi is a Tohoku University Global COE Research Assistant.

[†] Mami Tanaka is a Tohoku University Global COE Member.

the form of a very brief potential wave at the onset of the applied force, and a similar brief wave at its termination. This signal increases with applied force and drops to zero when the force remains constant. Furthermore, there is no response during the stationary plateau of the applied stimulus, and the voltage drops to a negative peak as pressure is removed, and subsequently decays to zero. This response is quite similar to the output signal from the Pacinian corpuscle in human skin, the sensory receptor in the dermis [1]. Tanaka developed a haptic sensor for monitoring skin conditions using PVDF film. This study shows that PVDF film is useful as the receptor of a tactile sensor [2].

In this study, a tactile sensor system that can measure the quality of detergents is developed. In the experiment, a ceramic plate coated with various detergents is exposed to flowing water, and the plate is rubbed by a subject's finger. The times at which subjects obtained a QQ feeling were investigated as the sensory test. A tactile sensor using PVDF film as the receptor was fabricated. In experiments, the sensor scans the surface of the plate at regular intervals, with a similar procedure to the sensory test. To compare the results with those of the human sensory test, a threshold of integrated sensor output was set, and the time to reach the threshold was investigated. The times were compared with the results of the sensory test.

2. Sensory Test of QQ Feeling

The sensory test of QQ feeling was conducted with four subjects using four kinds of detergent. During the action of washing out each detergent, the times until subjects felt QQ were measured.

A detergent on a ceramic plate was used as a measurement object. The plate was washed with the same detergent before use to strip off any left-over material on the plate and to keep the same condition on the surfaces of the plates. A 3% solution of each detergent was prepared. These 3% samples were poured onto the centers of plates in 30-ml volumes and spread uniformly over the plates, which had a diameter of 10 cm. Plates were rinsed with water flowing from a tube positioned 1 cm above the center of the plate. The flow speed of the rinse water was 1.6 l/min. During rinsing, the subject rubbed the surface of the plate with a finger. The time until the subject felt QQ was measured. Each sample was evaluated five times by four subjects. The rinse water was 25°C.

The results are presented in Table 1. The values in Table 1 are the average times from five trials with each sample evaluated by each subject. To reduce differences in sensitivity among individual subjects, the average time until subjects felt QQ for each sample was calculated and listed as the 'Average' in the table. The results show that all subjects felt QQ earlier using samples B and C than with the other samples, and that sample A required the most time for subjects to feel QQ.

Table 1. The time when subjects feel QQ feeling for each sample.

Subject / Sample	Time[s]				
	a	b	c	d	**Average**
A	17.59	19.67	19.63	20.26	**19.29**
B	9.30	14.68	10.85	8.80	**10.91**
C	3.82	8.72	6.62	9.26	**7.10**
D	12.55	16.54	15.73	18.06	**15.72**

3. Tactile Sensor System and Signal Processing

The geometry of the tactile sensor is presented in Fig. 1. The sensor has a layered structure. The base of the sensor is an aluminum plate; a polyurethane rubber block and PVDF film, which are waterproofed with tarpaulin, are stacked in sequence. These are covered with rubber film with an uneven surface to increase the sensitivity of the sensor. All layers are bonded to each other by double-sided tape.

The experimental set up and measurement conditions are shown in Fig. 2 and Table 2, respectively. A ceramic plate was fixed on the Z axial slide table, and the sensor was fixed to the arm on the X axial slider. To contact the sensor,

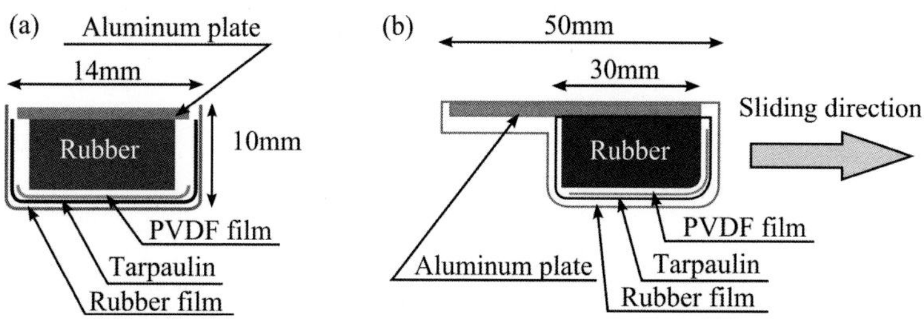

Figure 1. Geometry of the sensor: (a) Front view and (b) side view. Sliding direction of this senor is right in (b).

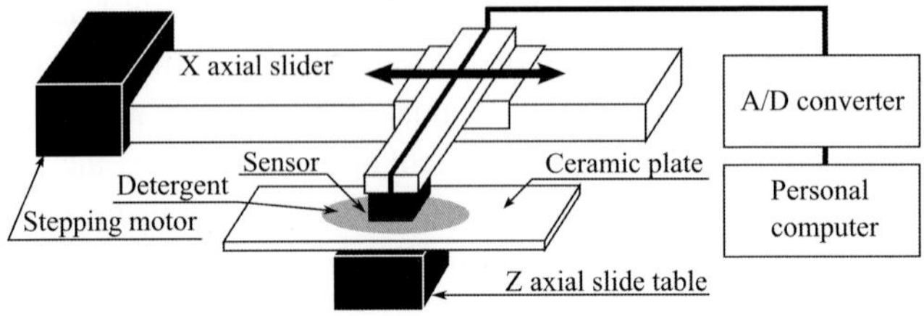

Figure 2. Experimental set-up.

Table 2. Measurement conditions.

Sampling frequency	5 kHz
Measurement term	2 s (including 0.5 s before and after sliding)
Sliding speed	50 mm/s
Sliding time	1 s

the plate was lifted up by the Z axial slide table, lifted a further 2 mm after the plate makes contact with the sensor to press the sensor onto the surface of the plate. The sliding direction of the sensor was parallel to the surface of the plate. The sensor slides on the surface of the plate using a stepping motor. The piezoelectric output of the PVDF film was digitized and stored by a personal computer.

Typical sensor outputs are presented in Fig. 3(a). In this figure, there are two peaks, one at the initiation of slipping and another at the end of slipping on the plate.

The equivalent circuit for a PVDF film is depicted in Fig. 4. It is known that the open-circuit voltage (V_i in Fig. 4), which is immeasurable, is proportional to the pressure on the PVDF film and that the closed-circuit voltage

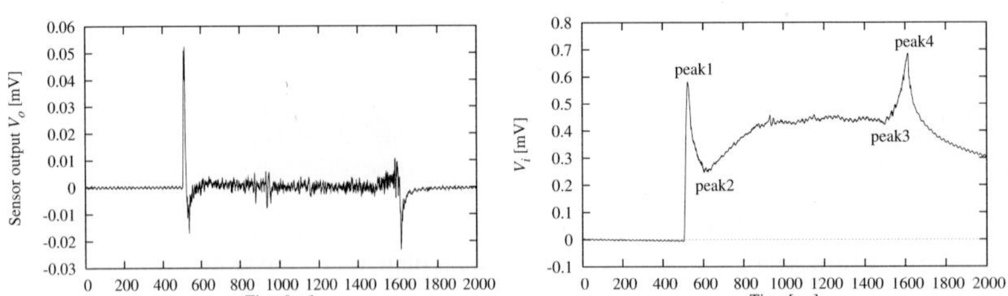

(a) An example of sensor output V_o. (b) An example of integrated sensor output V_i.

Figure 3. Examples of sensor output.

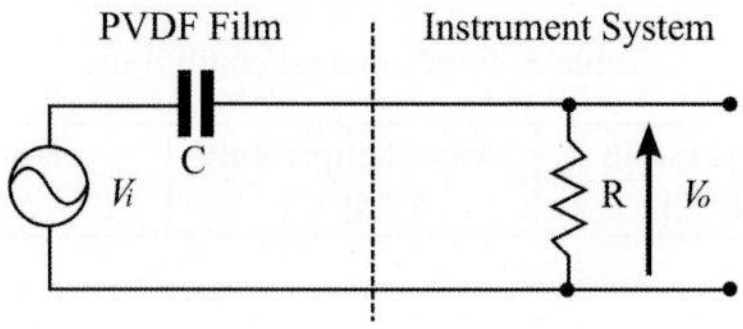

Figure 4. Equivalent circuit for PVDF film.

(V_o in Fig. 4), which is measurable, is a differential transform of the open-circuit voltage. It follows that the open-circuit voltage, which depicts pressure on the PVDF film [3], can be obtained by applying an integrated transform to the measured sensor outputs. Considering Fig. 4, V_i can be written as:

$$V_i(t) = V_o(t) * \left\{ \delta(t) + \frac{1}{RC} u(t) \right\} \tag{1}$$

Here, $\delta(t)$ is a delta function, $u(t)$ is a Heaviside unit function, $*$ is the convolution integral, and t is the time from the start of measurement. In the experiment, V_o is measured in the form of discrete signals, as shown in Fig. 3(a). Therefore, V_i can be calculated from the following discrete form of Eq. (1).

$$V_i(n) = V_o(n) + \frac{T_s}{RC} \sum_{i=1}^{n} V_o(i) \tag{2}$$

Here, T_S is the sampling period (0.2 ms), R is the resistance (1 MΩ), C is the capacitance of the PVDF film (1.38 nF) and n is the number of data points from the start of measurement. Applying Eq. (2) to the experimental data V_o, V_i is calculated and presented in Fig. 3(b). Considering that V_i is proportional to the pressure on the PVDF film, the value of peak 1 shown in Fig. 3(b) signifies the pressure on the PVDF film at the initiation of slipping. The value of peak 1 is defined as V_1, and the difference between peak 1 and peak 2 is defined as V_2. V_1 and V_2 were used to evaluate detergents as described below.

4. Sensor Measurement

4.1. *Experiment 1*

In experiment 1, the relationship between the amount of detergent remaining on a plate and the sensor output was investigated using four detergents (detergents A, B, C and D).

Table 3. Experimental conditions.

Air temperature	Water temperature	Humidity
17.9[°C]	17.5[°C]	52[%]

A ceramic plate was washed with the same detergent before use to strip off any left-over material on the plate and keep same condition of the surface of the plate. A 3% solution of each detergent was prepared, poured on the center of a plate by in a 30-ml volume, and spread uniformly over the plate making a diameter of 10 cm. The sensor was put on the plate 30 mm from the center, and sliding was commenced to collect sensor output. At the same time, the plate was rinsed with water from a tube positioned 1 cm above the center of the plate. The flow speed of the rinse water was 1 l/min. Measurement of sensor output was conducted every 10 s for 2 min. Samples were measured in triplicate, and the same procedure was repeated for all samples (sample A, B, C and D). The experimental conditions are shown in Table 3.

An example of sensor output is presented in Fig. 5. Each line in Fig. 5 is a time variation of the integrated sensor output. The result shows that V_1 and V_2 increase with time. To evaluate how V_1 and V_2 increase using each sample, thresholds were introduced. The threshold for V_1 was 0.8 and the threshold for V_2 was 0.5. The times T_1 and T_2 until V_1 and V_2 reached each threshold were measured for each sample. T_1 and T_2 values are shown in Table 4. T_1 and T_2 were small when using samples B and C, but large when using sample A. Additionally, the coefficient between T_1 and T_2 is 0.953, T_1 and T_2 indicate a similar tendency.

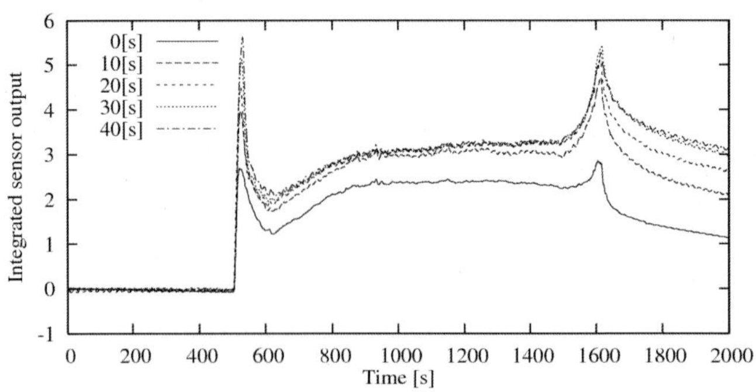

Figure 5. An example of integrated sensor output V_i. Each line is the time variation of every 10 s.

Table 4. T_1 and T_2, These are the times until V_1 and V_2 reached each threshold.

Sample / Time[s]	A	B	C	D
T_1	60	10	10	30
T_2	50	10	20	40

T_1 and T_2 were compared with the results of the sensory test described in section 2. The coefficient for the correlation between T_1 and the result of the sensory test was 0.95, and the coefficient for the correlation between T_2 and the result of the sensory test was 0.87. Thus there are strong correlations between the results of sensory tests by subjects and sensory experimental results.

4.2. *Experiment 2*

To more accurately determine T_1 and T_2, the measurement intervals applied in the experiment described in Sec. 4.1 were narrowed. From the results described in Sec. 4.1, T_1 and T_2 are in the time ranges shown in Table 5. The sensor scanned every 1 s during the intervals examined in this experiment.

Table 5. Time taken for V_1 and V_2 to reach each threshold using each detergent.

Detergent	A	B	C	D
Time[s]	40-60	0-10	0-10	20-40

The sensor slid on the surface of a plate every 10 s up to 2 min, and in the measurement time shown in the Table 5, the sensor scans every 1 s. The preparation and procedure for this experiment were the same as in experiment 1 in Sec. 4.1 except for the measurement intervals and times. Here, this sensor system can't finish a series of operations within 1 s. To solve this problem, the flow of rinse water is stopped every 1 s, and the sensor scans again the surface of the plate. The experimental conditions are shown in Table 6.

Table 6. Experimental conditions.

Air temperature	Water temperature	Humidity
18.0°C	14.9°C	52%

Table 7. T_1 and T_2 when the sensor scans every 1 s. Hyphen means no data (did not reach threshold).

	Time [s]							
	Threshold of V_1				Threshold of V_2			
Detergent	0.70	0.75	0.80	0.85	0.35	0.40	0.45	0.50
A	42	42	42	43	41	42	42	42
B	3	3	3	5	3	3	3	3
C	4	5	6	6	6	6	6	6
D	22	22	22	22	23	24	-	-

Table 8. Coefficients of the correlations of T_1 and T_2 with the results of sensory tests for each threshold.

	Threshold of V_1				Threshold of V_2			
	0.70	0.75	0.80	0.85	0.35	0.40	0.45	0.50
Coefficient correlation	0.92	0.93	0.93	0.95	0.94	0.93	0.92	0.92

To determine T_1 and T_2, six thresholds were added, and eight thresholds were set in total because V_1 reaches roughly 0.85 at the maximum. Four values were set as the thresholds of V_1 from 0.70 to 0.85. V_2 is less than 0.55, and some V_2 values measured using some samples were larger than 0.30 at the first scan. Thus four values were set as the thresholds of V_2 from 0.35 to 0.50. The set thresholds are shown by the top line in Table 7.

The results of this experiment are shown in Table 7. There are slight differences between the results using each threshold. However, T_1 and T_2 correlate strongly with the results of the sensory tests described in Sec. 2 (Table 8).

In addition, the value of peak 4 and the difference between peak 3 and peak 4 in Fig. 3(b) are defined as V_3 and V_4, respectively. The times until V_3 and V_4 reach the maximum are defined as T_3 and T_4, respectively. T_3 and T_4 were compared with the results of the sensory test. The coefficient of the correlation between T_3 and the result of the sensory test was 0.97, and the coefficient for the correlation between T_4 and the result of the sensory test was 0.98. There was also a strong correlation between V_3/V_4 and the results of sensory tests. These results show strong correlations between the QQ feeling and integrated sensor output, regardless of the value of the threshold.

5. Conclusion

An experimental study on the development of a sensor for evaluation of detergents has been presented. The obtained results can be summarized as follows.

1. A sensor for the evaluation of detergents was fabricated using PVDF film as the sensory material, and its validity for the sensor system was proven experimentally by using four detergents.
2. The time for the integrated sensor output to reach the threshold was found to have a high correlation with the results of sensory tests.

Acknowledgments

The authors acknowledge the support of Tohoku University Global COE Program "Global Nano-Biomedical Engineering Education and Research Network Centre".

References

1. G. M. Shepherd, *Neurobiology*, Oxford University Press, New York (1994).
2. M. Tanaka, Development of tactile sensor for monitoring skin conditions. *Journal of Materials Processing Tech.* **108**, 253–256 (2001).
3. T. Ikeda, *Fundamental of Piezoelectricity*, Oxford University Press, New York (1990).

Section 4

Nano-Biointervention

NONCONTACT MANIPULATION OF MICRO-NANO SCALE BIOLOGICAL OBJECTS IN A CHIP USING INTEGRATED OPTICAL TWEEZERS AND MICROTOOLS

FUMIHITO ARAI [*], HISATAKA MARUYAMA

Department of Bioengineering and Robotics, Tohoku University, Aramaki Aza Aoba 6-6, Aoba-ku, Sendai, Miyagi 980-8579, Japan

TOSHIO FUKUDA

Department of Micro-Nano Systems Engineering, Nagoya University, Furo-cho, Chikusa-ku, Nagoya, Aichi 464-8603, Japan

On-chip single cell analysis is an important approach for research and development in life science, pharmaceutical industry, livestock agriculture, and so on. Optical tweezers are well known as one of the noncontact manipulation methods used in a closed space and much has been reported. Although optical tweezers are suitable for manipulating a single micro to nano scaled particle, direct laser manipulation is not recommended for manipulation. Microtools offer the following advantages; (1) avoiding direct laser irradiation, (2) varying the trap force depending on the microtool, and (3) performing dexterous manipulation of micro-nano targets. We employed two different multi-beam optical tweezers such as time-shared scanning method and generalized phase contrast method and developed integrated optical tweezers. We fabricated various functional microtools such as a thermosensitive hydrogel microbead for detachable manipulation, a pH sensing gel-tool for local environment measurement, and an SU-8 microtool with an arbitrary shape. Recent research attention has been gathered dexterous manipulation and single cell measurement.

1. Introduction

Recently, manipulation and measurement of a single cell have become important for detailed investigation of cell properties. Although many studies analyzing cell properties analysis have been conducted using micromanipulators under an optical microscope, this method requires skillful operators and damage the cell. Moreover, environmental disturbance and contamination are also serious problems. Recently, on-chip cell analysis for this purpose has been receiving significant attention. Experiments on a microchip, which is a closed space, can avoid disturbance and contamination. However, micromanipulators can not be

[*] Fumihito Arai is a Tohoku University Global COE Member.

employed on a microchip. Noncontact manipulation methods such as optical tweezers are suitable for on-chip cell manipulation [1]. Although optical tweezers enable micromanipulation of individual cells, it may damage the cell by direct irradiation of a focused laser. Dexterous manipulation such as precise attitude control is also difficult by direct laser manipulation. For safe and dexterous manipulation of a cell, we proposed indirect laser manipulation of the cell using microtool [2]. Microtools are defined as microobjects having various functions, for example dexterous manipulation and local measurement. Figure 1 shows the history of microtools for optical tweezers. We succeeded in three-dimensional 6-DOF manipulation [3], DNA manipulation without any modification [4] and local environment measurement using microtool [5]. Recent research attention has been gathered dexterous manipulation.

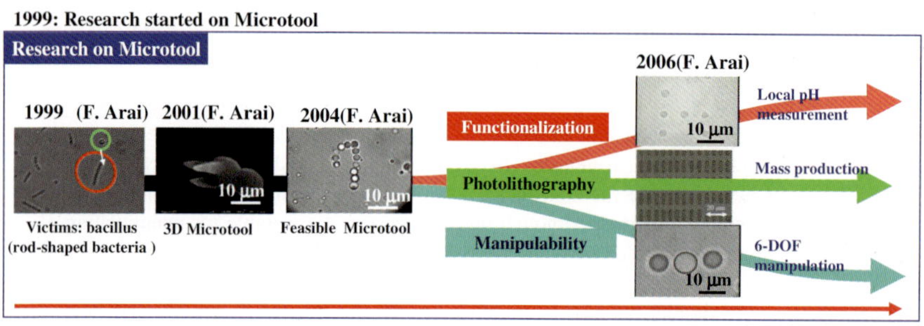

Figure 1. History of microtool for optical tweezers.

2. Optical Tweezers

We developed the time-shared scanning (TSS) method for the trajectory control of the multiple targets independently by the single laser to accomplish this [6]. We employed a laser with a wavelength of 1064 nm, because this wavelength is relatively safe to cells. Our TSS system is different from existing laser scanning micromanipulation. In our system, we can manipulate multiple targets independently by a single laser by changing the discrete laser scanning pattern as shown in Fig. 2(a). Multiple microtools are manipulated stably along with each designed trajectory by the proposed method as shown in Fig. 2(b).

We also employed a generalized phase contrast (GPC) method for laser manipulation of multiple objects. TSS involves high-speed laser scanning using galvano mirrors. GPC uses laser phase control. TSS is suitable for manipulating a few objects at a high speed, whereas GPC is suitable for trapping many objects. Recently, we have proposed integrated optical tweezers (IOT), which is integrated TSS and GPC because of its fast response [7].

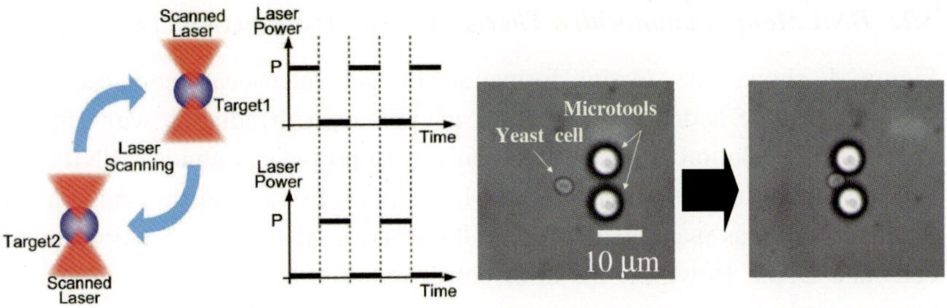

(a) Schematic diagram of TSS (b) Cell manipulation using two microtools

Figure 2. Time-shared scanning method for manipulating multiple objects.

3. Dexterous Manipulation with Microtool

3.1. *6-DOF Manipulation of Microobject [6]*

6-DOF manipulation was performed by 3D-TSS. 3D-TSS enables 3D control with 6DOF (position and orientation) of a microtool or a cell. Three-dimensional orientation control is effective in improving the cell surgeries such as gene injection. Figure 3(a) shows a schematic diagram of the optical system used in 3D-TSS. An inverted microscope is used as a platform. The laser is scanned in a 2D observation plane by controlling the angle of a pair of galvanometer mirrors using DC motors. A piezo stage was used to drive an objective lens for scanning the laser along the Z axis. These actuation units are controlled by a computer. Using the galvanometers and the piezo stage simultaneously, 3D laser scanning and 6-DOF manipulation are achieved as shown in Fig. 3.

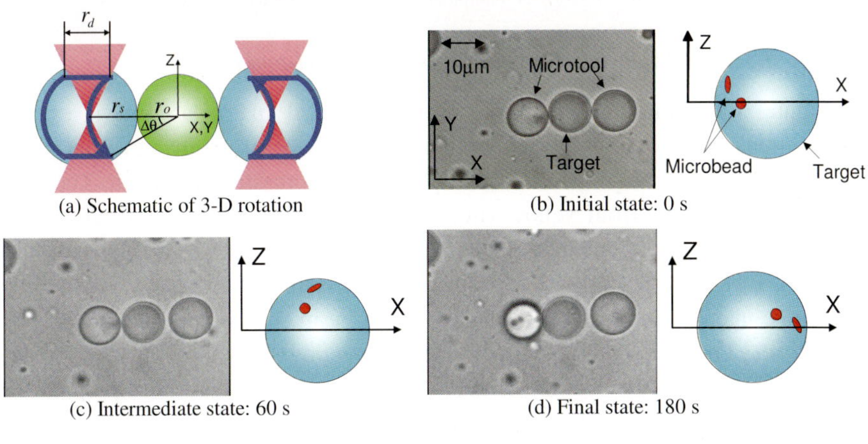

Figure 3. 3-D rotation of a microsphere using 6-DOF manipulation.

3.2. *DNA Manipulation with a Thermosensitive Hydrogel-tool [4]*

Figure 4 shows the *in-situ* formation of gel microbead composed of a thermoreversible hydrogel for indirect manipulation. Irradiation with 1064 nm laser to a solution mixed with poly-N-isopropylacrylamide (PNIPAAm) generated a hydrogel-tool at the laser focus point by heating the solution. The hydrogel-tool was manipulated by the laser and used for micromanipulation of cells and DNA. However, the laser power required to form this hydrogel-tool was generally too strong to perform manipulation by the conventional method. We reduced the laser power for the formation of the tool using the PNIPAAm aqueous solution was achieved by addition of additives was developed. YPD Broth (containing yeast extract, peptone, and dextrose) were selected as an additive and measured the relationship between the laser power, irradiation time, and diameter of the hydrogel-tool. Manipulations of cells and DNA using the laser-trapped hydrogel-tool were performed successfully.

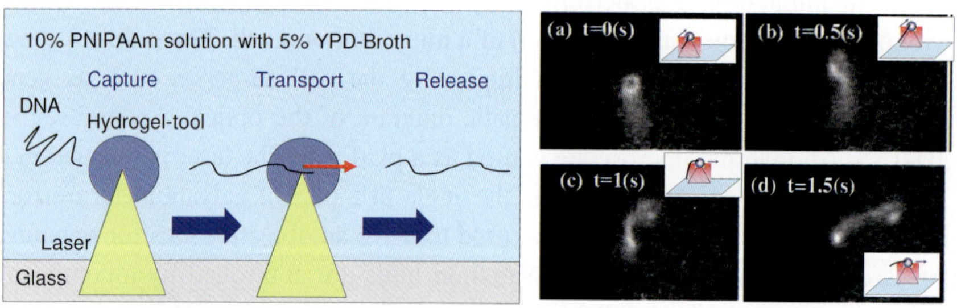

(a) Schematic of hydrogel-tool (b) DNA extension

Figure 4. Dexterous DNA manipulation using thermosensitive hydrogel-tool.

3.3. *Local pH Measurement with Functional Gel-tool [5]*

Figure 5(a) shows a schematic of on-chip local pH measurement using a functional gel-tool impregnated with the pH indicator Bromothymol Blue (BTB). The gel-tool is composed of a hydrophilic photo-crosslinkable resin. The color of BTB changes from yellow (in an acidic solution) to blue (in an alkaline solution). The pH sensing gel-tool was produced by stirring a mixture of the resin, BTB, and an 8.0 wt% phosphate dipotassium salt solution. The gel-tool can be manipulated using optical tweezers on a microchip and fixed on a glass surface. The pH value is measured by detecting the color of the tool and calibrated using the tool's red color difference (Cr). There is a proportional relationship between pH and Cr. The measurement precision of pH is about 0.2.

Figure 5(b) shows the pH measurements of the boundary domains of two solutions with differing pH values (pH 6 and pH 9). The local pH value in the boundary domain was measured by scanning the pH sensing gel-tool (size: about 5 μm) using optical tweezers. Stationary pH measurement is also possible by arranging the gel-tool in an arbitrary pattern on the glass surface. This sensing method can be employed not only for pH measurement but also for temperature measurement by utilizing different indicators.

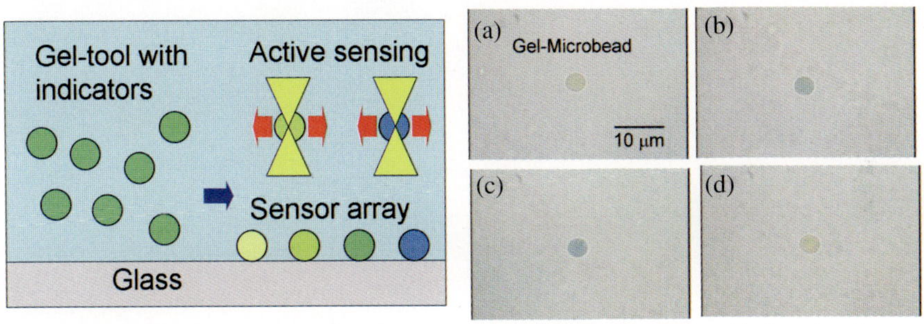

(a) Schematic of pH sensing gel-tool

(b) pH measurement with the gel-tool

Figure 5. Local pH measurement using pH sensing gel-tool.

4. Summary

Dexterous manipulation in microchip and microtool with various functions play an important role in the development of advanced methods and devices for cell surgery, cell measurement, and other biomedical engineering. For future improvement, it is obvious that micro-nanotechnology plays an important role. Our proposed methods enable advanced cell measurements and are anticipated to make great contributions to cell biology.

Acknowledgments

Work partially supported by 2007 Tohoku University Global COE Program Global Nano-Biomedical Engineering Education and Research Network Centre.

References

1. F. Arai, A. Ichikawa, M. Ogawa, T. Fukuda, K. Horio and K. Itoigawa, *ELECTROPHORESIS* **22**, 283 (2001).
2. F. Arai, H. Maruyama, T. Sakami, A. Ichikawa and T. Fukuda, *IEEE/ASME Transactions on Mechatronics* **8**, 3 (2003).
3. F. Arai, T. Endo, R. Yamauchi and T. Fukuda, *Journal ref: Journal of Robotics and Mechatronics*, **18**, 153 (2006).

4. A. Ichikawa, F. Arai, K. Yoshikawa, T. Uchida and T. Fukuda, *Appl. Phys. Lett.* **87**, 191108 (2005).
5. H. Maruyama, F. Arai and T. Fukuda, *Lab Chip*. **8**, 346 (2008).
6. F. Arai, K. Yoshikawa, T. Sakami and T. Fukuda, *Appl. Phys. Lett*. **85**, 4301 (2004).
7. F. Arai, K. Onda, H. Matsumoto and R. Iitsuka, *Proc. of μTAS 2008* **1**, 396 (2008).

ON-CHIP CELL MANIPULATION WITH MAGNETICALLY DRIVEN MICROTOOLS

FUMIHITO ARAI [*], YOKO YAMANISHI

Department of Bioengineering and Robotics, Tohoku University, 6-6-01 Aramaki-Aza-Aoba, Aoba-ku, Sendai, Miyagi, 980-8579, Japan

We describe a novel non-contact sorting system that uses an in-chip, magnetically driven micro-tool (MMT), rather than conventional biomanipulation by hand, which has a higher risk of contamination, a lower success rate and lower repeatability. A MMT can sort particles individually, and can be installed directly in a microchannel, unlike other conventional cell sorting systems based on field gradients or electrostatic force. The proposed method can be applied to other micro-objects independently of their dielectric properties and without the complexity and power requirements associated with optical systems. The drive unit was significantly downsized by amplifying the density of magnetic flux with a permanent magnet. They also require less power because no energy is required to keep the MMT fixed in a particular position. Sorting was performed using real-time sensing images of microbeads, and the system successfully sorted beads of different sizes at switching speeds up to 18 Hz. The MMT developed in this study is unique with respect to its flexibility and biocompatibility. In addition, since the PDMS (Polydimethylsiloxane) microchip is disposable, it can be applied to cell sorting with no risk of contamination. Potential fields of application of this technology include cloning techniques, which require sorting of oocytes, with and without nuclei.

1. Introduction

Microchip production has progressed remarkably in the fields of chemistry and biology with the development of micro and nano fabrication techniques. A microchip that can culture cells has been developed—it maintains the desired environment by controlling the oxygen, nutrients, and waste materials in the microchannel. In such microchannels, the transportation medium can be a single- or multiphase flow, typically containing microbeads, chemicals, or cells. The ability to manipulate microparticles in a microchannel offers great potential for biology, chemistry, and environmental analysis.

The most popular microactuators for application in the confined space of microchannels are electrostatic microactuators, optical tweezers, and magnetic microactuators. Coulomb force has often been used for manipulating cells on the order of 10 μm, whereas it is necessary to apply a high voltage to manipulate

[*] Fumihito Arai is a Tohoku University Global COE Member.

particles on the order of 100 μm (*e.g.* oocyte). But heat generated by high voltage can damage cells. The dielectrophoretic force can be adjusted by varying the square of the gradient of the electrical field, however, it is controllable only in the limited region adjacent to the electrodes, and requires higher voltages to sort larger objects. Optical tweezers can manipulate cells indirectly by non-contact actuation of microtools, thus reducing the risk of damaging cells, however, the generated force is on the order of several pN, which is not suitable for manipulating cells on the order of 100 μm. Magnetic manipulation, on the other hand, offers limited risk of cell contamination, and has been used in many studies because of its low cost [1-4].

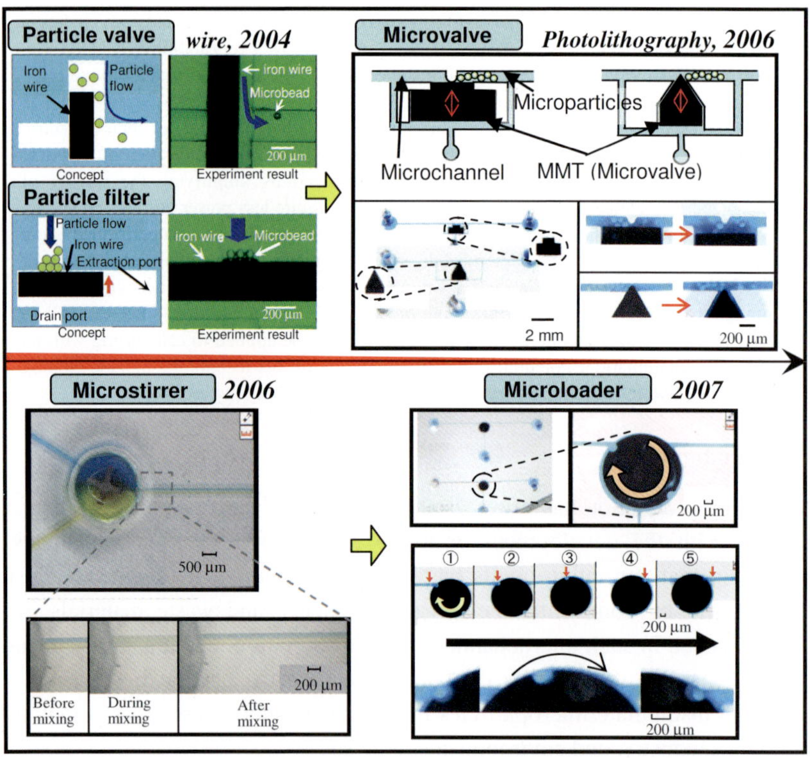

Figure 1. History of magnetically driven microtools.

In previous research we developed magnetically driven microtools (MMT). MMTs can be fabricated with high accuracy, mass produced, and the tools can be fabricated in arbitrary shapes. The technique provides many microchannel functions, such as a valve, stirrer, and loader [5]. Microdevices actuated by magnetic force have a simple structure, are low cost, and easy to integrate into a microchip. We recently developed an MMT and appropriate actuation system for sorting. The developed MMT is flexible and biocompatible and, by fabricating the microchannel from a disposable material, can be applied to cell sorting applications where there is a risk of contamination. By analyzing the image around the separation point, particles flowing in the microchannel were sorted individually and in real time, based on particle size. While developing the MMT drive unit, it was difficult to downsize the electromagnetic coil because of the large number of windings required to produce sufficient actuation power. Permanent magnets have been reported to have significant advantages in miniaturizing systems [6], therefore, we used a permanent magnet that operated indirectly, and were able to downsize the drive unit significantly. An automated sorting system, for different sizes of polystyrene beads, was also developed, using a real-time image processing system.

2. Fabrication and Actuation

2.1. *MMT Fabrication*

Figures 2(a)-(e) show the MMT fabrication process, which may be summarized as follows: (a) A thick negative tone resist (KMPR1050, KAYAKU MicroChem, Japan.) was spread over a silicon substrate, and an MMT mold was produced by photolithography; (b) a mixture of PDMS and magnetite (Fe_3O_4, av. diameter of 200 nm, 50 wt%) was spread over the patterned mold and baked in an oven (110°C, 10 min); and, (c) and (d) the MMT was obtained after treatment with a stripper liquid (KMPR Remover PG, 70°C, KAYAKU MicroChem, Japan.). The surface of the MMT was Teflon coated with CF_4 gas by plasma ashing (discharge power of 130 W) for 30 minutes to avoid any stiction in the microchannels. After fabrication (Fig. 2(e)), the MMT was directly installed in a microchip, which was fabricated by SU-8 photoresist (Figs. 2 (f)-(h)).

Magnetic powder in the polymer has a great advantage for fabricating arbitrary shapes. The baked PDMS (the catalyst was evaporated) is biocompatible and widely used as a biochip material. The magnetite is also biocompatible and widely used in drag delivery systems. In the case of ferromagnetic materials, it is difficult to fabricate fine and complicated shapes. To manipulate cells, the actuator shapes to be complicated to provide accurate manipulation while protecting the sensitive material of cell. This advantage overcomes the disadvantage of the induction of smaller forces compared to that with ferromagnetic materials.

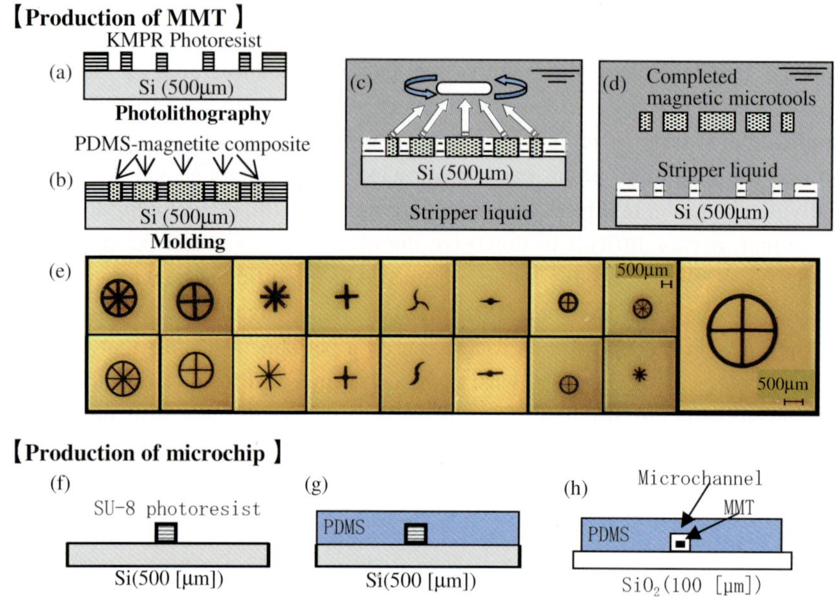

Figure 2. Fabrication processes of the MMT and microchannel.

2.2. *Actuation of the MMT*

Figure 3 shows the configuration of a MMT with vertical or lateral motion that consists of two modules—an upper module containing a disposable microchannel and a lower actuation module containing a drive unit. The density of magnetic flux generated by the electromagnetic coil is amplified by the permanent magnet (neodymium) unit mounted between the microchannel and the magnetic circuit, and the MMT is moved by non-contact actuation. The direction of current in the coil can be switched to reverse the electromagnet's polarity, resulting in translatory motion of the permanent magnet. The density of magnetic flux generated by the electromagnetic coil is amplified, and an adequate density of magnetic flux (maximum = 316 mT, with 1.5 V applied to the electromagnet) is transmitted to actuate the MMT in the microchannel.

The magnetic flux density for actuation was approximately 110 times that of a conventional setup without a permanent magnet. It is important to note that the MMT remains in position even after the magnetic circuit is switched off, enabling reduced power consumption, because no energy is consumed in keeping the MMT fixed in place. However, there might be a problem in that the magnetic field could be actuated by different electromagnets. Recently, the MMT was effectively magnetized, significantly reducing the interaction region, and thus, the entire chip system has been downsized remarkably.

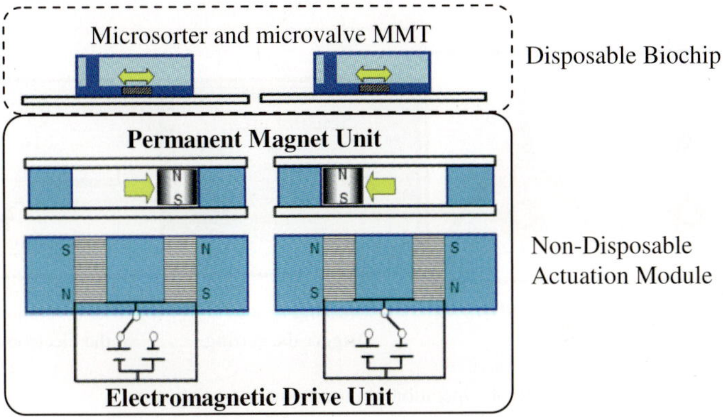

Figure 3. The MMT actuation module.

3. Operation of the MMT Sorting System

3.1. *Sort Automation with Image Processing*

The microsorter was successfully operated with lateral motion, as shown in Fig. 4(a), using the actuation module shown in Fig. 3. Also, it was confirmed that the edge of the MMT can be used to switch the microchannel. The maximum operating speed of the MMT was about 20 Hz. We successfully automated the sorting system using an image sensing system mounted upstream of the edge of the MMT. The system is first calibrated with an image of representative particles. The particles can then be sorted using a threshold, based on the calculation of the coefficient of correlation between the CCD image data and the sample image data. A sensing area is then set in the CCD image. A current, amplified by a power amplifier, changes the magnetic polarity of the electromagnet coil, and the output of a D/A (digital to analogue) circuit is varied. This process is controlled by the results of the image processing within the sensing area. Figure 4(b) shows a 100-μm bead passing through the sensing area (dashed square), and triggering the sorting module. The bead was successfully transported to the lower microchannel by moving the MMT to the upper wall of the microchannel. Similar results were observed for the 50-μm particles. We thus succeeded in automating the sorting of beads by size. The total sample size for the experiment was 20 and all particles were successfully sorted. The sorting rate in the current experiment, which is restricted by the image acquisition systems, was 10 Hz, and which can be improved to 10 times faster enough to catch up to the switching speed of the sorter (18 Hz for the current version).

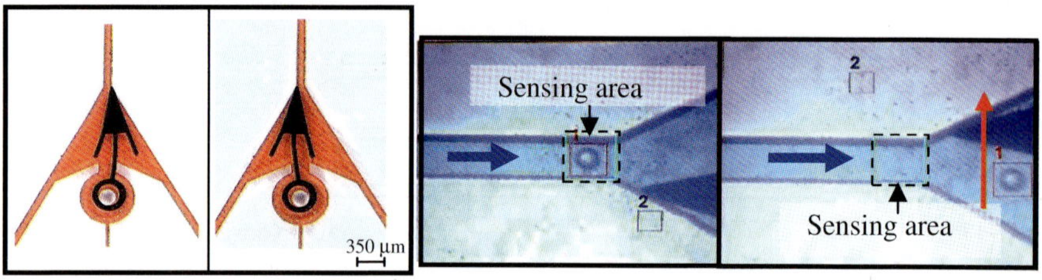

(a) Operation of MMT sorter (b) Sensing area to trigger the sorting (c) After reversing the polarity of the electromagnet

Figure 4. Operation of the automatic sorting system.

Acknowledgments

This work is supported by the Research and Development Program for New Bio-industry Initiatives and the Ministry of Education, Culture, Sports, Science, and Technology of Japan (Grants-in-Aid for Scientific Research 17040017 and 19016004), and the Global COE Program, "Global Nano-Biomedical Engineering Education and Research Network Center."

References

1. K. W. Oh and C. H. Ahn, *J. Micro. Microeng*. **16**, R13–R39 (2006).
2. H. Maruyama, F. Arai and T. Fukuda, *μ-TAS2005*, 1422–1424 (2005).
3. K. S. Ryu, K. Shaikh, E. Goluch, Z. Fan and C. Liu, *Lab Chip* **4**, 608–613 (2004).
4. M. Barbic, J. J. Mock, A. P. Gray and S. Schults, *App. Phys. Lett.* **79**, 1399–1401 (2001).
5. Y. Yamanishi, S. Sakkuma and F. Arai, *J. Robo. Mech*. **20**, 273–279 (2008).
6. O. Cugat, J. Delamare and G. Reyne, *IEEE Trans. Mag*. **39**, 3607–3612 (2003).

ANALYSIS OF TUMOR SUPPRESSOR GENE USING MOLECULAR IMAGING FOR PERSONALIZED MEDICINE

NATSUKO CHIBA *, LEIZHEN WEI †

Department of Molecular Immunology, Institute of Development, Aging and Cancer, Tohoku University, 4-1 Seiryo-machi Aoba-ku, Sendai, Miyagi, 980-8575, Japan

Breast cancer susceptibility gene 1 (*BRCA1*) is a prominent gene responsible for hereditary breast and ovarian cancers. Although increasing evidences indicate that BRCA1 participates in multiple cellular processes, its precise mechanisms in these processes remain unclear. To elucidate the molecular function of *BRCA1* as a tumor suppressor gene, we analyzed the various functions of BRCA1, such as apoptosis, DNA repair, and cell cycle regulation. We recently analyzed the cellular responses of BRCA1 to various DNA damages using a molecular imaging technique. In particular, we examined the *in situ* response of BRCA1 accumulation at laser micro-irradiation–induced double-strand breaks (DSBs) and single-strand breaks (SSBs). BRCA1 accumulated at DSBs via its amino (N)- and carboxyl (C)-termini. N-terminal BRCA1 rapidly accumulated at DSBs in a Ku80-dependent manner. In contrast, BRCA1 accumulated at SSBs via its C-terminus and with different kinetics from its accumulation at DSBs. BRCA1 accumulation at SSBs was abolished by caffeine treatment of the cells, which inhibits the PI3-kinases ataxia telangiectasis mutated (ATM) and ataxia telangiectasis mutated-related (ATR). To analyze the response of BRCA1 to UV irradiation, we generated local UV irradiation in cell nuclei using an isopore membrane filter. Interestingly, BRCA1 accumulation was dependent on a protein that is involved in the nucleotide excision repair pathway. Furthermore, BRCA1 immediately accumulated at the locally UV-irradiated sites in a transcription-dependent manner. These results suggest that BRCA1 is involved in the transcription-coupled repair (TCR) process for UV irradiation–induced DNA damage. Furthermore, we have identified a new BRCA1-associated protein by proteomics, and are analyzing the properties of this protein. These precise analyses of BRCA1 and its related proteins will contribute to the development of novel molecular diagnoses and therapies for the personalized medicine of breast and ovarian cancers.

1. Introduction

Approximately 5% of all breast cancers show a familial pattern of occurrence. In 1990, genetic studies provided the first evidence that the risk of breast cancer development is linked (in some families) to mutations in chromosome 17q21.

* Natsuko Chiba is a Tohoku University Global COE Member.
† Leizhen Wei is a Tohoku University Global COE Fellow.

Breast-cancer susceptibility gene 1 (*BRCA1*) was identified in 1994 by positional cloning [1], and our research has focused on the function of this gene. *BRCA1* encodes 1863 amino acids (a.a.), and more than 200 different *BRCA1* germ-line mutations associated with cancer susceptibility have been identified. Carriers of *BRCA1* mutations also have increased susceptibility to ovarian cancers; the risk of breast cancer development is about 50-80%, and that of ovarian cancer development is 12-60%, in *BRCA1* mutation carriers. *BRCA1*-related breast cancers are characterized by an early onset, a high frequency of bilateral involvement, and a poor prognosis.

Somatic *BRCA1* mutations are rarely observed in sporadic breast cancers. However, *BRCA1* mRNA and protein expressions are downregulated in ~ 30% of sporadic breast cancers and 70% of ovarian cancers. This may be due to non-mutational mechanisms, such as acquired promoter methylation or malfunctions in upstream pathways that regulate *BRCA1* expression. Therefore, the reduced expression or function of *BRCA1* is thought to be an important contributing factor in sporadic cancers.

BRCA1 contains a RING domain in its amino (N)-terminus and two tandem BRCT domains in its carboxy (C)-terminus. The BRCT domain is frequently found in DNA repair proteins, and functions as a binding module for phospho-serine peptides. BRCA1 also has a nuclear export signal (NES), a nuclear localization signal (NLS), and a DNA-binding region. After DNA damage, BRCA1 is phosphorylated by Chk2 or ATM (Fig. 1). The N-terminal region of BRCA1 directly interacts with BARD1, thereby enhancing the ubiquitin polymerase activity of BRCA1. Like BRCA1, BARD1 has a RING domain at its N-terminus and two BRCT domains at its C-terminus.

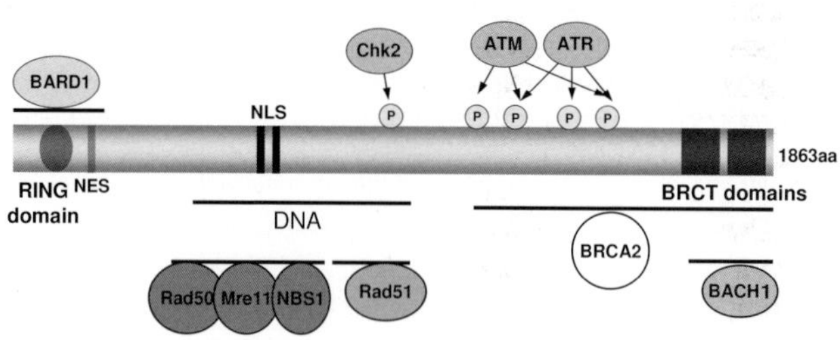

Figure 1. Structure of the BRCA1 protein. BRCA1 contains an N-terminal RING domain, a nuclear export signal (NES), a nuclear localization signal (NLS), a DNA binding region, and two C-terminal BRCT domains. BRCA1 is associated with a number of proteins, as indicated. Sites phosphorylated by Chk2 or ATM are indicated by arrows.

BRCA1 has been implicated in a number of cellular processes, including the regulation of DNA damage repair, transcription, the cell cycle, chromatin remodeling, and apoptosis. To elucidate the molecular mechanisms of the tumor suppressive functions of BRCA1, we have focused on analyzing the various functions of BRCA1:

(1) Using biochemical techniques, we resolved four distinct BRCA1-containing 6.3mm.
(2) We found that BARD1 is an integral component of the RNA polymerase II (RNAPII) holoenzyme (holo-pol). Furthermore, N-terminal BRCA1 was important for its association with holo-pol and for its localization at nuclear foci in the S-phase of the cell cycle [3].
(3) The expression of N-terminal–deleted BRCA1 revealed a powerful growth suppressive effect that was associated with increased apoptosis [4].
(4) We identified RNAPII as a substrate for the ubiquitinative activity of BRCA1. After UV irradiation, a subpopulation of RNAPII was phosphorylated and recognized by the C-terminus of BRCA1 as a substrate. BRCA1/BARD1 ubiquitinated the phosphorylated RNAPII, which was subsequently degraded at the proteasome. BRCA1/BARD1 could then recruit other DNA repair factors to the site of DNA damage. We presume that BRCA1 is involved in TCR processes by the mechanism shown in Fig. 2 [5].
(5) Finally, we have developed a functional assay for the molecular diagnosis of BRCA1 by its ubiquitin ligase activity and subcellular localization in the S-phase and after DNA damage.

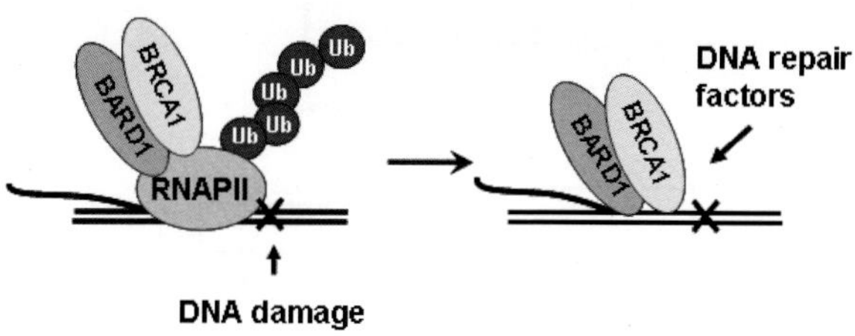

Figure 2. After DNA damage, a subpopulation of RNAPII is phosphorylated. BRCA1/BARD1 ubiquitinates the phosphorylated RNAPII, which is subsequently degraded at the proteasome. BRCA1/BARD1 then recruits other DNA repair factors to the site of DNA damage.

Recently, our research has focused on the DNA repair functions of BRCA1. Among the various functions of BRCA1, its tumor suppressor activity is thought to be primarily attributable to its involvement in DNA repair. Clinically-observed missense mutations often result in non-functional BRCA1 proteins that cannot repair DNA damage. However, the precise mechanism(s) of BRCA1 in the DNA repair pathway are not fully understood. Therefore, we collaborated with Dr. Akira Yasui of the Department of Molecular Genetics, Institute of Development, Aging, and Cancer, Tohoku University, to analyze the cellular responses of BRCA1 to various DNA damages using a molecular imaging technique

2. BRCA1 Response to Various DNA Damage Induced by Laser Micro-irradiation

2.1. *Laser Micro-irradiation System*

To elucidate the molecular mechanisms of BRCA1 in the DNA repair processes following various DNA damages, we analyzed BRCA1 accumulation at the site of DNA damage induced by laser micro-irradiation in living cells. Our experimental system consisted of a laser micro-irradiation apparatus combined with a confocal microscope. We used a 365-nm pulse laser system to irradiate cells in the epi-fluorescence path of the microscope system. The irradiation dose was modulated by passing lasers through distinct filters in front of the lens. A 405-nm pulse laser system was also used. The power of the laser scan was controlled by the number of scans (Fig. 3). Using this system, various types of DNA damage, such as single-strand breaks (SSBs), double-strand breaks (DSBs), and oxidative base damage, were produced at restricted nuclear regions of the cells [6,7].

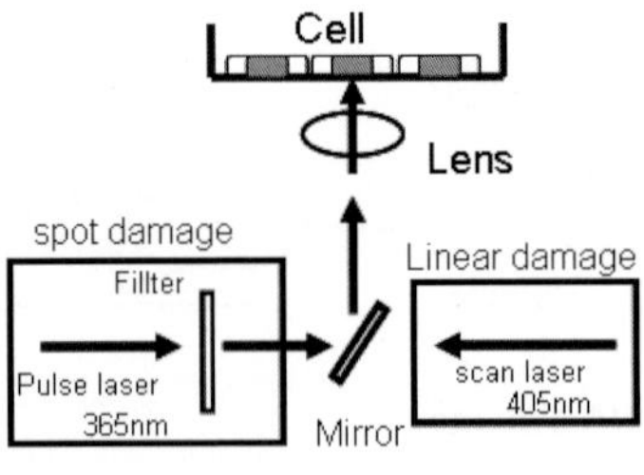

Figure 3. Laser micro-irradiation systems. The left column shows the 365-nm pulse laser irradiation system, producing low dose or high dose irradiation regulated by a filter in front of the mirror. The right column shows the 405-nm laser system. Fluorescent images were obtained and processed using an FV-500 confocal scanning laser microscope system. A 405-nm scan laser system for the irradiation of cells in the epifluorescent path of the microscope system was used. One scan of the laser light at full power delivers ~ 1600 nW.

2.2. *BRCA1 Accumulation at the Laser-irradiated Site is a Response to DNA DSBs*

We analyzed the *in situ* response of BRCA1 at DSBs produced by laser micro-irradiation. The N- and C-terminal fragments of BRCA1 accumulated at the DSBs independently and with distinct kinetics. N-terminal BRCA1 fragments accumulated at DSBs immediately after laser irradiation and dissociated rapidly. In contrast, C-terminal BRCA1 fragments accumulated at DSBs more slowly, but remained at the sites. Interestingly, the rapid accumulation of N-terminal but not C-terminal BRCA1 at DSBs depended on the Ku80 protein that functions in non-homologous end-joining (NHEJ). Furthermore, this accumulation was independent of the BARD1 protein that binds to N-terminal BRCA1. Two small regions in the BRCA1 N-terminus independently accumulated at DSBs and interacted with Ku80. Missense mutations found within N-terminal BRCA1 in cancers significantly altered the accumulation kinetics at DSBs. One such mutation abolished the association of BRCA1 with the Ku80 protein. These findings might provide a molecular basis for BRCA1 involvement in the NHEJ pathway of the DNA DSB repair process.

2.3. *BRCA1 Accumulation at the Laser-irradiated Site is a Response to DNA SSBs*

2.3.1. *BRCA1 accumulation at the laser-irradiated site is a response to DNA SSBs*

After SSBs were introduced into the nuclei of human Saos-2 cells by laser micro-irradiation, the cells were fixed and processed for immunofluorescence. Endogenous BRCA1 clearly accumulated at the irradiated sites and colocalized with XRCC1, which rapidly accumulates at SSBs and functions in SSB repair processes [6-8]. We then examined the real-time accumulation of GFP-tagged BRCA1 in living cells after laser-irradiation. GFP-BRCA1 was transfected into Saos-2 cells, the cells were laser-irradiated, and the mean intensities of GFP-BRCA1 were quantified. GFP-BRCA1 clearly accumulated at the irradiated sites (Fig. 4). This accumulation was slow and gradual, indicating that the accumulation kinetics of overexpressed GFP-BRCA1 mimic those of endogenous BRCA1.

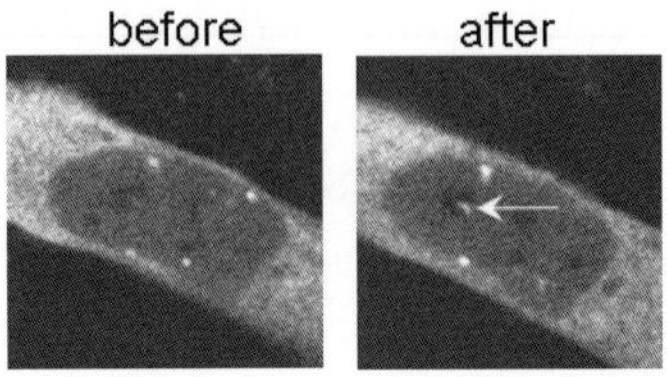

Figure 4. GFP-tagged BRCA1 accumulates at the SSB sites induced by laser-irradiation in Saos-2 cells. Arrow indicates the laser-irradiated sites.

We next identified the BRCA1 regions that mediate BRCA1 accumulation at sites of irradiation. Several GFP-tagged deletion mutants of BRCA1 were constructed. The full-length BRCA1, as well as the deletion mutants Δ305-770, Δ775-1292, Δ1-302, and Δ1527-1863, were transfected into Saos-2 cells. All of the BRCA1 deletion mutants except Δ1527-1863 accumulated at SSBs. The C-terminal a.a.1527-1863 fragment accumulated at the laser-irradiated sites. These results suggest that the BRCA1 C-terminus is important and sufficient for BRCA1 accumulation at SSBs. We then quantified the mean intensity of GFP-BRCA1 at the accumulation sites, and examined the accumulation kinetics of full-length BRCA1 and BRCA1 fragments for 600 s after irradiation. The BRCA1 fragments slowly and gradually accumulated at the SSBs, reaching a plateau after 420 s.

2.3.2. *BRCA1 accumulation at SSBs is completely abolished by caffeine treatment*

After DNA damage, many proteins, including BRCA1, are phosphorylated by PI3-kinases, such as ATM and ATR. We examined whether BRCA1 accumulation at SSBs was affected by treatment of the cells with caffeine, which inhibits ATM and ATR. When Saos-2 cells were incubated with caffeine before laser-irradiation, the accumulation of C-terminal BRCA1 at SSBs was completely abolished. This result suggests that the BRCA1 accumulation at SSBs is dependent on PI3-kinases.

2.3.3. *Requirement of other DNA repair factor for the BRCA1 accumulation of at SSBs*

We next examined whether BRCA1 accumulation at SSBs was dependent on specific DNA repair factors by using factor-deficient cells. GFP-tagged C-terminal BRCA1 fragments were transfected into Chinese hamster ovary (CHO)-derived *XRCC1*-deficient cells. The C-terminal fragment accumulated at laser-induced SSBs in *XRCC1*-deficient cells. In contrast, GFP-tagged DNA polymerase β did not accumulate at the laser-induced SSBs. Therefore, XRCC1 is not required for the accumulation of BRCA1 at irradiated sites.

2.3.4. *BRCA1 is important for resisitance to SSBs*

Finally, we performed a colony formation assay to examine whether C-terminal BRCA1 is important for the resistance of cells to SSB formation. We transfected HCC1937 cells with plasmids to express HA-tagged BTCA1 or control vector, and then assessed the colony forming ability of these cells after treatment with the SSB-inducing compound methyl methanesulfonate (MMS). HA-BRCA1 expression was confirmed by Western blotting using an antibody against BRCA1. HCC1937 cells expressing HA-tagged full-length BRCA1 were more resistant to treatment with MMS, suggesting that MMS resistance depends on the C-terminus of BRCA1.

2.3.5. *Discussion*

In this study, we laser micro-irradiated living cells and demonstrated that endogenous and GFP-tagged BRCA1 proteins accumulated at SSBs. Using various deletion mutants of BRCA1, we found that BRCA1 was slowly recruited at laser irradiation–induced SSBs via its C-terminal regions. The kinetics of BRCA1 accumulation at SSBs were quite distinct from the kinetics of accumulation at DSBs. Accumulation of C-terminal BRCA1 at SSBs was abolished by the treatment of cells with caffeine, suggesting that BRCA1 accumulation at SSBs depends on the phosphorylation of BRCA1 or other proteins by ATM and/or ATR. Interestingly, although C-terminal BRCA1 accumulated slowly at DSBs similar to its accumulation at SSBs, BRCA1 accumulation at DSBs was not diminished by caffeine treatment. Taken together, these results suggest that BRCA1 accumulates at SSBs and DSBs through different mechanisms.

We previously reported on the accumulation kinetics of proteins involved in the repair of SSBs [6]. Compared with other SSB repair proteins, BRCA1 displays accumulation kinetics that are similar to those of PCNA and CAF1-p150. PCNA is involved in a late step of the long-patch base excision repair pathway and CAF1 mediates chromatin assembly in the late steps of SSB repair [9,10]. BRCA1 might also be involved in the late processes of SSB repair, together with PCNA and/or CAF1-p150.

3. BRCA1 Response to Local UV Irradiation

In the field of DNA repair, BRCA1 is engaged in DNA DSB repair pathways, homologous recombination (HR), and NHEJ. BRCA1 is also involved in TCR processes [9], and BRCA1-deficient cells are defective in a process in which oxidative base damages are removed preferentially from the transcribed DNA strand [10]. After UV irradiation, BRCA1 localizes to nuclear foci and is phosphorylated at Ser1423 and Ser1524 by ATR. Previously, we and other groups reported that BRCA1 ubiquitinates RNAPII after UV irradiation [5,11,12]. The predominant DNA damages produced by UV irradiation are cyclobutane pyrimidine dimers and 6-4 photoproduct adducts. These lesions are removed by nucleotide excision repair (NER), which eliminates a wide variety of bulky helix–distorting DNA lesions. NER operates via two pathways, global genome repair (GGR) and TCR. GGR can repair DNA lesions at any location in the whole genome, while TCR selectively removes DNA lesions on the transcribed strands of expressed genes.

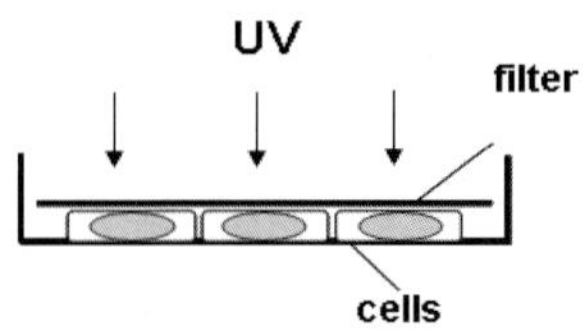

Figure 5. Schematic of cellular exposure to local UV irradiation using a membrane filter.

To gain further insights into BRCA1 regulation of the TCR pathway, we induced UV damage in small restricted areas of cell nuclei by using an isopore membrane filter. Cells monolayered in 35-mm glass-bottom dishes were treated with or without a transcriptional inhibitor, were covered with a polycarbonate isopore membrane filter (3-μm diameter pores), and were UV-irradiated (254-nm) with a dose of 60 J/m^2 (Fig. 5). The polycarbonate blocked the 254-nm UV light, and cells were exposed only though the filter pores. Immunostaining with an antibody against BRCA1 revealed that BRCA1 immediately accumulated at the site of local UV irradiation and colocalized with cyclobutane pyrimidine dimmers (CPD), which was the predominant form of UV-induced DNA damage (Fig. 6). This accumulation was abolished by a transcriptional inhibitor, and was dependent on the presence of a protein that is involved in the NER pathway. Furthermore, BRCA1 was associated with this repair factor. Our data reveal the mechanism of BRCA1 involvement in the TCR pathway at lesions induced by UV irradiation.

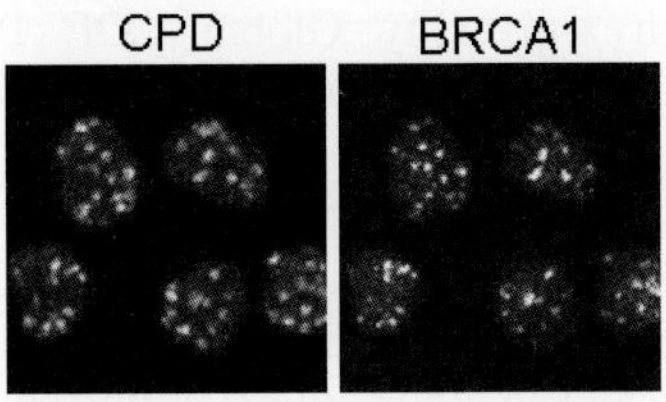

Figure 6. BRCA1 accumulation at sites of local UV irradiation. Saos-2 cells were fixed after UV irradiation through membrane filters, and were stained with antibodies against CPDand BRCA1.

4. Identification of BRCA1 Associated Molecules

We also identified a new BRCA1-associated protein by a proteomic study, and are currently examining its function. Interestingly, this protein localizes at the centrosome, suggesting that it functions in mitosis with BRCA1.

5. Conclusion

It was recently reported that the inhibition of poly(ADP-ribose) polymerase 1 (PARP1), a critical enzyme in the SSB repair pathway, leads to a severe, highly-selective toxicity in *BRCA1*-deficient cells [13]. This seems to be because PARP inhibition results in unrepaired SSBs, giving rise to DSBs. These provide new concepts for the development of potential cancer treatments. The reduced expression or function of BRCA1 is observed in sporadic cancers, suggesting that the precise analysis of BRCA1-related proteins will contribute to the discovery of novel molecular targets for cancer chemotherapy.

In cancer treatment, it is important to select a treatment method and predict its effectiveness in individual cancer patients. Molecular targeting therapies are focus of investigation with a view to developing novel approaches to cancer control. New biomarkers are similarly essential for improved cancer diagnostics. Our research has focused on the identification and functional analysis of critical proteins mediating carcinogenesis and on the development of diagnostic biomarkers and molecular targets for therapies for personalized medicine. Molecular imaging techniques will be useful tools for these aims.

Acknowledgments

This study was supported by Grants-in-Aid from the Ministry of Education, Science, Sports and Culture of Japan. N. Chiba and L. Wei also acknowledges the support of Tohoku University Global COE Program "Global Nano-Biomedical Engineering Education and Research Network Centre".

References

1. Y. Miki, J. Swensen D. Shattuck-Eidens, P. A. Futreal, K. Harshman, S. Tavtigian, Q. Liu, C. Cochran *et al.*, A strong candidate for the breast and ovarian cancer susceptibility gene BRCA1. *Science* **266**, 66–71 (1994).
2. N. Chiba and J. D. Parvin, Redistribution of BRCA1 among four different protein complexes following replication blockage. *J Biol Chem.* **276**, 38549–38554 (2001).
3. N. Chiba and J. D. Parvin, The BRCA1 and BARD1 association with the RNA polymerase II holoenzyme. *Cancer Res.* **62**, 4222–4228 (2002).
4. F. You, N. Chiba, C. Ishioka and J. D. Parvin, Expresssion of an amino-terminal BRCA1 deletion mutant causes a dominant growth inhibition in MCF10A cells. *Oncogene.* **23**, 5792–5798 (2004).
5. L. M. Starita, A. A. Howitz, M-C. Keogh, C. Ishioka, J. D. Parvin and N. Chiba, BRCA1/BARD1 ubiquitinate phosphorylated RNA polymerase II. *J Biol Chem.* **280**, 24498–24505 (2005).
6. L. Lan, S. Nakajima, Y. Oohata, M. Takao, S. Okano, M. Masutani, S. H. Wilson and A. Yasui, *In situ* analysis of repair processes for oxidative DNA damage in mammalian cells. *Proc.Natl.Acad.Sci.USA* **101**, 13738–13743 (2004).
7. L. Lan, S. Nakajima, K. Komatsu, A. Nussenzweig, A. Shimamoto, J. Oshima and A. Yasui, Accumulation of Werner protein at DNA double-strand breaks in human cells. *J Cell Sci.* **118**, 4153–4162 (2005).
8. W. W. Au and B. R. Henderson, The BRCA1 RING and BRCT domains cooperate in targeting BRCA1 to ionizing radiation-induced nuclear foci. *J Biol Chem* **280,** 6993–7001 (2005).
9. D. W. Abbott, M. E. Thomson, C. Robinson-Benion, G. Tomlinson, R. A. Jensen and J. T. Holt, BRCA1 expression restores radiation resistance in BRCA1-defective cancer cells through enhancement of transcription-coupled DNA repair. *J Biol Chem.* **274**, 18808–18812 (1999).
10. F. Le Page, V. Randrianarison, D. Marot, J. Cabannes, J. Feunteun and A. Sarasin, BRCA1 and BRCA2 are necessary for the transcription-coupled repair of the Oxidative 8-Oxoguanine lesion in human cells. *Cancer Res.* **60**, 5548–5552 (2002).
11. F. E. Kleiman, F. Wu-Baer, D. Fonseca, S. Kaneko, R. Baer and J. L. Manley, BRCA1/BARD1 inhibition of mRNA 3' processing involves

targeted degradation of RNA polymerase II. *Genes & Development* **19**, 1227–1237 (2005).
12. W. Wu, H. Nishikawa, R. Hayami, K. Sato, A. Honda, S. Aratani, T. Nakajima, M. Fukuda and T. Ohta, BRCA1 ubiquitinates RPB8 in response to DNA damage *Cancer Res.* **67**, 951–958 (2007).
13. H. Farmer, N. McCabe, C. J. Lord, A. N. Tutt, D. A. Johnson, T. B. Richardson, M. Santarosa, K. J. Dillon, I. Hickson, C. Knight, N. M. Martin, S. P. Jackson, G. C. Smith and A. Ashworth, Targeting the DNA repair defect in BRCA mutant cells as a therapeutic strategy. *Nature* **434**, 917–921 (2005).

DEVELOPMENT OF A TRANSCUTANEOUS ENERGY TRANSMISSION SYSTEM FOR ADVANCED MEDICAL APPLICATIONS

KENTARO KATO [2)], KOTA SHINOHE [1)], TETSUYA TAKURA [1)],
FUMIHIRO SATO [1,2)], HIDETOSHI MATSUKI [2)*]

1) Department of Electrical and Communication Engineering, Tohoku University, 6-6-05, Aoba, Aramaki-aza, Aoba-ku, Sendai, Miyagi 980-8579, Japan

2) Department of Biomedical Engineering, Tohoku University, 6-6-05, Aoba, Aramaki-aza, Aoba-ku, Sendai, Miyagi 980-8579, Japan

In this paper, we report the progress of several studies conducted in the Matsuki laboratory, including the developments of implanted functional electrical stimulation (FES) systems, functional hyperthermia, transcutaneous power-signal supply systems for an implantable total artificial heart, power supply systems for a rechargeable cardiac pacemaker, and temperature control for an SMA artificial anal sphincter. In the section on the implanted FES system, we describe the safety and high-performance rehabilitation of lost motor function. Duplex communication between a mounted system and implanted devices is an essential technology in meeting this goal. In the section on the real-time internal radiation dose measurement system, we describe the wireless communication system, and a circuit that obtains radiation dose data.

Keywords: contactless energy transmission, magnetic core, high frequency, magnetic flux density.

1. Introduction

We are examining the *in vivo* application of electromagnetic fields with regard to biomagnetism and energy management. From the perspective of electrical engineering, the equipment power source is an important issue for a majority of medical devices. Power savings (*i.e.*, low power consumption) and wireless capability are keys to promoting the widespread use of medical equipment in a variety of areas, including social welfare and nursing. The next generation of electrical equipment need not be connected directly to a power source — contactless power transfer will benefit medical equipment. Being able to use equipment with no internal energy source should dramatically improve the envisioned health care system. We expect such equipment to contribute greatly

[*] Hidetoshi Matsuki is a Tohoku University Global COE Member.

to medical treatment in the 21st century. Based on this viewpoint, we conducted various studies in the developments of implanted functional electrical stimulation (FES) systems, functional hyperthermia systems, power-signal supply systems for implantable total artificial heart and rechargeable cardiac pacemaker, and temperature control for an SMA artificial anal sphincter. This paper shows the progress of several studies in the Matsuki laboratory: an implanted FES system, and a real-time internal radiation dose measurement system.

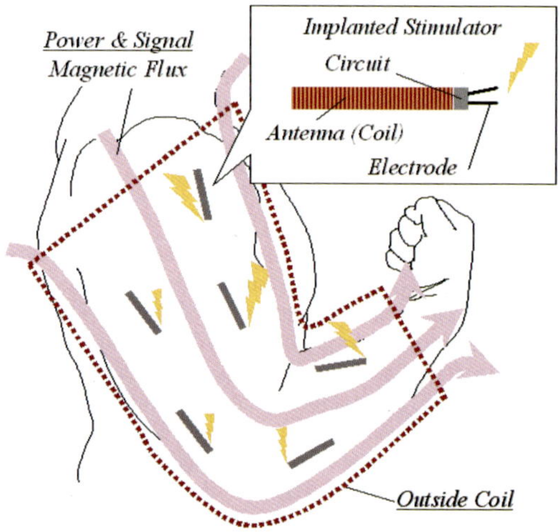

Figure 1. Direct Feeding Method.

2. The Implanted FES System

2.1. *FES*

When people suffer spinal code injures in an accident, the control signals from their brain are not transmitted to their muscles, and hence, they suffer paralysis. In some cases, however, their muscles can be constricted by applying electrical stimulation because the muscles retain their excitability. A control pattern of these electrical stimulations enables rehabilitation of their motor functions. This therapy is called FES for patients with paralysis. The implanted direct feeding method (DFM) is one FES technique, in which small stimulators are implanted near the target points, and electrical stimuli are applied to them (Fig. 1). Implanted stimulators have an antenna, which consists of coils and capacitors. The driving and stimulus energy and stimulus pattern signals are applied through mounted coil using magnetic coupling. This allows high-precision stimulation without percutaneous points and long wires. In the implanted DFM, feedback information from the implanted stimulating system is necessary to confirm system operation and its safe usage, or to enable the DFM of FES to rehabilitate more complex motor functions. To date, however, we have not developed an implanted unit that has such a function. In this study, we show an implanted antenna that was invented for duplex communications. We report on the proposed advanced antenna and experimental communication results to verify its capability.

2.2. *Principle of MCDR*

We proposed a magnetic connective dual resonance (MCDR) antenna for the implanted duplex communication antenna. The MCDR antenna design is shown in Fig. 2. It has two resonance circuits connected magnetically to allow it to function as in duplex communications.

When the MCDR antenna sends a signal, the current, which is proportional to the signal voltage, flows in Lp, generating magnetic flux. The signal voltage is not sufficiently large for the magnetic flux from Lp to transmit a signal, therefore, the magnetic flux from Lp then enters the serial resonant circuit Rs, which resonates the induced current. Meanwhile, at the receiving end, capacitor Cp resonates with the circuit Rp and a large signal voltage is applied to Cp.

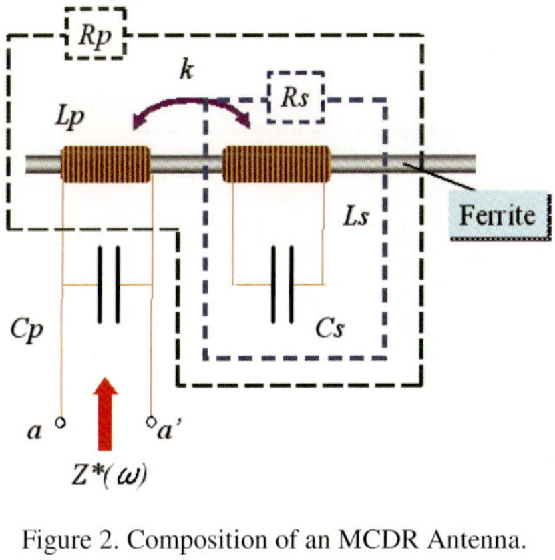

Figure 2. Composition of an MCDR Antenna.

2.3. *Impedance and Optimal Coupling Coefficient*

The impedance at the view points of a-a' is given by Eq. (1), where the resistances of coils have not been considered. The antenna can transmit or receive signals at the frequency which the impedance becomes zero or infinity. Therefore, Eq. (1) gives the transmitting frequency fio and receiving frequency foi. Because these frequencies are real numbers, the MCDR antenna can operate as sending and receiving modes. The capacitance values are determined by the following equations.

$$Z^*(\omega) = j\omega Lp \frac{1-\omega^2 CsLs(1-k^2)}{1-(CpLp+CsLs)\omega^2 + CpLpCsLs(1-k^2)\omega^4} \quad (1)$$

$$f_{io} = \frac{1}{2\pi\sqrt{LsCs(1-k^2)}} \quad (2)$$

$$f_{oi} = \frac{1}{2\pi}\sqrt{\frac{CpLp+CsLs \pm \sqrt{(CpLp-CsLs)^2+4CpLpCsLsk^2}}{CpLpCsLs(1-k^2)}} \quad (3)$$

If the coupling coefficient between Lp and Ls is too small, the MCDR antenna cannot send the signal. When it is larger, a larger magnetic energy is transmitted from Lp to Ls, but if the coupling is too strong, it increases the mutual inductance between Lp and Ls, and slows the performance. Therefore we analyzed the relationship between the coupling coefficient and received voltage. We analyzed the voltage at R_{load} while varying the coupling coefficient between Lp and Ls. Figure 3 shows the simulated circuit. The resonant frequencies are aimed at below; fio = 2.0 MHz and foi = 1.6 MHz. The inductance and resistance values are determined by experimentation Cp and Cs depend on Eqs. (4) and (5).

$$Cs = \frac{1}{(2\pi f_{io})^2 Ls(1-k^2)} \quad (4)$$

$$Cp = \frac{1}{(2\pi f_{oi})^2 Lp}\frac{1-(2\pi f_{oi})^2 LsCs}{1-(2\pi f_{oi})^2 LsCs(1-k^2)} \quad (5)$$

The result is shown in Fig. 4. Communication is the easiest when k = 0.45 (transmission from the implant) and k = 0.20 (reception at the implant). Although both sending and receiving showed higher performance at about 0.33, we selected 0.45 because of the requirement for lower energy in implanted devices than those outside the body.

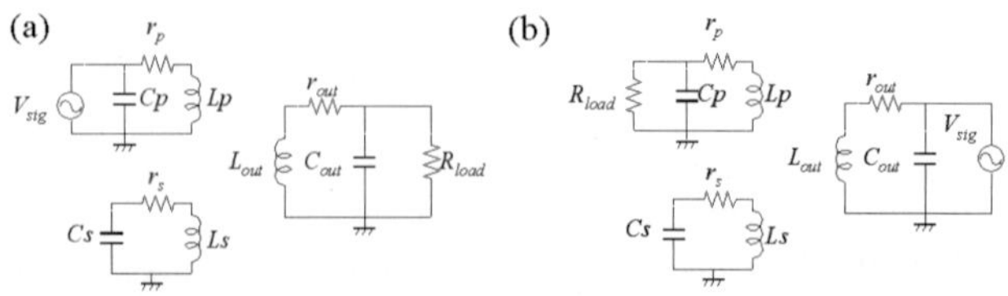

Figure 3. Simulated Circuit; (a) Sending (Inside to Outside), and, (b) Receiving (Outside to Inside).

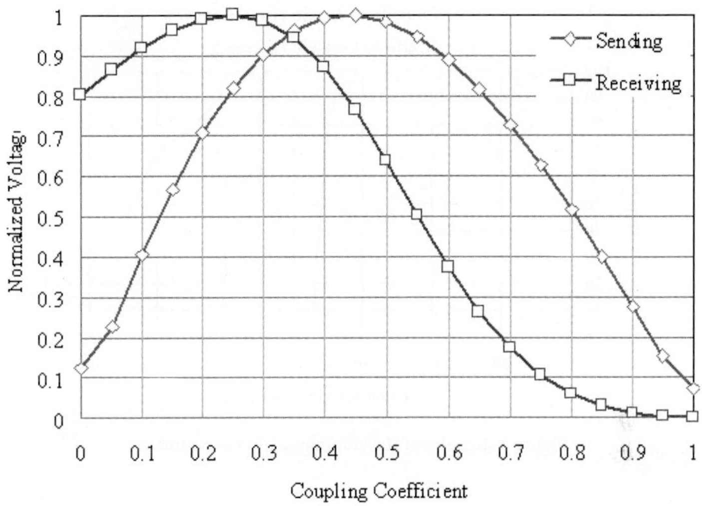

Figure 4. The optimal coupling coefficient.

2.4. *Experiment and Results*

We made an MCDR antenna using the above criteria. The capacitors were mounted on the back of the board, and the ferrite core (N-120L) was 0.7 × 0.7 × 10 mm. The distance between Lp and Ls was 2 mm and the coupling coefficient was 0.465. We experimented with reception at nearly 1.62 MHz and transmission at nearly 2.02 MHz.

To demonstrate the communication ability of the MCDR antenna, we used a flat solenoid coil as the mounted coil [4]. The MCDR antenna was set at a location and angle at which the communication is the easiest. Although there are some problems with displacement or rotation of implanted elements, these are solved by the mounted coil and we did not evaluate them. While alternating the distance between the MCDR antenna and the flat solenoid coil, we transmitted 10000 bits and measured the bit error rate (BER) for each distance. The modulation method was phase shift keying (PSK), the driving voltage was 3.3 V, and the baud rate was 115.2 kbps.

The results are shown in Fig. 5. In this experiment, we obtained the result that the optimum receiving distance was 90 mm and the transmission distance was 60 mm. In the clinical usage, implanted elements are buried at a depth of 10 to 20 mm. Therefore, these antennae are suitable for the DFM for FES.

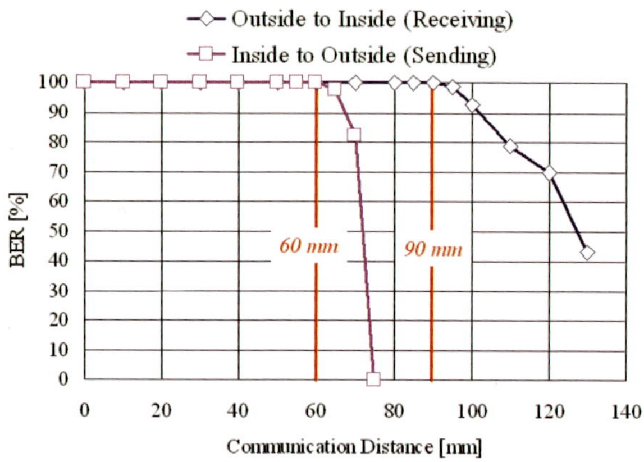

Figure 5. Results of the communication experiment.

3. Real-time Internal Radiation Dose Measurement System

3.1. *Purpose and Outline*

We propose a the real-time internal radiation dose measurement system, as shown in Fig. 6. This system uses a CdTe detectors, which is a semiconductor detector, as an X-ray detector. The CdTe detector has a Schottky barrier diode structure, which provides good energy resolution. Ge and Si detectors are also available (other semiconductor detectors), however, they must be used at extremely low temperature to reduce leakage current and noise, whereas a CdTe detector can be used at room temperature. When X-rays are radiated to a CdTe detector, electrons and holes are generated and move to electrodes by potential difference. In other words, a CdTe detector outputs a charge as the radiation dose data. We measured output of the circuit when radiating the detector with X-rays at a distance of 100 cm, while changing the radiation dose rate from 1 to 6 Gy/min. Figure 7(a) and (b) show the output of the circuit. Figure 7(b) shows an expanded view of one current pulse shown in Fig. 7(a). From Fig. 7(a), we found that the output current waveforms were periodic pulses. Using the output voltage and circuit resistance, the maximum value of the current was determined to be 60 μA.

The real-time internal radiation dose measurement system consists of an implantable dosimeter and a wireless communication system. When a tumor is radiated with X-rays by an external radiation machine, such as LINAC (Medical Linear Accelerator), the radiation dose is measured by an implanted dosimeter with a CdTe detector. The radiation dose data is converted into digital data and transmitted outside the body by the wireless communication system inside the body. The measurement and transmission of the internal radiation dose data are carried out in real time during radiation therapy. This feedback of the transmitted data from a control room to an external irradiation machine allows exact irradiation. An implantable dosimeter includes a CdTe detector, an internal coil and circuit. It is injected near the tumor with a syringe or a catheter. After radiation therapy, it is not removed—it is injected so that it does not task the body of a patient, and surgery is needed to remove it. The wireless communication system should operate over a range of at least 200 mm.

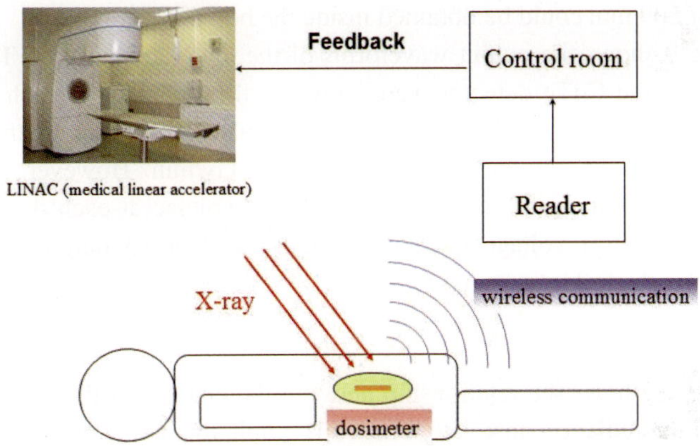

Figure 6. Real-time internal radiation dose measurement system.

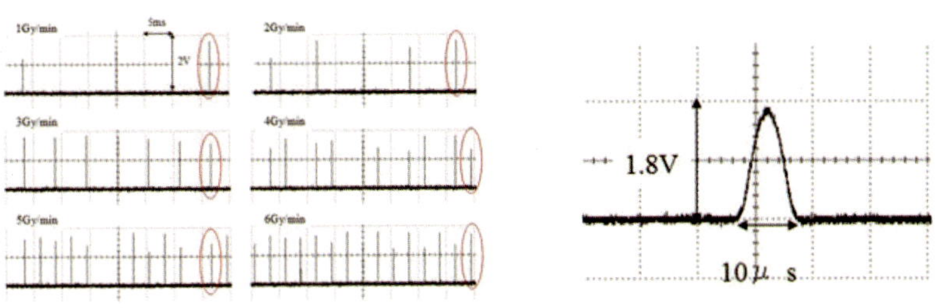

(a) Current waveforms at each dose rate. (b) Expanded view of one current pulse.

Figure 7. Current waveform from the CdTe detector.

3.2. *Experiment and Results*

Figure 8 shows the communication coverage area. When the external coil was located at the origin, that is to say, the dislocation in the directions of X- and Y-axis was zero, the communication distance was 300mm along the Z-axis. When the communication distance was the required 200 mm, the permissible dislocation was 90 mm along the X-axis, and was 200 mm along the Y-axis. For rotation, communication was possible when the angle was 90°. Considering symmetry along the X- and Y-axis indicates that communication was possible through any angle. When immersed in the physiological salt solution, the communication area was same as that in the air, and communication by magnetic fields was not affected. In summary, we confirmed that the communication distance of 200 mm could be obtained inside the body.

Figure 9 shows the output waveforms of the integrating circuit. The output charges from the CdTe detector were stored in the capacitor at each dose rate and all stored charges were discharged by reset pulses. For example, three pulses were integrated for a radiation dose rate of 2 Gy/min. However, there is a constant offset voltage of 250 mV in the output voltage at each dose rate. By removing the offset voltage outside the body, the correct output charges are calculated by the following equation.

$$Q_{out} = C(V_{out} - V_{offset}) \tag{6}$$

Figure 10 shows the relationship between the charges of the CdTe detector output and the radiation dose rate. From this graph, the output charges change in proportion to the radiation dose rate. The radiation dose rate delivered to the tumor is determined by the output charges from the CdTe detector.

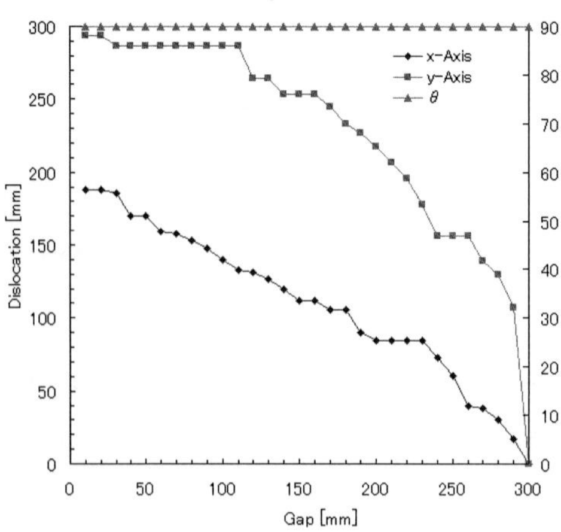

Figure 8. Communication area (characteristized by dislocation and rotation).

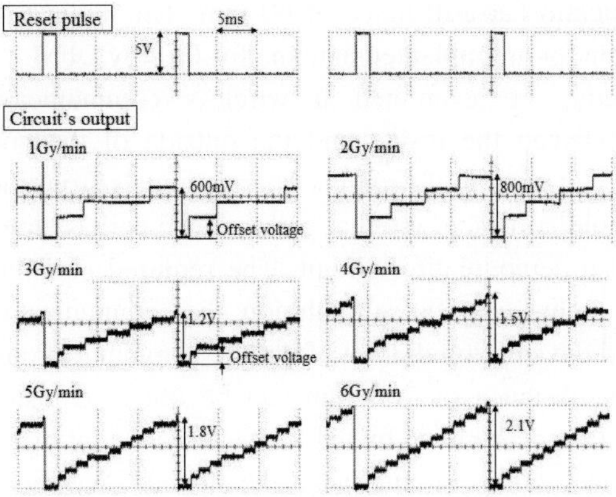

Figure 9. Output waveforms of the integrating circuit at each dose rate.

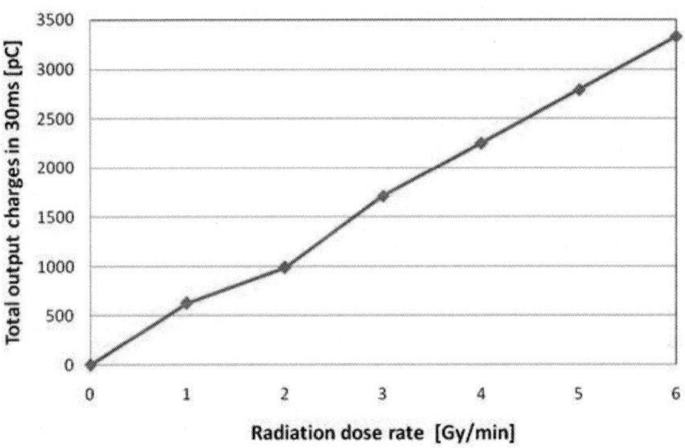

Figure 10. Relationship between the charges from the CdTe detector output and the radiation dosage rate.

4. Conclusion

In the DFM for FES, a duplex communication function of the implanted elements is absolutely necessary for efficient rehabilitation and safe therapy. To realize this function, we proposed the MCDR antenna, which has two resonant circuits connected magnetically. We analyzed and found the optimum coupling coefficient for duplex communication. In addition, by analyzing the impedance of the MCDR antenna, we obtained the design specifications. With this design, we made the MCDR antenna and conducted a communication experiment, which verified the duplex communication ability of the MCDR antenna and achieved duplex communication at a distance of 60 mm. This indicates that the MCDR antenna is suitable for an implanted antenna for FES by DFM.

In this study, we examined a wireless communication system that communicates between the inside and the outside of the body, and a signal amplification circuit to amplify minute charge signals from the CdTe detector. We used a solenoid coil as an internal coil, a plane coil as an external coil, PSK modulation, and a demodulation circuit. The required communication distance was realized. In a physiological salt solution, the communication area was same as that in air, and the solution did not affect the communications. In other words, a communication distance of 200 mm can be obtained inside the body. The radiation dose rate is determined by the total charge output by the CdTe detector; therefore, an integrating circuit was developed. We confirmed that the integrating circuit amplified the minute charge signals and the radiation dose rate was determined from the output of the circuit. In the future, the power transmission system to drive the internal circuits needs to be developed, and the influence of the linkage by the power transmission coils on the signal transmission coils should be examined.

Acknowledgments

This work was supported in part by the Global COE Program "Global Nano-Biomedical Engineering Education and Research Network Centre" in Tohoku University.

References

1. D. Prodanov, E. Marani, J. Holsheimer, Functional ElectricStimulation for sensory and motor functions: Progress and Problems, http://www.diagnosticarea.com/publications/FES_Review.html (2008).
2. F. Rattay, *Electrical nerve stimulation: theory, experiments and applications,* Springer-Verlag, Wien (1990).
3. F. Sato, T. Nomoto, H. Matsuki and T. Sato, A new Contactless Power-Signal Transmission Device for Implanted Functional Electrical Stimulation. *IEEE Trans. Magn.* **40** (2004).
4. N. Jinguuji and H. Matsuki, Transcutaneous Signal-Transmission System Using Cross Coils. *J.Magn. Soc. Jpn.* **22**, 745–748 (1998).
5. M. Takahashi, K. Watanabe, F. Sato and H. Matsuki, Signal Transmission System for High Frequency Magnetic Telemetry for an Artificial Heart. *IEEE Trans. Magn* **37**, 2921–2924 (2001).
6. T. Somekawa, T. Takura, F. Sato, H. Matsuki, and T. Sato, Prototype of Full-duplex Communication for Implantable Signal Transmission System of Magnetic Eight-Figure Coil. *J. Magn. Soc. Jpn.* **31** (2007).

DEVELOPMENT OF BIO-IMAGING WITH FUNCTIONAL NANO-OBJECTS

NORIAKI OHUCHI [1,2]*, MASAAKI KAWAI [1], YUU SAKURAI [1], HIDEO HIGUCHI [3], YOSHIO KOBAYASHI [4], KOHSUKE GONDA [2], MOTOHIRO TAKEDA [2]

1) Division of Surgical Oncology, Graduate School of Medicine, Tohoku University, Sendai, Japan

2) Division of Nano-Medical Science, Graduate School of Medicine, Tohoku University, Sendai, Japan

3) Department of Physics, Graduate School of Science, The University of Tokyo, Tokyo, Japan

4) Department of Biomolecular Functional Engineering, College of Engineering, Ibaraki University, Hitachi, Japan

Nano-materials are expected to be used for medical applications such as research on pharmacokinetics and novel therapeutic agents. A high-resolution *in vivo* 3D imaging system with very high-spatial accuracy was established, which enables visualization at the single molecular level. The movement of fluorescent nano-particles in a tumor interstitium and the membrane dynamics were measured in metastatic tumor cells using the *in vivo* imaging.

A thorough understanding of the functional vascular pore size and interstitial pathway to the tumor in a single molecule level is necessary to design to optimize the delivery of macromolecular drugs. Nano-particles with different sizes were injected into mice with a human breast cancer xenograft and the movement of the particles was tracked in the tumor interstitium. The diffusion coefficient and velocity of the particle movements were inversely related to the drug size and position of the particles. The movement of the complexes at each stage was "stop-and-go", due to the convection current and Brownian motion. The image analysis of the delivery processes of single nano-particles with different sizes *in vivo* provides valuable information for therapeutic macromolecules which will be useful to increase therapeutic efficacy.

The dynamics of cell membranes based on their fluidity and morphology are crucial issues for cell motility during the spread of cancer. A metastasis-activating factor in the cell membrane in stromal tissue of a living tumor in mice was labeled with antibody-conjugated quantum dots. The cells located away from vessels exhibited very slow diffusion velocity and no membrane protrusion was seen. In contrast, in the cells neighboring the vessels, the diffusion velocity was 10-300 times faster than that in cells far from the vessels. Moreover, the cells formed an invadopodia, a kind of pseudopodia, in the direction to the vessel. These results show that the cancer cells dramatically change their membrane fluidity during the metastatic process and form invadopodia to migrate directionally, suggesting potential targets for drugs to treat invasive tumors.

---

* Noriaki Ohuchi is a Tohoku University Global COE Member.

A novel contrast medium for X-rays and its distribution in the tumor was also examined. Nano-sized silica coated silver iodide beads exhibited contrast enhancement in the tumor for over one hour. Sentinel node navigation is an important clinical application for fluorescence nano-particles and silver iodide beads. The results in animal models demonstrate the potential of fluorescence and X-ray measurement as alternatives to existing tracers.
Future innovation in cancer imaging by nano-technology will provide extensive pharmaceutical and biological information, not only in clinical fields, but also in basic medical science.

1. Introduction

Nano-materials are widely applied to medical diagnoses and treatments such as research on pharmacokinetics and novel therapeutic agents. Functional nano-particles include quantum dots (QDs), nano-micelle, nano-sized gold particles, silica coated silver iodide beads and oligo-dendrimers. They have unique properties and are potentially useful in medicine. The systemic distribution of the nano-materials is an important issue that should be determined.

Recent studies have proven the specific influence of particle size in malignancies because the tumor vessels have immature pores and increased permeability [1]. This phenomenon is called extended permeability and retention (EPR). So the size of nano-particles is an important parameter in the design anticancer drugs. The size of nano-particles is also important for sentinel node navigation surgery [2].

In cancer metastasis, cells first detach from the parent tumor, migrate and invade into surrounding connective tissue and blood vessels and then spread to other organs through the blood stream [3,4]. To move within tumor tissues, the cancer cells are thought to form the pseudopodia termed filopodia, lamellipodia and invadopodia, driven by actin polymerization, in the direction of cellular migration and invasion [5-9], thus indicating that the membrane dynamics play a fundamental role in cancer metastasis. To determine the membrane dynamics during metastasis, the understanding of *in vivo* membrane fluidity and morphology is very important. A trajectory analysis of membrane protein movement is one of best approaches to estimate membranous fluidity. The membrane protein randomly moves by passage between the compartments (40-700 nm) partitioned by actin filaments underneath the cell membrane [10-12]. The diameter of a filopodia is 100-200 nm [8]. Therefore, to intrinsically understand the membranous fluidity and morphology *in vivo*, imaging with at least several dozen nanometers-spatial resolution is essential.

Previous *in vivo* imaging using luciferase- or GFP-expressing cancer cells clarified the behavior of metastatic cancer cells in living mice [13-16]. However, since the spatial resolution in the imaging was at the micrometer-level, the details of the membrane dynamics in cancer cells *in vivo* are unknown. The bright and photostable QDs, fluorescence nanocrystals are useful for detecting the fluorescent signal with high-spatial resolution [17-19]. Previous studies succeeded in tracking anti-HER2 (human epidermal growth factor Receptor 2) monoclonal antibody-conjugated QDs *in vivo* [20].

This study was designed to clarify three applications of functional nano-particles. The first is to detect *in vivo* movement of nano-particles in the interstitium of a tumor, because the size may affect not only the accumulation of nano-particles but also the movement in the interstitium. The utility of functional nano-particles was established along with the development of novel optical measurement methods for both the nano-scale and macro-scale. The second goal was to develop a method to image tumor cell's membrane protein using antibody-conjugated QDs and visualize the details of membrane fluidity and morphology during metastasis with high-spatial resolution *in vivo*. The third application was to perform contrast enhancement of rat and tumor using a novel X-ray contrast media composed of silica coated silver iodide beads.

2. Materials and Methods

2.1. *Nano-scale Measurement of Fluorescent Nano-particles in the Interstitium*

2.1.1. *Fluorescent nano-particles*

'FluoSpheres®', which are polystyrene microspheres manufactured by Invitrogen, Molecular Probe (Oregon, USA) provided the particles 40 and 100 nm. The Quantum dot 705 ITK kit (Quantum Dot Corp., Hayward CA) was used for particles of 20 nm.

2.1.2. *Cell culture and animal preparation*

A human breast cancer cell line KPL-4 was kindly provided by Dr. J. Kurebayashi (Kawasaki Medical School, Kurashiki, Japan) [2]. KPL-4 cells were cultured in Dulbecco's modified Eagle's medium (DMEM) supplemented with 5% fetal bovine serum under 5% CO_2/air.

A suspension of KPL-4 (1.0 x 10^7 cells / 100 μl DMEM) cells was transplanted subcutaneously into the dorsal skin of female Balb/c nu/nu mice at 5 - 7 weeks of age (Charles River laboratories Japan, Yokohama, Japan). Several weeks after tumor inoculation, mice with a tumor volume of 100 - 200 mm^3 were used for the study. The mice were placed under anesthesia by ketamine and xylazine.

2.1.3. *Measurement setup and detection of single particle movement*

The optical system consisted of an epi-fluorescent microscope, Nipkow lens type confocal unit and an electron multiplier type CCD camera (Fig. 1). This system can capture images of single nano-particles at a rate of 33 ms/frame. The xy-position of the fluorescent spot was calculated by fitting to a 2D-Gaussian curve. A single fluorescent nano-particle could be identified by the fluorescence intensity. In addition, quantitative and qualitative information such as velocity, directionality and transport mode was obtained using the time-resolved trajectories of nano-particles.

The movement of nano-particles in the interstitium of the tumor was detected by the optical system after intravenous administration of fluorescent nano-particles.

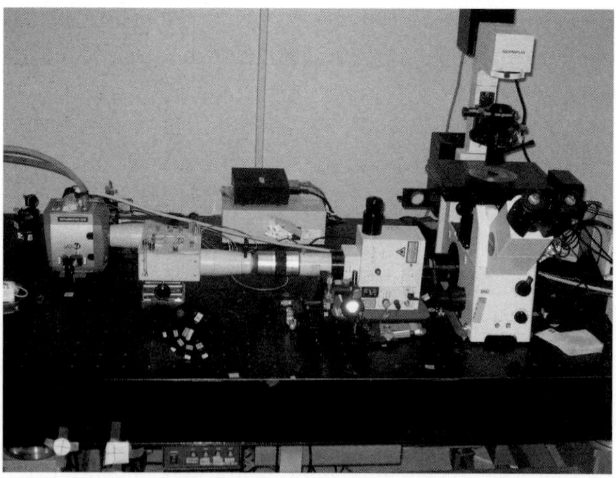

Figure 1. The optical system with a confocal microscope.

2.2. *Imaging of Cancer Metastasis in Living Tumor*

2.2.1. *Membrane protein-antibody-conjugated QDs*

Membrane protein-antibody-conjugated QDs were prepared using a Qdot® 705 Antibody Conjugation Kit (Invitrogen) where the number indicates the emission wavelength.

2.2.2. *Cell culture*

The metastatic human cancer cell line was cultured in DMEM (GIBCO) containing 10% FBS under 5% CO_2/air.

2.2.3. *Optical system with a confocal microscope*

The optical system for observations of the fluorescence of QDs consisted primarily of an epi-fluorescent microscopy (IX-71, Olympus) with modifications, a Nipkow disk type confocal unit (CSU10, Yokogawa) and an electron multiplier type charge-coupled device camera (EM-CCD, Ixon DV887, Andor Technology; Fig. 1). The objective lens (X60, PlanApo, 1.40 NA, Olympus) was used for *in vivo* imaging. QDs were illuminated by a green laser (532 nm wavelength, CrystaLaser). The laser-excited fluorescence was filtered with a 685-725nm band-pass filter or >580nm long-pass filter. Images were taken at the rate of 5 frames per second. Moreover, in order to remove the oscillation of the heartbeat and respiration during the observation, a special stage was originally developed and attached to the above microscopy system.

Cancer cells (1 X 10^6) were suspended in 100μl L-15 (GIBCO) containing 10% FBS and transplanted subcutaneously into the dorsal skin of female SCID mice at 5-7 weeks of age (Charles River). Five to ten weeks after the transplantation, membrane protein-antibody-conjugated QDs were injected into tail vein of the mouse (Final concentration in blood is 5 nM). The mouse was placed under anesthesia with a mixture of ketamine and xylazine and the anesthetized condition was maintained within the course of the imaging session. Minimal surgery was performed to expose the living tumor by removing the skin, with as little damage to surrounding blood vessels as possible. The polyvinyl chloride plate (0.5 mm thickness) containing a small window (10 mm X 10 mm) was mounted on the exposed tumor and both were bonded by adhesive. The tumor-bearing mouse was fixed on a special stage, removing the oscillation, by fixation between the plate and stage and then observed. The position of the QDs on the tumor cell membrane was tracked using a previously described single molecule tracking method [19].

2.3. *Contrast Enhancement of a Tumor by Novel X–ray Contrast Media*

2.3.1. *Preparation of silica coated silver iodide beads*

The silica-coated silver iodide beads (AgI beads) of 40-80 nm were produced by the Stöber method. Potassium iodide and $AgClO_4$ was used for the formation of the silver-iodide core. Silica-coating of the fluorescent microspheres was carried out with ammonia-catalyzed reaction of tetraethylorthosilicate (TEOS) in ethanol–water solution. The ethanol solution of TEOS was added to aqueous PVP solution with stirring after addition of the suspension of the fluorescent microspheres. The hydrolysis of TEOS was initiated by addition of the aqueous ammonia solution to form a silica shell on the microspheres for 24 hr with stirring. Silica coated silver iodide nano-particles have a diameter of 40-80 nm [21].

2.3.2. *Cell culture and animal preparation*

A saline suspension of the silica-coated AgI beads was intravenously administered to rats with xenografts of human breast cancer (KPL-4). The CT images were sequentially acquired at 15, 30 and 60 min after intravenous administration.

These studies were carried out in accordance with the Institutional Animal Use and Care Regulations of Tohoku University, after receiving approval of the Committee on Animal Experiments.

3. Results

3.1. *Nano-scale Measurement of Fluorescent Nano-Particles in Tumors*

The movement of single particles was observed in the perivascular, interstitial and intracellular regions of tumors. The particles could be successfully detected using the measurement system. The HER2 protein expressed in the cancer cells and its dynamics *in vivo* were visualized by QDs and fluorescence polystylene beads on a nano-scale. Fluorescent nano-particles of various sizes were also detected by the optical system. The perivascular region, interstitium and intercellular region in the interstitium of the tumor were investigated.

In the perivascular area, the 20 nm particles exhibited almost uni-directed movement by interstitial flow. In contrast, the 40 and 100 nm particles showed uni-directed movement with increased random diffusion.

In the tumor interstitial region, the movement of the 20 nm consisted of uni-directed movement and random diffusion. The 40 and 100 nm particles showed movement affected only by diffusion.

In the intercellular region, the 20 nm particle movement appeared to be highly restricted. The movement of the 40 and 100 nm particles appeared to represent only diffusion. That indicated Brownian motion.

3.2. *Imaging of Cancer Metastasis in Living Tumor*

The goal was to understand the membrane dynamics during cancer metastasis. The phase contrast image and differential interference image could not be used *in vivo*. Therefore, the labeling of membrane protein with QDs was useful. To investigate whether this method allows visualization of the membrane microarchitecture in cells, the cultured cells were stained with membrane protein antibody-conjugated QDs. The fluorescent image of QDs showed spotty distribution in a single image with a 200 ms-exposure time. In the image, the outline of cell is not clear. However, since the membrane proteins move randomly on the cell membrane, the outline image of cell membrane could be obtained by making a superexposed image (200ms X 25 frames). When the fluorescent superexposed image and phase contrast image in the same cell were compared, the filopodia, a typical component of the cell microarchitecture, was observed in both images (Fig. 2), indicating this method is useful for visualization of membrane morphology *in vivo*.

Next, to image the membrane dynamics of metastatic cancer cells *in vivo*, a tumor-bearing mouse was prepared by transplantation of metastatic human breast cancer cells. Five to ten weeks later after transplantation, the membrane protein antibody-conjugated QDs was injected into tail vein of the mouse and then living tumor cells were observed following cancer metastasis: (1) in cells far from a blood vessel and (2) cells near a blood vessel.

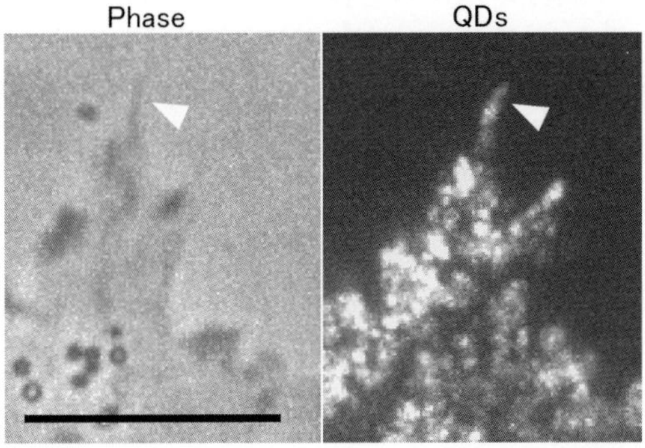

Figure 2. Cultured cell labeled with membrane protein antibody-conjugated QDs. White arrow heads show a filopodia in a cell. Excitation was 532 nm. Emission was >580nm. Superexposured time was 5 s. Bar, 10μm.

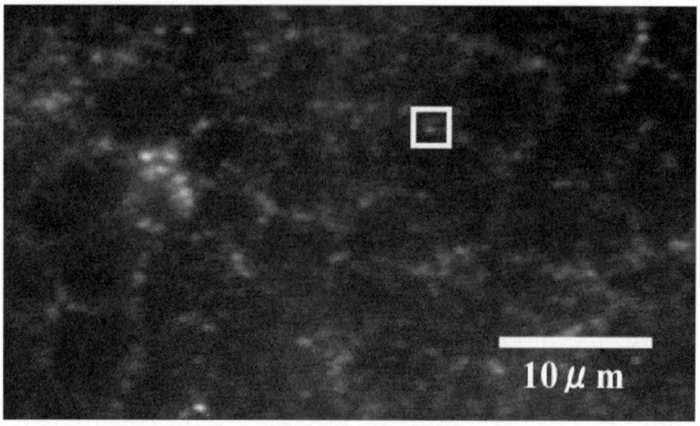

Figure 3. Cells far from the blood vessels labeled with membrane protein antibody-conjugated QDs. A white square shows a representative image of QDs bound to membrane proteins on the edge of tumor cell. Excitation was 532 nm. Emission was >580nm. Exposure time, 0.2s.

(1) Cells far from vessels were visualized. In the cells, little active membrane dynamics was seen. The diffusion velocity of the membranous protein as typified by the white squares (Fig. 3) was 70 nm^2/s, which was 700-1500 times slower than that of the cells *in vitro* (data not shown). In Fig. 3, when a minimally-mobile particle was tracked, the standard deviation of the QDs-position was 17.2 nm for the X axis and 18.2 nm for the Y axis, indicating that this imaging method achieves < 18 nm-spatial resolution *in vivo*. In addition, in all of the *in vivo* imaging analyses, the most lateral QDs in the cell image superexposed over several seconds was defined as the QDs on the cancer cell membrane.

(2) Cells near a blood vessel were visualized. The cell formed an invadopodia, which plays a critical role in migration within the stromal connective tissue. The diffusion velocity of the membranous protein in the cells was 10-300 times faster than that of cells far from the vessels (Fig. 3). Moreover, in the cells near a vessel, the diffusion velocity on the invadopodia was 30 times faster than that on the other membranous regions in the same cell. These results suggest that, in the cells near vessel, faster diffusion velocity of membrane protein is involved in the invadopodial formation.

3.3. *Contrast Enhancement of a Tumor by Novel X–ray Contrast Media*

Enhancement of a tumor was visualized after administration of AgI beads. The Hounsfield number of the tumor gradually increased and achieved a maximum value in 60 min.

4. Discussion

This study captured the specific movement of nano-particles in various sizes in live mice after nano-particles had been injected into the tail vein of mice. In the former study six stages of drug delivery are observed, 1) vessel circulation, 2) extravasation, 3) movement into the extracellular region, 4) binding to HER2 on the cell membrane, 5) movement from the cell membrane to the perinuclear region after endocytosis and 6) in the perinuclear region [4]. That study suggests that there are many stages to be considered for better drug delivery. The interstitium plays a very important role for drug delivery, because it is the scaffold of the tumor cells and the structure of the interstitium that defines the interface of tumor cells to drugs. The movement of the particles in each process was also found to be "stop-and-go", *i.e.*, sometimes the particles stay within a highly restricted area and then move suddenly after release. This indicates that the movement was promoted by a motive power and constrained by both the 3D-structure and protein-protein interactions. The motive power of the movement was produced by the diffusion force driven by thermal energy [2,21,22], interstitial lymph flow and active transport by motor proteins [23]. The cessation of movement is most likely induced by a structural barricade such as a matrix cage [2,21,22] and/or a specific interaction between proteins, *e.g.*, an antibody and HER2 [22], motor proteins and rail filaments such as actin filaments and microtubules [23]. That study strongly suggests that the size of drug carrier can be an important drug delivery modifier. If a carrier in size of 20 nm is used, it may reach cells soon and immediately diminish. Larger size can stay in the interstitium longer. Smaller particles can be used for drugs for treatment to allow drugs to access cancer cells. Larger particles can be used for the localization of a malignant lesion for surgical or radiological treatment.

There have been many different approaches to tumor-targeting "nanocarriers" including anti-cancer drugs, for passive targeting such as Myocet [23], Doxil [24] and for active targeting such as MCC-465 [25], anti-HER2 immuno-liposome [26]. The biological behavior of nano-carriers, including such crucial features as their transport in the blood circulation, interstitial behavior, cellular recognition, translocation into the cytoplasm and the final fate in the target cells, still remains poorly understood. These results suggest that the transport of nano-carriers can be quantitatively analyzed in the tumors of living animals by the present method.

The results of imaging of the cellular movement showed the cells far from vessels did not show active membrane dynamics. The reason for this might be that the amount of a membrane dynamics- activating stimulant, which is derived from blood component, is very low and of the region contains closely-attached cells and extracellular matrix that inhibit the membrane fluidity [27]. The cells near the vessel were attracted to vessel by the stimulant, formed a invadopodia in the direction of vessel. The concentration gradient of the stimulant through stromal extracellular matrix might induce formation of local invadopodia. Membrane fluidity on the invadopodia was 30 times faster than other membrane region. Since the diffusion velocity of membrane protein was controlled by compartmentalization by the actin filaments, the increased activity of actin dynamics in the invadopodia might decrease the compartmentalization and accelerate the diffusion velocity of the membrane protein.

The *in vivo* imaging in this study will allow researchers to achieve an intrinsic understanding of the membrane fluidity and morphology of metastatic cancer cells. These findings could be never presumed from only *in vitro* data. Therefore, this study opens new avenues for examination of biological phenomenon *in vivo*.

Silica coated silver iodide beads had contrast enhancement in a xenograft of breast cancer. In an earlier study, the slow increase in contrast enhancement of the liver and spleen was attributed to sedimentation of silica coated silver iodide beads in the capillary system. Enhancement of the tumor may therefore be attributed to EPR other than the sedimentation of silica coated silver iodide beads in the capillary system.

5. Conclusions

In vivo fluorescent imaging of nano-particles and antibody conjugated QDs was performed at a nano scale in mice. In addition, sentinel nodes were detected by the silica coated fluorescent beads and contrast enhancement of the xeno-grafts of the tumor by silica coated silver iodide beads for X-CT in rats at a macro scale. Fluorescent measurement at the nano scale provides essential information for the development of drug delivery systems. In addition, CT images of tumor were acquired with novel contrast media with a unique advantage. Nano-materials for medicine would be a great aid to improve the made to order medicine by their highly sensitive and super selective property for diagnoses. Advanced sensing technologies such as single molecular imaging techniques are also required to make the best use of the functional nano-materials to achieve hyper sensitive and super selective imaging. These novel advanced nano-technologies may therefore revolutionize medicine in the near future.

Acknowledgments

This work is supported by Global Center of Excellence (GCOE) Program "Global Nano-Biomedical Engineering Education and Research Network Centre", the Japan Society for the Promotion of Science. This work was also supported by Grants-in-Aid for Scientific Research in Priority Areas from the Japan MEXT (H.H.), a Grants-in-aid for Research Project, Promotion of Advanced Medical Technology (H18-Nano-001) from the Ministry of Health, Labor and Welfare of Japan (N.O.), Special Coordination Funds for Promoting Science and Technology of JST (K.G. and H.H.), and CREST of JST (H.H.).

References

1. M. R. Dreher, W. Liu, C. R. Michelich *et al.*, *J Natl Cancer I* **98**, 335–344 (2006).
2. J. Kurebayashi, T. Otsuki, C. K. Tang *et al.*, *Br J Cancer* **79**, 707–717 (1999).
3. A. F. Chambers, A. C. Groom and I. C. MacDonald, *Nature Rev. Cancer* **2,** 563–572 (2002).
4. P. Schedin, *Nature Rev. Cancer* **6,** 281–291 (2006).
5. T. D. Pollard and G. G. Borisy, *Cell* **112**, 453–465 (2003).
6. E. Sahai, J. Wyckoff, U. Philippar, J. E. Segall, F. Gertler and J. Condeelis, *BMC Biotechnol.* **5**, 14 (2005).
7. H. Yamaguchi, M. Lorenz, S. Kempiak *et al.*, *J. Cell Biol.* **168,** 441–452 (2005).
8. P. K. Mattila and P. Lappalainen, *Nature Rev. Mol. Cell Biol.* **9,** 446–454 (2008).
9. C. Sarmiento, W. Wang, A. Dovas, *J. Cell Biol.* **180,** 1245–1260 (2008).
10. Y. Sako and A. Kusumi, *J. Cell Biol.* **125,** 1251–1264 (1994).
11. N. Morone, T. Fujiwara and K. Murase, *J. Cell Biol.* **174,** 851–862 (2006).
12. N. L. Andrews, K. A. Lidke, J. R. Pfeiffer *et al.*, *Nature Cell Biol.* (2008).
13. W. Wang, J. B. Wyckoff, V. C. Frohlich *et al.*, *Cancer Res.* **62**, 6278–6288 (2002).
14. J. Condeelis and J. E. Segall, *Nature Rev. Cancer* **3**, 921–930 (2003).
15. R. M. Hoffman, *Nature Rev. Cancer* **5**, 796–806 (2005)
16. D. E. Jenkins, Y. S. Hornig, Y. Oei *et al.*, *Breast Cancer Res.* **7**, R444–R454 (2005).
17. W. C. Chan and S. Nie, *Science* **281**, 2016–2018 (1998).
18. D. S. Lidke, P. Nagy, R. Heintzmann *et al.*, *Nature Biotechnol.* **22**, 198–203 (2004).
19. S. Li-Shishido, T. M. Watanabe and H. Tada, *Biochem. Biophys. Res. Commun.* **351**, 7–13 (2006).

20. H. Tada, H. Higuchi, T. M. Wanatabe and N. Ohuchi, *Cancer Res.* **67**, 1138–1144 (2007).
21. T. M. Watanabe and H. Higuchi, *Biophys. J.* **92**, 4109–4120 (2007).
22. A. Kusumi, C. Nakada, K. Ritchie *et al.*, *Annu. Rev. Biophys. Biomol. Struct.* **34**, 351–378 (2005).
23. M. Kross, B. Niemann, U. Massing *et al.*, *Cancer Chemother Pharmacol* **54**, 514–24 (2004).
24. M. E. O'Brien, N. Wigler, M. Inbar *et al.*, *Ann Oncol* **15**, 440–449 (2004).
25. T. Hamaguchi, Y. Matsumura, Y. Nakanishi *et al.*, *Cancer Sci.* **95**, 608–13 (2004).
26. J. W. Park, D. B. Kirpotin, K. Hong *et al.*, *J Control Release* **74**, 95–113 (2001).
27. H. Yamaguchi, J. Wyckoff and J. Condeelis, *Curr. Opin. Cell Biol.* **17**, 559–64 (2005).

DEVELOPMENT OF THE VARIOUS KINDS OF ARTIFICIAL ORGANS AND CLINICAL APPLICATION OF THE NEW DIAGNOSIS TOOL

TOMOYUKI YAMBE [*]

Department of Medical Engineering and Cardiology, Institute of Development, Aging and Cancer, Tohoku University,4-1 Seiryo-machi, Aoba-ku, Sendai, 980-8575, Japan

Aim of the research in the Department of Medical Engineering and Cardiology is the development of the new diagnosis tool and therapeutic tools for the medical fields and their clinical application. Some examples of the research theme are an artificial internal organs based on Nano-technology. And clinical application of a new diagnosis tool for the evaluation of the baroreflex system was carried out in our laboratory. In this paper, recent progress in therapeutic tool and diagnosis tools iswas introduced.

1. Introduction

The aim of the research in the Department of Medical Engineering and Cardiology is the development of new diagnosis and therapeutic tools for medical fields and their clinical application. One example of the research theme in our laboratory is the development of various kinds of artificial internal organs based on nanotechnology, such as an artificial sphincter, myocardium, esophagus, stent, and machine that stops convulsions. As for implantable artificial organs, the space for implantation is restricted. Therefore, micro-device development is indispensable. Nanotechnology and micromachining technology development are very important. At Tohoku University, the development of various artificial organs has been furthered according to the tradition of nano-machine micro-machine development. Various artificial organ projects being conducted at Tohoku University are introduced.

In addition, clinical application of a new diagnosis tool for the evaluation of the baroreflex system was carried out in our laboratory. In this paper, recent progress in therapeutic and diagnosis tools is introduced.

[*] Tomoyuki Yambe is a Tohoku University Global COE Member.

2. Artificial Internal Organs

2.1. *Completely Implantable Artificial Sphincter*

After surgery for colon cancer, several patients must have stoma, and thus, the patients cannot control defecation. This is not very good for their quality of life (QOL). The artificial sphincter that we invented makes it possible for a patient to control defecation. Figure 1 is an illustration of the artificial sphincter.

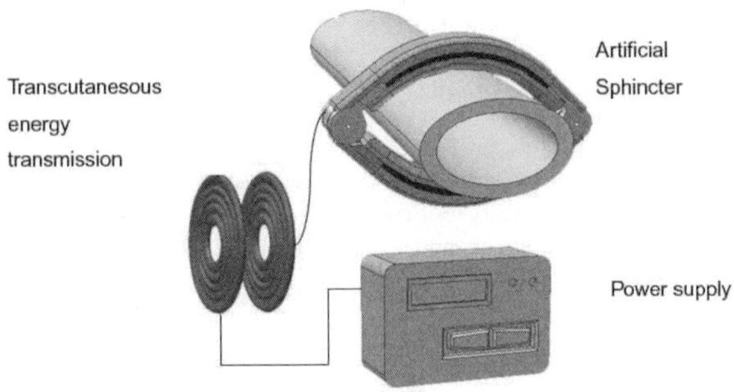

Figure 1. Schematic illustration of the completely implantable artificial sphincter

We used a shape memory alloy. Two shape memory alloy boards were combined. Energy was transmitted by the transcutaneous energy transmission system (TETS). Figure 2 is a photograph of the animal model of the stoma. When a patient goes to the toilet, the patient brings the TETS. An artificial sphincter opens and enables the patient to defecate. The patient can control defecation if this system is used. Thus, a patient's QOL will be greatly improved.

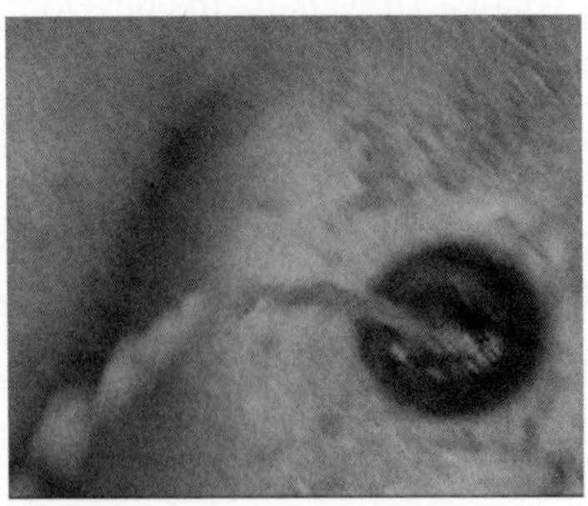

Figure 2. Photograph of an animal experiment of the artificial sphincter opening

2.2. *Artificial Esophagus*

About 10,000 Japanese per year die due to cancer of the esophagus. It is well-known that operations performed for esophagus cancer are difficult, because reconstruction of the esophagus is required. Operating is easy if there is an artificial esophagus. The esophagus moves food by peristalsis. A simple pipe cannot be substituted for an esophagus.

We invented an artificial esophagus in which peristalsis is possible. The artificial esophagus consists of macromolecule material and artificial peristalsis muscles. We used a Gore-Tex artificial vascular graft because this material is suitable for a living body. A human esophagus swallows by peristalsis. For peristalsis to occur, the shape memory alloy ring was used. The coil was made from the fiber of the shape memory alloy, which improved durability by nanotechnology molecular crystal arrangement. The ring of the coil contracted, and peristalsis occurred, pushing the contents of the esophagus through the intestinal tract. An animal experiment using a goat of the same weight as Japanese people was tried. The artificial esophagus replaced the goat's excised esophagus. The peristalsis motion had occurred by the artificial esophagus in the body of the goat. By using an artificial esophagus, operating on the esophagus because of cancer becomes easy. In the future, we can perform operations using an endoscope and replace the diseased esophagus with an artificial esophagus. Since there is little invasion, operations on the elderly will become possible. It is expected that the invention of an artificial esophagus with peristalsis movement will create great progress in esophagus cancer surgical therapy.

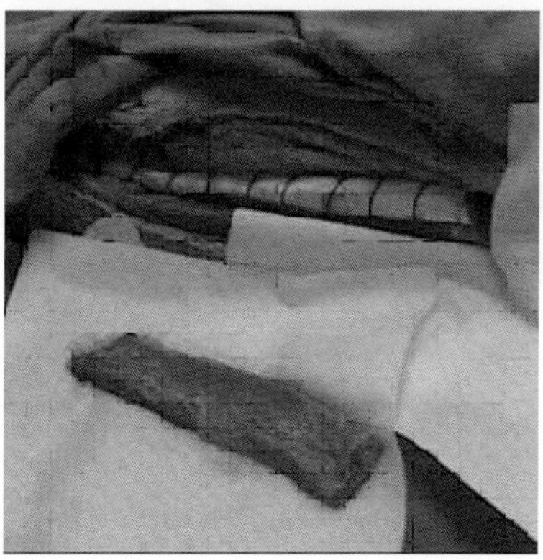

Figure 3. Photograph of an animal experiment using the artificial esophagus

2.3. *Peristalsis Stent with Hyperthermia Function*

Operating on the esophagus due to cancer is one of the most difficult operations even now, although medicine has progressed. One of the most important points is the difficulties of esophagus reconstruction. In an operation, since the stomach and intestines are used as a substitute, the invasion becomes large, and operating on the elderly becomes difficult. Although improvement in a patient's prognosis is expected if the cancer is resectable, there are many patients whose esophageal cancers are too advanced for surgery at the time of diagnosis.

Thus, a peristalsis stent with hyperthermia function for terminal esophageal cancer patients, for whom an operation cannot be conducted, was invented. The peristalsis stent with hyperthermia function has three characteristics: 1) completely noninvasive, 2) hyperthermia on the carcinoma tissue, and 3) peristalsis function. This stent could be an alternative to terminal esophagus cancer therapy.

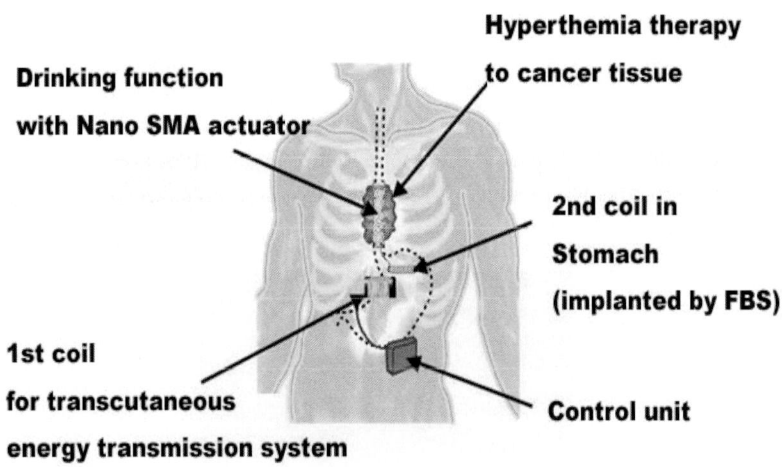

Figure 4. Schematic illustration of the peristalsis stent with hyperthermia

2.4. *Undulation Pump VAD Project*

Rotary blood pumps (RP) are effective as small ventricular assist devices. They may be still more effective if pulsation is possible. We studied an undulation pump (UP), which is a small RP, which can produce pulsation. In Japan, projects to develop an implantable-type UP ventricular assist device (UPVAD) and UP total artificial heart (UPTAH) are underway.

Six universities and several companies are working together in charge of this five-year development project.

Tohoku University is investigating the influence of the UP under development on internal-organ circulation. Goats with the same weight as Asians were used for the experiment. The UP was implanted using the left-heart bypass system. Because of the animal experiments, the blood circulation dynamic state was stabilized, and a sufficient left-heart-supporting effect was observed. We observed a tendency in which myocardial blood flow and kidney blood flow changed due to varied UP mode support. The problem of multiple organ failure is important during the clinical application of a ventricular assist device. Assistance for internal-organ circulation is important in preventing multiple organ failure. It has been concluded that UP may be useful in preventing multiple organ failure.

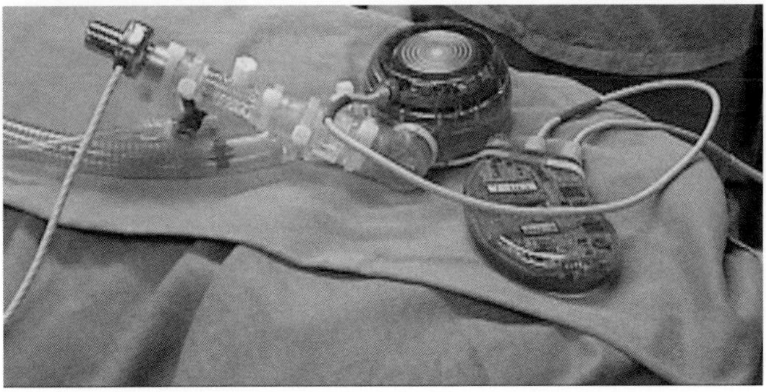

Figure 5. Photograph of an undulation pump VAD

2.5. *Long-term Animal Experiments of the Evaheart Rotary Blood Pump*

Rotary blood pumps are the most desirable VAD system when we consider clinical application for small people, including children. The Eva heart is the rotary blood pump developed by the Sun Medical Co., and chronic animal experiments that tested endurance and anti-thrombogenicity were performed at Tohoku University.

A record for longest survival was obtained: 823 days of survival is the world record for a completely implantable ventricular assist device in a chronic animal experiment.

Figure 6. World record for longest survival of an animal with implantable VAD

2.6. *Artificial Myocardium Using Nanotechnology*

The purpose of this research is to develop a nano artificial myocardium. Therefore, nano sensor and nano control chip development has been performed. An artificial myocardium actuator was also developed using nanotechnology. A control-objectives value setup that imitated the baroreflex system was tried using a nano sensor and a micro control chip. In this research, a nano thin film sensor that adapted diamond-like carbon (DLC) was developed, and a patent was applied for (application for patent 2003-317956).

Outstanding organism affinity can be expected, and application for artificial organs in all fields can be expected. Furthermore, a nano sensor that adapted an optical fiber was also developed, and the animal experiments were successful. Since information, such as each ventricle, can be evaluated simultaneously, the optimal drive of an artificial myocardium is possible. Sunagawa *et al.* reported a new bionic medical treatment method that can increase a heart failure patient's probability of survival by applying biological information and performing circulation dynamic state control. Heart failure medical treatment by pacing both ventricles is the established methodology. A hypothesis is drawn from these two reasons. Hemodynamics are checked by the nano sensor, and it is expected that a patient's prognosis is sharply improvable by controlling an artificial myocardium optimally. The actuator in which micro-machining is also possible is used for the artificial myocardium that this research developed.

An artificial myocardium is a system that assists the pulsation of the heart. The external surface of the heart is equipped with an artificial myocardium. Therefore, similar to the conventional artificial heart, there is no risk of a thrombus, and the artificial myocardium does not have the problem of the durability of an artificial valve. When there is no necessity, an artificial myocardium does not operate, and since circulation is performed only with the heart, improvement in the durability of an artificial-myocardium system is expected.

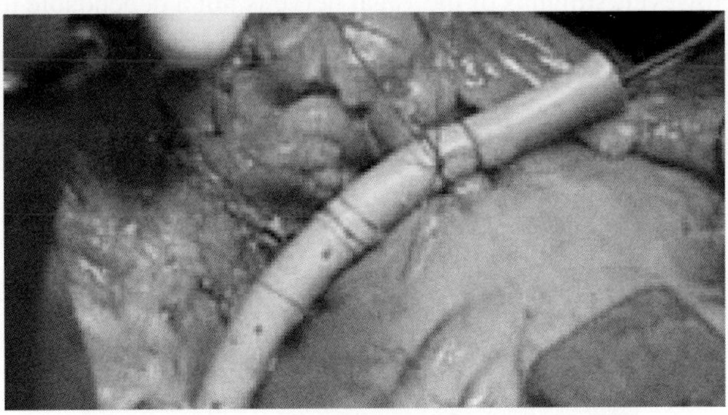

Figure 7. Photograph of the completely implantable artificial myocardium

Currently, artificial myocardium research is under promotion using an electro-hydraulic system in chronic animal experiments.

Furthermore, the molecular crystal arrangement of a shape memory alloy is prepared by applying nanotechnology. Development of the artificial muscles that have improved durability and the speed of response progresses by reducing a hysteresis. Now, cultivation of a heart muscle cell has attracted attention with reproduction medical treatment. Although it is easy to create a cell sheet, 3-dimensional construction is very difficult. Of course, reconstruction of the coronary arteries is needed, and thrombus formation is also a concern.

On the other hand, the nano-level actuator developed with nanotechnology is important. If this succeeds, it is expected that such an actuator will become applicable to various artificial organs.

A transcutaneous energy transmission system using nanotechnology enables the creation a smaller, completely implantable device. Animal experiments were performed, and the support effect was obtained by mechanical assistance.

An artificial myocardium may become a candidate for mechanical assistance in the future.

2.7. *Brain Function Control Units*

About 0.5% to 1% of the general population has epilepsy (NIH Consensus Panel, 1990). Seizures cannot be controlled by medication in 20% of epilepsy patients. A third of these patients need surgical resection of the epileptic focus to control the attacks. However, surgical resection of a part of the brain has a risk of complications. Some patients feel a symptom before suffering epileptic seizures. When a symptom is felt, there is a treatment, which performs vagal nerve stimulus. However, stimulus of the vagal nerve is not a dependable treatment.

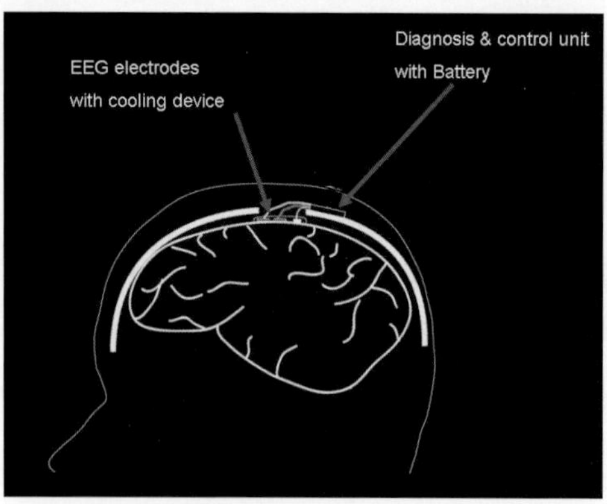

Figure 8. Epilepsy attack control machine

We invented the cerebral partial cooling method and applied for a patent (2004-304964).

Using this method, if a patient feels a symptom of an impending epileptic attack, the patient pushes a switch. The control switch is implanted under the skin. With the switch on, the epileptic focus is cooled, and convulsions are prevented beforehand. A battery is used in patients who suffer only a few attacks. A transcutaneous energy transmission system (TETS) is used in patients who suffer frequent attacks. In this system, if a patient feels a symptom, the TETS will be used. The outer coil is attached to the inner coil under the skin. Energy is supplied, and the focus is cooled. We are currently investigating a prediction algorithm for epilepsy. If prediction is possible, automatic control of epilepsy will become possible. This will be good news for patients with epilepsy.

3. Diagnose the Baroreflex Sensitivity of an Artery

A baroreflex function is the one of the most important factors in patients with hypertension. The heart rate (HR) falls reflexively when blood pressure (BP) rises. In addition, an artery is extended reflexively, when blood pressure rises. Blood pressure returns to a normal value because of heartbeat reduction and blood vessel extension.

The sensitivity of the baroreflex system is calculated at the heartbeat reaction to the change in blood pressure. However, even if the reaction of a heartbeat is measured, the baroreflex function of an artery is not known. Thus, we invented a new method of diagnosing the baroreflex sensitivity of an artery, using the pulse wave velocity (PWV).

Since PWV correlates with the elasticity of an artery, PWV has been used for atherosclerosis diagnosis until now. Therefore, if PWV is used, a change in the elasticity of an artery can be diagnosed. The reaction of an artery can be measured if the elasticity of an artery is diagnosed. Therefore, if PWV is measured, the baroreflex function of an artery can be diagnosed. However, there have been no reports of a method that measures baroreflex sensitivity by PWV, until now.

Therefore, this report is the first in the world. In ten healthy objects, the time series of heart rate variability, fingertip pulse change, and blood pressure were recorded. The obtained data were inputted in a PC using the AD converter.

Preventive medicine is a priority for most governments because of increasing medical and health care expenses. In Japan, metabolic syndrome has recently been identified as playing an important role in the pathophysiology of many diseases. Hypertension, hyperlipidemia, diabetes mellitus, and obesity are important consequences of metabolic syndromes. Among these, hypertension is paramount when considering the disruption of organ function. Thus, prevention of hypertension is very important.

The baroreflex system is a key indicator of hypertensive pathophysiology [12-14]. When the BP increases, the HR decreases, and there is peripheral arterial dilation [12,14]. By decreasing the cardiac output and peripheral arterial resistance, BP returns to normal. Hypertension is a concern in the young as well as the elderly [12-14]. Baroreflex sensitivity is reduced in younger hypertensive patients [12]. However, currently there is no simple and sensitive diagnostic method for measuring arterial behavior in the baroreflex system.

This study describes the development and clinical application of a novel baroreflex diagnosis machine and offers a preliminary consideration of its clinical applicability.

3.1. *Diagnosis of Arterial Baroreflex Sensitivity*

Every medical student studies the baroreflex system as a typical example of homeostasis [12-14]. When blood pressure increases, baroreceptors in the carotid arteries and aortic arch sense the increase in the baroreflex sensitivity. When this information is transmitted to the central nervous system, the HR decreases, and the arteries dilate. These reactions restore normal BP. Baroreflex sensitivity is evaluated by measuring the HR response to the BP changes. The slope of the linear regression line demonstrates the sensitivity of the baroreflex system of the heart.

While the HR response in the baroreflex system can be monitored, no method currently exists to evaluate arterial baroreflex function, possibly because of the difficulty in evaluating vascular tone during wakefulness.

Recently, new methodologies—brachial ankle pulse wave velocity (baPWV) and cardio-ankle vascular index (CAVI)—have been developed to evaluate human arterial stiffness [17-20]. These methodologies noninvasively evaluate arterial wall stiffness using the pulse waveform of the brachial and ankle arteries. These methodologies are based on the premise that pulse wave velocity (PWV) is correlated with arterial wall stiffness. Thus, PWV increases when the arterial wall becomes harder and decreases when the arterial wall softens.

In the baroreflex system, the arterial wall softens in response to an increase in BP, thereby decreasing the vascular resistance. BP will return to normal because of the decrease in resistance. The softness of an arterial wall can be measured by PWV.

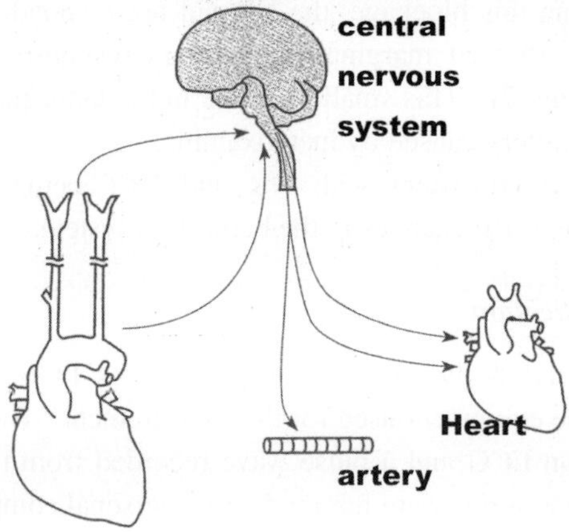

Figure 9. Schematic representation of the baroreflex system

The PWV value could possibly be used to quantitatively measure the baroreflex sensitivity of the arterial wall. PWV can be calculated from the pulse wave transmission time (PTT) and distance. Thus, measurements of PTT and BP permit an evaluation of arterial baroreflex response.

3.2. *Animal Experiments*

To evaluate the autonomic responses of the PTT or PWV, chronic animal experiments were carried out using healthy adult goats. The goats used in the experiment weighed almost the same as the average Japanese person. All experiments were approved by the Ethical Committee of the Institute of Development, Aging and Cancer, Tohoku University. Under anaesthesia, the chest cavity of the goat was opened in the fourth intercostal space. Implants, including an electromagnetic flow meter, electrodes for electrocardiogram (ECG), and a catheter-tip pressure sensor, were inserted into the femoral artery, and a fluid-filled catheter was inserted into the left ventricle. After the chest was closed, the goats were moved to their cages. Measurements were taken when the goats were conscious in the chronic stage.

After intravenous injection with methoxamine, BP suddenly increased. HR reduced in response to the increase in systolic pressure, and PTT increased in response to the BP change. Prolongation of the PTT indicated the softening of the artery. Therefore, PTT and PWV were thought to be indicative of the autonomic response of an artery. To demonstrate the autonomic nervous control of the HR and PTT, the autonomic nerves were blocked using atropine and propranolol. During this blockage, the HR did not respond to the BP change; however, the PTT showed marginal decrease in response to the increase in systolic pressure (Fig. 7). This small decrease in the latter might have been due to hardening of the artery caused by methoxamine.

The results are consistent with HR and PTT being indicative of the autonomic response to BP changes in the baroreflex system.

3.3. *Clinical Application*

PTT and PWV were easily measured by monitoring ECGs and pulse waves. Figure 13 shows the equipment used for the measurement. The newly developed system used only an ECG and a pulse wave recorded from the radial artery or fingertip. These time series were inputted into a personal computer and analyzed quantitatively using custom-developed software. HR was calculated from the reciprocal of the inter-R-wave interval of the ECG signal. PTT was defined as the time interval from the peak of the R-wave to the point at which the pulse wave signal began to increase. HR and PTT were interpolated by cubic spline functions to continuous-time functions, and were resampled every 0.5 s. The utility of this system for the quantitative diagnosis of the baroreflex sensitivity of an artery was recognized by the patent application.

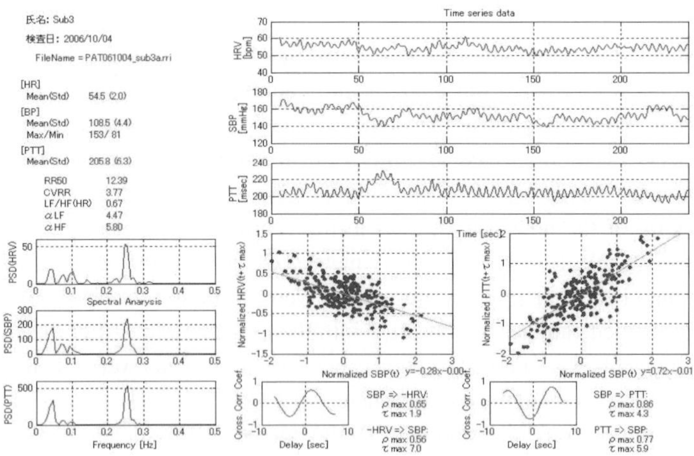

Figure 10. Clinical report of the baroreflex system

The arterial responses were measured in terms of the PWV calculated from the PTT from the heart to an artery. In this system, the HR change corresponding to the BP change in the time series sequence was observed. Delay time was measured by the cross-correlation function. The slope in the changes in BP and HR indicates the sensitivity of the baroreflex system of heart. Furthermore, this system could also measure the sensitivity of the baroreflex system of an artery.

Figure 10 displays an example of a patient report. Upper tracings of the HR, BP, and PTT are shown, and their spectral analysis data are displayed on the left. In the lower right-hand part of the report, cardiac and arterial baroreflex sensitivity is provided along with an analysis of the cross-correlation function.

The clinical research in our study began after ethical committee approval was obtained. So far, the results have shown that our system can successfully detect decreased sensitivity of the baroreflex system in hypertensive patients. We are now analyzing various data of patients with hypertension.

Further examination will be needed using more cases. This new method may be useful for following up patients with hypertension.

Acknowledgments

This work was partly supported by support from the Tohoku University Global COE Program "Global Nano-Biomedical Engineering Education and Research Network Centre."

References

1. Jpn.Pt.Appl. 3910020, Artificial sphincter
2. Jpn.Pt.Appl. 2006-141881: Peristaltic conveying devaice
3. Jpn.Pt.Appl. 2005-335435: multi functional stent
4. Jpn.Pt.Appl. 2005-349025: regenerative scaffold
5. WO2004/112868, artificial myocardium
6. Jpn.Pt.Appl. 2007-82898: transcutaneous energy transmission system
7. China patent 01121110.5, Artificial Sphincter
8. Jpn.Pt.Appl. 2006-142040: doppler flow meter evaluating flow of the bypass graft
9. Jpn.Pt.Appl. 2006-271105: diagnosis system to evaluate baroreflex function
10. Jpn.Pt.Appl. 2006-281664: control acupuncture
11. Jpn.Pt.Appl. 2006-141993: cooling device
12. Z. Novakova, E. Zavodna, H. Hrstkova and P. Honzik, Overweight and decreased baroreflex sensitivity as independent risk factors for hypertension in children, adolescents, and young adults. *Physiol Res.* (2007 May 30, Epub ahead of print).

13. Q. Fu, N. E. Townsend, S. M. Shiller, E. R. Martini, K. Okazaki, S. Shibata, M. J. Truijens, F. A. Rodriguez, C. J. Gore, J. Stray-Gundersen and B. D. Levine, Intermittent hypobaric hypoxia exposure does not cause sustained alterations in autonomic control of blood pressure in young athletes. *Am J Physiol Regul Integr Comp Physiol.* **292**, R1977–R1984 (2007), (2007 Jan 4, Epub).
14. N. Honzikova, Z. Novakova, E. Zavodna, J. Paderova, P. Lokaj, B. Fiser, P. Balcarkova and H. Hrstkova, Baroreflex sensitivity in children, adolescents, and young adults with essential and white-coat hypertension. *Klin Padiatr.* **218**, 237–242 (2006).
15. T. Yambe, Y. A. Kovalev, I. A. Milyagina, V. A. Milyagin, Y. Shiraishi, M. Yoshizawa, Y. Saijo, T. Yamaguchi, M. Shibata and S. Nitta, A Japanese-Russian collaborative study on aging and atherosclerosis. *Biomed Pharmacother.* **58** (Suppl 1), S91–S94 (2004).
16. A. Yamashina, H. Tomiyama, T. Arai, Y. Koji, M. Yambe, H. Motobe, Z. Glunizia, Y. Yamamoto and S. Hori, Nomogram of the relation of brachial-ankle pulse wave velocity with blood pressure. *Hypertens Res.* **26,** 801–806 (2003).
17. T. Yambe, X. Meng, X. Hou, Q. Wang, K. Sekine, Y. Shiraishi, M. Watanabe, T. Yamaguchi, M. Shibata, T. Kuwayama, M. Maruyama, S. Konno and S. Nitta, Cardio-ankle vascular index (CAVI) for the monitoring of the atherosclerosis after heart transplantation. *Biomed Pharmacother.* **59** (Suppl 1), S177–S179 (2005).
18. T. Yambe, M. Yoshizawa, Y. Saijo, T. Yamaguchi, M. Shibata, S. Konno, S. Nitta and T. Kuwayama, Brachio-ankle pulse wave velocity and cardio-ankle vascular index (CAVI). *Biomed Pharmacother.* **58** (Suppl 1) S95–S98 (2004).
19. H. Kumagai, H. Suzuki, M. Ryuzaki, S. Matsukawa and T. Saruta, Baroreflex control of renal sympathetic nerve activity is potentiated at early phase of two-kidney, one-clip Goldblatt hypertension in conscious rabbits. *Circ Res.* **67**, 1309–1322 (1990).
20. A. D. Jose and R. R. Taylor, Autonomic blockade by propranolol and atropine to study intrinsic myocardial function in man. *J Clin Invest.* **48**, 2019–2031 (1969).

PULSE DIAGNOSIS MACHINE AND AUTOGENIC TRAINING

TOMOYUKI YAMBE *

Department of Medical Engineering and Cardiology, Institute of Development, Aging and Cancer, Tohoku University, 4-1 Seiryo-machi, Aoba-ku, Sendai, 980-8575, Japan

In traditional and alternative medicine, pulse diagnosis is one of the most important diagnostic methods. However, there have been no papers reporting medical evidence of a pulse diagnosis. We invented a quantitative diagnosis machine, which could carry out pulse diagnosis scientifically. This machine performs pulse diagnosis using three pressure sensors. The pulse diagnosis is reproduced by various pressures being applied to three pressure sensors on an external surface. Three experimental series, including experiments using a model circulation circuit, an animal experimental series, and a clinical application of the pulse diagnosis machine, were conducted. Chaos theory was used to evaluate the clinical application of the pulse diagnosis machine. For this part of the study, the data obtained using the machine are in agreement with findings from traditional Chinese medicine. Therefore, some aspects of Chinese medicine may have a scientific/medical basis. For the first time, scientific evidence could validate pulse diagnosis using this pulse diagnosis machine. In a clinical setting, we evaluated pulse diagnosis, baroreflex sensitivity and autonomic function during autogenic training. Altered autonomic function was observed, suggesting the usefulness of pulse wave analysis in clinical situations.

1. Introduction

There is a concept called "pulse diagnosis" in Oriental medicine (1-3). An Oriental medicine practitioner places three fingers over the radial artery of the left wrist and right wrist, as shown in Fig. 1. Varying pressure is applied by a finger and a pulse-wave form is sensed manually. Pulse diagnosis using this methodology is made to diagnose various diseases. Arrhythmia, flow rate and pressure information from the heart, for example, can be detected in the pressure wave form of a pulse wave.

The change in arterial elasticity that results from atherosclerosis can also be diagnosed using this technique. Information on the amount of circulating blood, reflecting renal function and viscosity of blood, as well as liver function, are also included in the pulse wave. The blood sugar level accompanying diabetes also influences viscosity. Information on central nervous system function, which

* Tomoyuki Yambe is a Tohoku University Global COE Member.

controls blood vessel movement, and information on changes of the nutrition state associated with alimentary canal disease may also be derived from these pulse waves.

However, the transfer of knowledge relating to the methodology of diagnosis is impaired if trainees do not follow instruction with experience. Despite the thousands of years of tradition behind Oriental medicine, any medical or scientific persuasive power is completely lost if it is promoted without objectivity. E ven if pathological conditions could be diagnosed using traditional medicine in the past, it becomes impossible to completely diagnose conditions without appropriate transfer of knowledge. The maintaining of traditional diagnosis methods relies on generalization, quantification and information sharing.

To continue the use of pulse diagnosis in the modern age, we developed a novel pulse-wave diagnostic machine.

This study aimed at the contemporary use of pulse diagnosis by learning from the past. If a pulse diagnosis machine is used, a scientific quantitative diagnostic method could be introduced into Oriental medicine for empirical use for the first time. Here, we describe the equipment and experiments to evaluate it, and describe the experimental results.

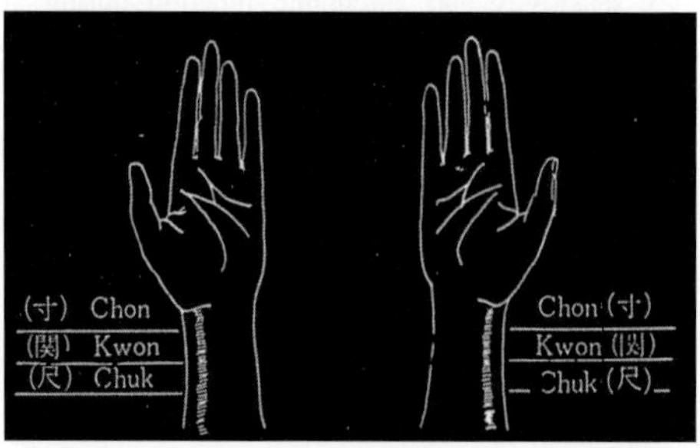

Figure 1. Concept diagram of the Pulse diagnosis

2. Arterial Pulse Waveform

Before we discuss the development of the pulse diagnosis machine, we would like to discuss the physiological meaning of an arterial pulse wave form. Information relating to the physiology of the human body is theoretically included in the pressure wave form of an artery. Therefore, theoretically, various quantitative diagnoses can be embodied in this wave form. The rising foot of an artery wave is generated about 100 ms after an ECG Q wave. It has been medically observed that this delay time is also applicable as an index of cardiac function. When the heart contracts, the aortic valve opens and blood flows into the ascending aorta from the left ventricle. The intra-aortic pressure increases quickly, simultaneously with blood inflow, and a pressure wave is formed. Therefore, important information about a cardiac function is included in the rising foot.

The wave of the first peak in an artery pressure waveform and the percussive wave in the first half of the two waves near the pulse-wave peak, broadly represent the peak of a blood-flow wave. Although this wave has the highest score in healthy young people, the subsequent tidal wave has the highest score in elderly people and atherosclerotic patients. The tidal wave or recoil wave is also referred to as the second wave, catacrotic wave, elastic wave or Spatsystolisher Gipfel. In many people, the main components of the tidal wave are the reflected waves from the peripheral circulation, and a shallow valley (mid-systolic dip) between the tidal wave and the percussive wave. This wave is small in young people and increases with atherosclerosis, and can also be a characteristic of aging or a rise in blood pressure. Moreover, it becomes smaller as the pulse wave proceeds along the peripheral artery.

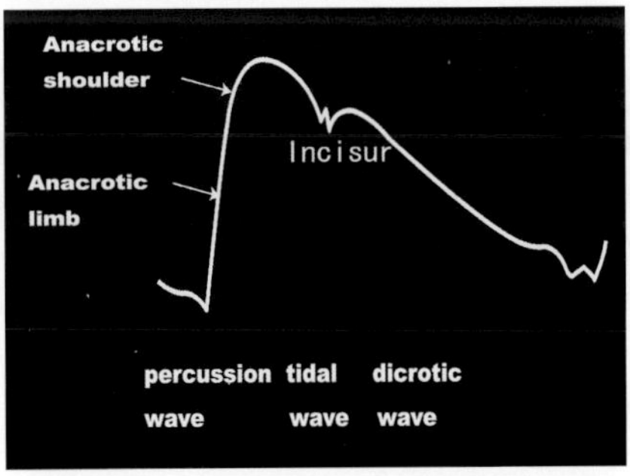

Figure 2. Time series data of the Arterial pulse pressure waveform

As for the tidal wave, Idiopathic Hypertrophic Subaortic Stenosis (IHSS) and aortic valve stenosis significantly influence the waveform. The contractile dynamic state and the left ventricle outflow also influence this wave formation.

The difference in height between the percussive wave and the tidal wave and the ratio of the full wave quantity are displayed as the augmentation index (AI= Δ p/A). This is broadly used to shows the elastic modulus of a blood vessel system. This has also been applied in commercial use as an index called blood vessel age.

A notch, known as the incisor or dicrotic notch, is formed in the downward foot of the main artery pressure waveform. This is a small vibration produced in a pulse wave simultaneously with the closing of an aortic valve. This represents the switch from the contraction phase to the diastolic phase. Since it is an index which represents closing of an aortic valve, it has been reported that this aspect of the waveform disappears when aortic valve closing is insufficient, as in aortic regurgitation (AR). With aging and hypertension, when compliance of the main artery becomes smaller, there is a tendency for the dicrotic notch to decrease.

The upheaval, which appears at an extended early stage is called the dicrotic wave, and originates as a reflection from the peripheral circulation of an artery pulse wave. Generally, the tidal wave and the dicrotic wave are inversely correlated. In young people, the dicrotic wave is large, whereas, in arteriosclerosis with an early advance of a reflected wave, the tidal wave becomes large. This dicrotic wave is important to maintain a coronary–arterial blood flow. The arterial pressure waveform in which this wave becomes large is called the dicrotic pulse. For example, in dilated cardiomyopathy , the tidal wave disappears and turns into the dicrotic wave. In contrast, in AR, the dicrotic wave fall is prominent.

The elasticity of arteries fluctuates in response to changes associated with atherosclerosis in addition to autonomic nerve tone. Therefore, in pulse waves, the control exerted by the central nervous system also becomes an important parameter. Furthermore, if the amount of circulating blood increases, theoretically, this will cause an increase in the entire pressure waveform time series curve. Similarly, blood viscosity changes will affect the whole waveform. Thus various parameters work mutually in the formation of the pulse wave.

Therefore, it should be possible to diagnose medical information from the wave form of the arterial pulse wave.

3. Overview of the Pulse Diagnosis Machine

Pulse diagnosis in Oriental medicine involves placing three fingers above the radial artery of both wrists, as shown in Fig. 1. Varying pressure is applied by a finger and the pulse-wave form is sensed manually. Pulse diagnosis is made using this methodology. Therefore, it is theoretically possible to measure pressure electronically using three precise pressure sensors, as shown in Fig. 3. If this machine is used, an objective and reproducible, quantitative method for pulse diagnosis is possible.

Figure 3. Sensors for the Pulse diagnosis machine

The pressure sensor was placed noninvasively on the skin surface above the radial artery, as is done in manual pulse diagnosis. This is represented in Figs. 3 and 4.

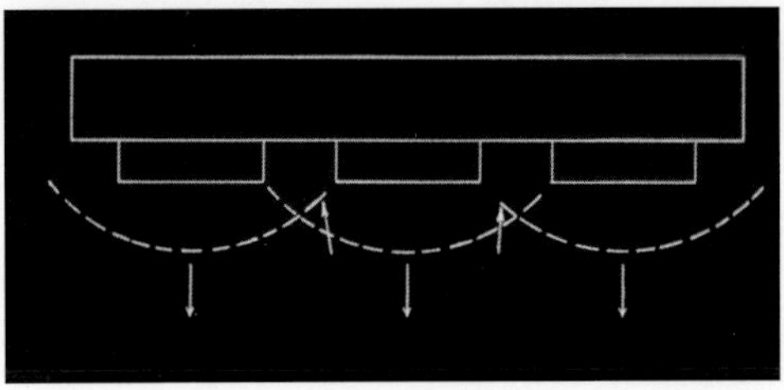

Figure 4. Concept diagram of three pressure sensors and Pulse diagnosis

The whole sensor was pressed in the direction of a perpendicular from external surface by the air pressure in which the quantitative evaluation is possible. The system digitizes the sensed pressure waveform of the pulse and records the data on a data recorder. Arterial pressure is detected by pressure transfer from the upper surface of the radial artery, as represented in Fig. 4. The basic performance of the model circulation circuit and safety and validity were evaluated and confirmed in animal experiments during the system development.

It is considered to be standard practice to evaluate clinical application last. Then, the model circulation circuit, which was based on the Windkessel theory of left-heart circulation, was prepared and equipped with pre-load, after load, and compliance using an artificial heart. The performance of this model was confirmed.

Figure 5 shows the pulse diagnosis sensor attached to the model circulation, which imitated the left-heart circulatory system.

Figure 5. A photograph of the pressure sensors for Pulse diagnosis and model circulation of a radial artery

The pulse diagnosis sensor was attached to model arteries of various elastic properties to imitate the radial artery. Various air pressures were applied to the external surface of the sensor, and quantitative pulse diagnosis according to Oriental medicine was tested.

By utilizing a reproducible methodology that used such model circulation, we were able to quantitatively check the measurement results.

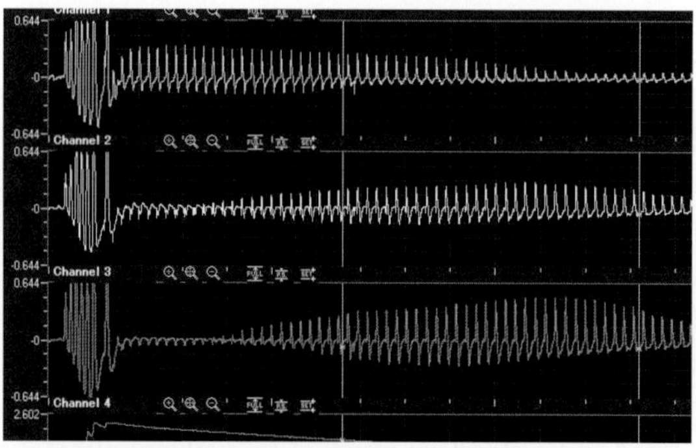

Figure 6. Alteration of an external surface pressures and pulse diagnosis sensor waveform. From upper tracing, waveform of the sensors of proximal, mid, and peripheral were shown

If model circulation is used, it is easy to change the contractility of the left-heart circulation. It is also simple to change the peripheral blood vessel resistance and vessel compliance. Therefore, factors determined for pulse diagnosis can be evaluated quantitatively using this model circuit.

The pressure results of the pulse diagnosis while gradually decreasing the outside pressure are shown in Fig. 6, which, from top, shows the proximal, mid side and peripheral regions of the pulse wave.

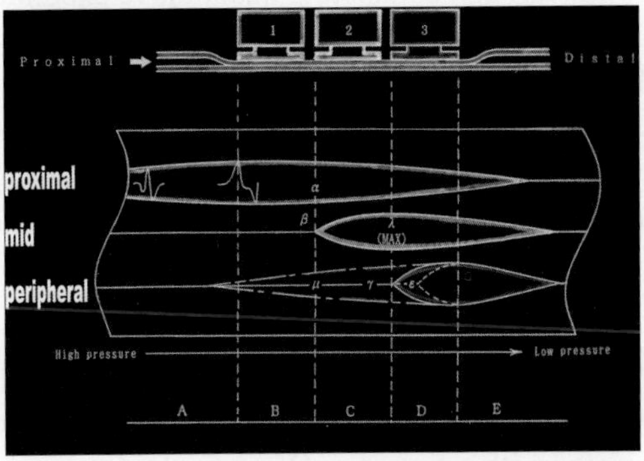

Figure 7. Theoretical background of the three pressure sensors waveform were shown

If it presses the external surface by high pressure and a radial artery is blocked by an external force, the pressure is undetectable in the peripheral regions. On the other hand, if the external pressure is gradually decreased, a gradually increasing sensing wave is detected in the central region and the peripheral region. A simulated pressure wave is shown in Fig. 7.

It is thought that the measurement result of having been good quantity has presented the height of the probability of scientific measurement like a theoretical background. It becomes indispensable experimentation after the scientific probate examination of a model circulation circuit according to a living body. Next, quantitative examination of the pulse-diagnosis machine was attempted using a goat of roughly the same body weight as a typical Japanese person.

The goat weighed about 60 kg. Its neck was cut open, the carotid artery was exposed, the sensor was placed *in situ*, and the pulse-wave sensing examination was tested.

The pulse diagnosis measurement was done as with the model circulation, and similar results were obtained. Data on the detection of the wave-like changes due to medication were also of interest. Unlike a human subject, in an animal experiment, it is also possible to manipulate the animal subject to illuminate the strange and laborious workings of the heart, such as the compliance of blood vessels, peripheral resistance, the amount of circulating blood, pre load, and after load .

Therefore, the concept of Oriental Medicine and the wave diagnosis of the pulse diagnosis technique, which have been in use since ancient times, can be given a scientific basis through conducting quantitative experimentation.

If this machine is used, the light of objective science can be applied to the world of Oriental Medicine for the first time.

Thus far, we have presented results of basic examination of the pulse diagnosis machine using a model circulation and in an animal model. Next, we checked the safety, quantitative use of the machine and the validity within clinical settings.

4. Methodology of Analyzing the Time Series Data

For pulse diagnosis in Oriental medicine, varying pressures are applied and the radial artery is palpated with a fingertip. Of particular interest is the fact that this method reflects hydrodynamics. Blood flow is first intercepted as the pressure of the radial artery by the side of the heart. After that, blood flow islowly resumes and the pressure pulse-wave in the peripheral region is generated gradually and can be detected. Much information, not only about the dynamic state of the circulation, such as a cardiac function, but also about the internal organs can be derived from this waveform. It is theoretically considered that such information is contained in the wave height and the waveform of a pulse-diagnosis wave so that it may be expressed in principle as blood flow resumption shown in Fig. 7.

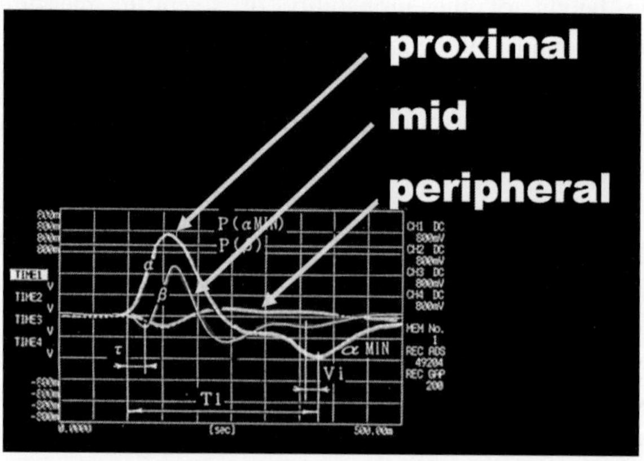

Figure 8. Three waves of the sensors from the Pulse Diagnosis Machine

The viscous factor of blood also affects the pulse wave. The pulse wave form is also affected by local factors in the skin acting on the blood vessel, such as NO, and by the action of the blood vessel at the time of re-reflux during transient ischemia. Therefore, functional diagnosis of the inner skin, which is important in advanced atherosclerosis, can also be determined.

The pulse diagnosis sensor detects three waves, as shown in Fig. 8, corresponding to the proximal, central and peripheral regions. The wave height declines in the peripheral region because pressure is applied from an external surface (Fig. 4).

As shown in Fig. 8, various parameters can be calculated from this wave and allows scientific/quantitative pulse diagnosis for the first time. Various fluctuation components exist within the time series information of the cardiovascular system, are represented by the wave of a pulse diagnosis, and is controlled by the autonomic nervous system. Therefore, autonomic nervous system diagnosis can also be done using a pulse wave.

Therefore, this system will offer significant clinical advantages, particularly because it is a non-invasive measurement system.

5. Time Series Data Analysis for the Evaluation of the Baroreflex

Every medical student studies the baroreflex system as a typical example of homeostasis. When blood pressure increases, baroreceptors in the carotid arteries and aortic arch sense the increase in the baroreflex sensitivity. When this information is transmitted to the central nervous system, the heart rate decreases and the arteries dilate. These processes restore the normal blood pressure. The baroreflex sensitivity can be evaluated by measuring the heart rate response to changes in blood pressure. The slope of the linear regression line indicates the sensitivity of the baroreflex system of the heart.

Although the heart rate response in the baroreflex system can be monitored, there is no method to evaluate arterial baroreflex function, possibly because of the difficulty in evaluating vascular tone while the subject is awake.

Recently, new methodologies, such as brachial ankle pulse wave velocity and cardio-ankle vascular index, have been developed to evaluate human arterial stiffness. These methodologies non-invasively evaluate arterial wall stiffness using the pulse waveform of the brachial and ankle arteries. These methodologies are based on the premise that pulse wave velocity (PWV) is correlated with arterial wall stiffness. Thus PWV increases when the arterial wall becomes harder and decreases when the arterial wall softens.

In the baroreflex system, the arterial wall softens in response to an increase in blood pressure, thereby decreasing the vascular resistance. The blood pressure subsequently returns to normal owing to the decrease in resistance. The softness of an arterial wall can be measured by PWV.

The PWV value could therefore be used to quantitatively measure the baroreflex sensitivity of the arterial wall. PWV can be calculated from the pulse wave transmission time (PTT) and distance. Thus measurements of PTT and blood pressure allow us to evaluate the arterial baroreflex response.

6. Autogenic Training

Autogenic training is a relaxation technique that was developed by the German psychiatrist Johannes Schultz. This technique involves daily practice sessions that last for around 15 minutes; the practitioner will repeat a set, which can induce a state of relaxation. This technique can be used to treat many stress-induced psychosomatic disorders. It is a method that influences one's autonomic nervous system.

In this paper, we evaluated the acute effect of autogenic training on autonomic nerve function during the first session.

Table 1. Effect of Autogenic training

	Before	During	After
HR	66.8±9.6	67.2±8.3	66.7±9.0
BP	119.5±8.2	120.2±10.7	117.8±12.0
PTT	144.0±15.6	144.7±16.4	144.6±16.0
SV	80.9±13.4	82.1±13.4	80.3±13.8
TPR	16.0±2.2	15.8±2.3	15.7±2.5

Table 1 shows the effects of autogenic training on hemodynamic parameters. There is no significant difference in between before, during and after training.

Table 2. Heart rate variability and baroreflex sensitivity in autogenic training

	Before	During	After
LF/HF	1.76±0.87	5.85±4.82	2.19±1.59
BP-HR	0.64±0.30	0.73±0.31	0.72±0.29
BP-PTT	0.30±0.16	0.43±0.22	0.38±0.15
BP-SV	0.22±0.02	0.02±0.01	0.02±0.01

Table 2 shows heart rate variability and the baroreflex sensitivity measured before, during and after the autogenic training session. During the session, a significant increase in LF/HF was noted, which may be because this was the first session. A significant increase in sympathetic tone was observed during the session.

These approaches using the newly developed pulse diagnosis machine and baroreflex evaluation may be useful when we consider the clinical use of various kinds of relaxation techniques in the future.

Acknowledgments

This work was partly supported by support of Tohoku University Global COE Program "Global Nano-Biomedical Engineering Education and Research Network Centre".

References

1. J. Lee, The systematical analysis of oriental pulse waveform: a practical approach. *J Med Syst.* **32**, 9–15 (2008).
2. M. Park, H. Kang, Y. Huh and K. C. Kim, Cuffless and noninvasive measurement of systolic blood pressure, diastolic blood pressure, mean arterial pressure and pulse pressure using radial artery tonometry pressure sensor with concept of Korean traditional medicine. *Conf Proc IEEE Eng Med Biol Soc. 2007,* 3597–3600 (2007).
3. A. Joshi, S. Chandran, V. K. Jayaraman and B. D. Kulkarni, Arterial pulse system: modern methods for traditional Indian medicine. *Conf Proc IEEE Eng Med Biol Soc. 2007,* 608–611 (2007).

DETECTION ALGORITHM OF FATAL ARRHYTHMIAS FOR IMPLANTABLE CARDIOVERTER DEFIBRILLATORS USING JOINT PROBABILITY *

MAKOTO YOSHIZAWA [1)†], HIROYUKI KINOSHITA [2)],
KAZUO SHIMIZU [3)], MASASHI INAGAKI [4)], KAZUNORI UEMURA [4)],
MASARU SUGIMACHI [4)], KENJI SUNAGAWA [5)]

1) Research Division on Advanced Information Technology, Cyberscience Center, Tohoku University, 6-6-05, Aoba, Aramaki, Aoba-ku, Sendai, Miyagi, 980-8579, Japan

2) Graduate School of Engineering, Tohoku University, 6-6-05, Aoba, Aramaki, Aoba-ku, Sendai, Miyagi, 980-8579, Japan

3) Olympus Corporation, 2-3-1 Nishi-Sinjuku, Shinjuku-ku, Tokyo, 163-0914, Japan

4) National Cardiovascular Center, 5-7-1 Fujishirodai, Suita, Osaka, 565-8565, Japan

5) Kyushu University, 3-1-1 Umade, Higashi-ku, Fukuoka, Fukuoka, 812-8582, Japan

A new algorithm based on a joint probability density function has been developed to detect fatal arrhythmias for implantable cardioverter-defibrillators (ICDs). In the algorithm, two sampled ECG signals measured in the right and the left ventricles are counted to create a two-dimensional frequency distribution. On the basis of the distribution, three kinds of indices, *i.e.*, mutual information, Pearson's χ^2-statistic, and Pearson's correlation coefficient, are calculated to assess the independence of the two ECG signals. The independence of the two signals differentiates sinus rhythm (SR)/atrial tachycardia (AT) from ventricular tachycardia (VT)/ventricular fibrillation (VF). In VT/VF, the two ECG signals are independent of each other while those in SR/AT are dependent on each other due to the common source of those signals: the His-Purkinje bundle. Validation of this method using a dog showed that the proposed method using the χ^2-statistic can accurately distinguish between supraventricular and ventricular arrhythmias without detection of the R-wave.

1. Introduction

Traditional implantable cardioverter-defibrillators (ICDs) use information from ECG signals to detect ventricular arrhythmias such as ventricular fibrillation (VF) and ventricular tachycardia (VT) as well as supraventricular arrhythmias such as atrial tachycardia (AT) [1-4]. It is difficult for the common algorithms

* This work is supported by Grant in Aid for Scientific Research of Ministry of Health, Labor and Welfare.

† Makoto Yoshizawa is a Tohoku University Global COE Member.

based on the R-R interval of the ECG signals to definitively distinguish among VF, VT, AT and sinus rhythm (SR). In such algorithms, it is necessary to adjust several parameters of the ICD specifically to each patient with the ICD. Furthermore, the accuracy of detecting arrhythmias depends strongly on the accuracy of detecting R-waves.

To develop a method which does not require precise recognition of R-waves for distinguishing between VF/VT and SR/AT, we have determined statistical independence between the right ventricular ECG and the left ventricular ECG at the coronary sinus. The proposed method has been evaluated using three indices of statistical independence on data obtained from a dog with an ICD.

2. Methods

2.1. *Outline of Conventional and Proposed Methods*

2.1.1. *Conventional detection methods*

Conventional methods used in ordinary ICDs for differentiating ventricular arrhythmia and supraventricular arrhythmia, detect R- and P-waves with electrodes placed in a ventricle and in an atrium, respectively, and their temporal information is analyzed to distinguish VT/VF and AT/AF. However, the conventional methods have the following problems:

1) Their accuracy depends strongly on the accuracy of detecting R-waves and P-waves, and it is difficult to maintain robust detection even when under-sensing or over-sensing happens [1].
2) It is difficult to adjust parameters (thresholds) optimally for many criteria of detection [3,4].
3) The specificity of detection tends to be low, and especially, there is a real possibility of miscategorizing supraventricular arrhythmia as ventricular arrhythmia.

One of the possible reasons for these problems is that the accuracy of detecting peaks of R-waves or P-waves must be high. This is because appropriate conditional branching based on precise peak recognition of the ECG signal is required for classifying arrhythmia according to a difference of the time delay in excitation between the atrium and the ventricle or a difference in irregularity between supraventricular arrhythmia and ventricular arrhythmia.

2.1.2. *Concept of proposed detection algorithm*

In this study, we have proposed an algorithm for detecting fatal arrhythmias for an ICD which does not depend heavily on the accuracy of peak detection in ECG signals by focusing on the independence of ECG signals from the right and left ventricles.

First, it can be assumed that SR/AT and VF/VT can be differentiated on the basis of independence of the right and left ventricular ECG signals as shown below.

In case of SR, cardiac electrical excitation is transmitted normally from the sinus node to the right and left ventricles via the His-Purkinje bundle, and thus the two ventricles are excited simultaneously. In case of AT, ectopic electrical excitation occurs in the atrium and is also transmitted through the His-Purkinje bundle to cause simultaneous excitation of the two ventricles.

On the other hand, VT/VF is caused by ectopic electrical excitation in the ventricles which is not transmitted through the His-Purkinje bundle. Therefore, excitation in the two ventricles is not synchronized. This means that the occurrence probabilities of the ECG signals of the two ventricles are dependent on each other in SR/AT but are independent of each other in VT/VF. Thus, a criterion of independence is a key to classification of arrhythmias.

2.2. *Indices of Independence*

In this study, mutual information (MI) and Pearson's chi-squared statistic (χ^2-statistic) have been adopted as mathematical indices of independence (or dependence) for the two ECG signals. Pearson's correlation coefficient (PCC) has also been calculated to compare with the above statistics.

2.2.1. *Mutual information*

Mutual information, MI is a statistical index which quantifies how much one random event can be predicted on the basis of another [5].

Consider two events A and B which are denoted discretely by $\{A \mid a_i; i = 1, \cdots, m\}$, and $\{B \mid b_j; j = 1, \cdots, n\}$, respectively. As shown in Table 1, assume that the two dimensional (joint) frequency of each element a_i and b_j is represented by k_{ij}, and let the sum of all elements be equal to N as follows:

$$\sum_{i=1}^{m}\sum_{j=1}^{n}k_{ij} = N \tag{1}$$

In this situation, the joint probability $p(a_i, b_j)$ of a_i and b_j can be estimated by:

$$p(a_i, b_j) \approx \frac{k_{ij}}{N} \tag{2}$$

By letting

$$k_{i\bullet} = \sum_{j=1}^{n} k_{ij},\ k_{\bullet j} = \sum_{i=1}^{m} k_{ij} \tag{3}$$

each occurrence probability $p(a_i),\ p(b_j)$ can be estimated by:

$$p(a_i) \approx \frac{k_{i\bullet}}{N} \tag{4}$$

$$p(b_j) \approx \frac{k_{\bullet j}}{N} \tag{5}$$

Mutual information $MI(A,B)$ between the events A and B is defined as:

$$MI(A,B) = \sum_{i=1}^{m}\sum_{j=1}^{n} p(a_i, b_j)\log\frac{p(a_i, b_j)}{p(a_i)p(b_j)} \tag{6}$$

Substituting Eq. (2-5) into Eq. (6), we have:

$$MI(A,B) = \log N + \sum_{i=1}^{m}\sum_{j=1}^{n} k_{ij} \log\frac{k_{ij}}{k_{i\bullet}k_{\bullet j}} \tag{7}$$

The smaller $MI(A,B)$ is, the lower the probability is that one event can be predicted on the basis of another. This implies that two events are more independent.

Table 1. Joint frequency distribution of events a_i and b_j.

	b_1	b_2	$\cdots$	b_n	Total
a_1	k_{11}	k_{12}	$\cdots$	k_{1n}	$k_{1\bullet}$
a_2	k_{21}	k_{22}	$\cdots$	k_{2n}	$k_{2\bullet}$
$\vdots$	$\vdots$	$\vdots$	$\ddots$	$\vdots$	$\vdots$
a_m	k_{m1}		$\cdots$	k_{mn}	$k_{m\bullet}$
Total	$k_{\bullet 1}$	$k_{\bullet 2}$	$\cdots$	$k_{\bullet n}$	N

2.2.2. χ^2-*statistic*

In general, Pearson's χ^2-statistic tests whether two random variables are independent of each other or not [6].

Assume that events A and B have the same frequency distribution as Table 1. Under the null hypothesis, H_0

H_0: A and B are independent of each other,

consider the χ^2-statistic T defined by:

$$T = \sum_{i=1}^{m}\sum_{j=1}^{n}\frac{(O_{ij} - E_{ij})^2}{E_{ij}} \tag{8}$$

$$= \frac{1}{N^2}\sum_{i=1}^{m}\sum_{j=1}^{n}\frac{(Nk_{ij} - k_{i\bullet}k_{\bullet j})^2}{k_{i\bullet}k_{\bullet j}} \tag{9}$$

where

$$O_{ij} = \frac{k_{ij}}{N} \text{ : observed joint frequency} \tag{10}$$

$$E_{ij} = \frac{k_{i\bullet}}{N}\cdot\frac{k_{\bullet j}}{N} \text{ : theoretical joint frequency under } H_0 \tag{11}$$

Eq. (8) or (9) estimates the statistical error between O_{ij} and E_{ij}. Under a certain condition, T would be approximated by the chi-square distribution $\chi^2_{(m-1)(n-1)}$ with $(m-1)(n-1)$ degrees of freedom [6]. Using a significance level α, if

$$T \geq \chi^2_{(m-1)(n-1)}(\alpha) \tag{12}$$

then the null hypothesis H_0 is rejected, and it can be estimated that the two events are not independent of each other. This means that the smaller T is, the higher the probability is that the two events are independent.

2.2.3. *Correlation coefficient*

To compare with the above statistical indices for independence (or dependence), Pearson's correlation coefficient (PCC) has been used.

Let x_k, y_k; $k = 1, \cdots, K$ denote two random variables which are the sampled values of the right and the left ventricular ECG signals in this study. The *PCC* is calculated from

$$PCC = \frac{\sum_{k=1}^{K}(x_k - \bar{x})(y_k - \bar{y})}{\sqrt{\sum_{k=1}^{K}(x_k - \bar{x})^2 \sum_{k=1}^{K}(y_k - \bar{y})^2}} \tag{13}$$

where $\bar{x}$, $\bar{y}$ are mean values of x_k, y_k, respectively. It means that the smaller *PCC* is, the lower correlation between the two variables.

2.3. *Signal Processing of Animal Experiment*

As shown in Table 2, we measured the right and left ventricular ECG signals in a dog when arrhythmias were induced with electrical stimulation. In this experiment, AT was simulated by right atrial pacing, and VT was simulated by right or left ventricular pacing. VF was induced by extrasystolic stimulation after the R-wave or in the middle of T-wave.

Each ECG signal was sampled at the rate of 250 Hz. The sampled values were segmented by a data window with the length of 1 s ($K = 250$) as shown in Fig. 1. The number of the segmented data classified into each class (each bin) was counted to create the joint frequency distribution as shown in Table 1. In the next step, the data window was shifted by 0.2 s (50 samples), and the same process was repeated. Each class of the distribution was obtained by segmenting the value between the minimum and the maximum values observed in the data window into $m = n = 5$ segments for each axis and then $m \times n = 25$ bins were prepared.

Table 2. Data set measured in a dog for evaluation.

Mode	Number of episodes	Length of observation (in seconds)
SR	9	241.5
AT	7	64.7
VT	15	172.1
VF	4	79.4

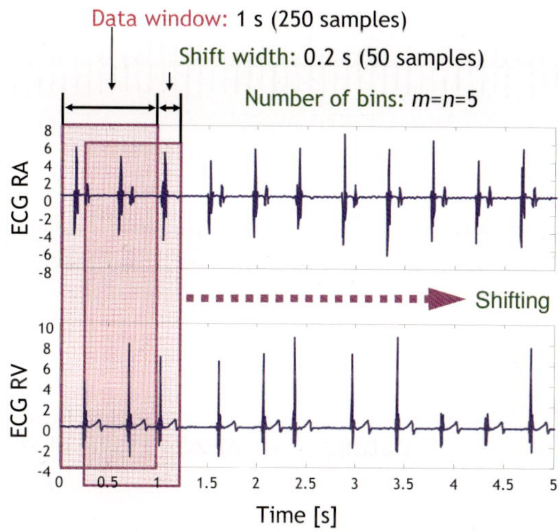

Figure 1. Data window used for creating a joint frequency distribution.

3. Results

3.1. *Time Course of Indices of Independence*

Examples of the time course of the three indices of independence, *i.e.*, mutual information *MI* corresponding to Eq. (7), the χ^2-statistic *T* corresponding to Eq. (9), and the correlation coefficient *PCC* corresponding to Eq. (13) are shown in Figs. 2-5.

These figures indicate that three indices were high in SR/AT but were low in VT/VF. It suggests that *PCC* was not stable, but *MI* and *T* were stable. In particular, the χ^2-statistic was stable and its response to the change from SR to VT/VF was quick and finished before 1 second.

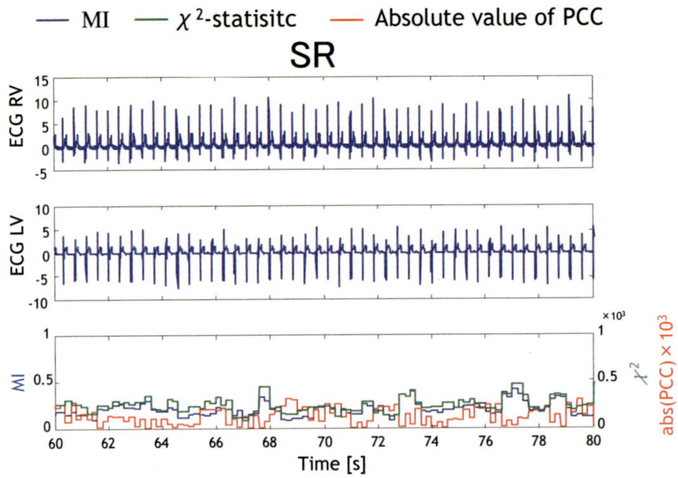

Figure 2. Independence indices in sinus rhythm (SR).

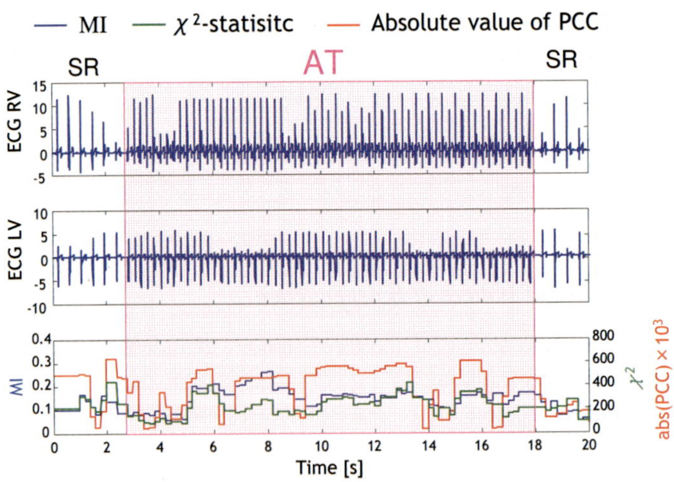

Figure 3. Independence indices in atrial tachycardia (AT).

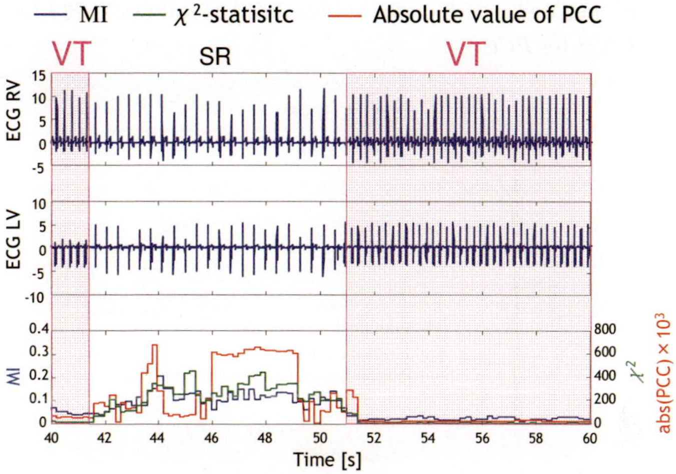

Figure 4. Independence indices in ventricular tachycardia (VT).

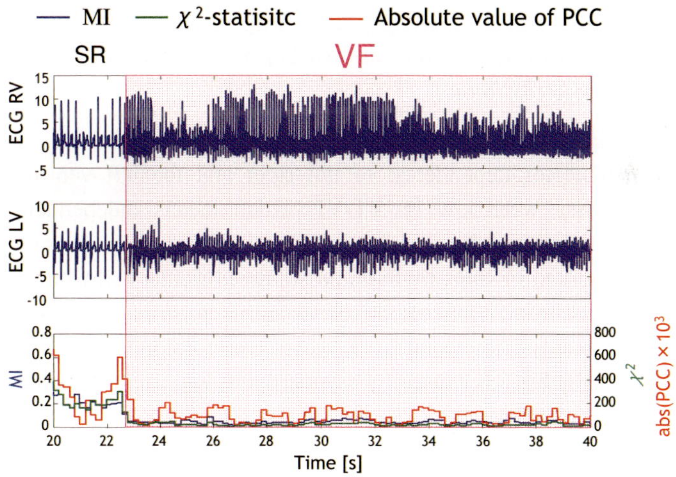

Figure 5. Independence indices in ventricular fibrillation (VF).

3.2. *ROC Curve and ROC Area*

For indices *T*, *MI* and *PCC*, the receiver operating characteristic (ROC) curve of VT/VF versus SR/AT can be depicted in Fig. 6. The ROC area, which is defined as the area under each curve, is calculated as follows: 0.987 for *T*, 0.978 for *MI*, and 0.898 for *PCC*.

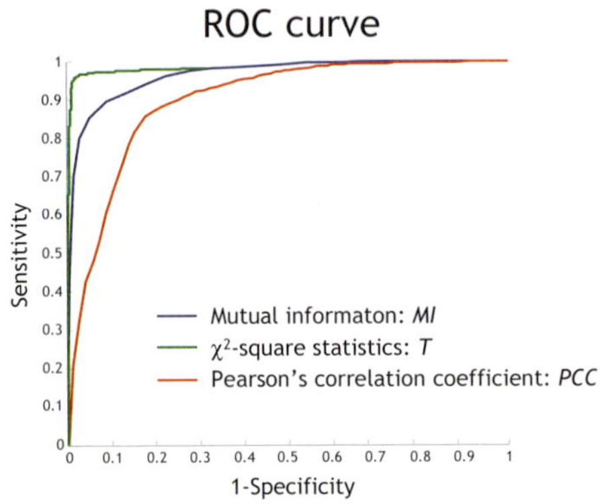

Figure 6. ROC curve of VT/VF versus SR/AT.

4. Discussion

4.1. *Difference in Independence between Supraventricular and Ventricular Arrhythmias*

As shown in Figs. 2-5, the three evaluation indices *MI*, *T*, and *PCC* were high in SR/AT, which means dependence of the right and the left ventricular ECG signals while they were low in VT/VF which means independence. This supports the predictions given in 2.1.2. That is to say, it is possible that the dependence observed in sinus rhythm and supraventricular arrhythmia is caused by electrical synchronization of the two ventricles through the His-Purkinje bundle, which mainly corresponds to R-wave, while the independence observed in ventricular arrhythmia is caused by asynchronous and irregular behavior of the two ventricles.

4.2. *Why Correlation Coefficient Was Unstable*

The reason why the behavior of correlation coefficient *PCC* was unstable in comparison with *MI* and *T* seen in Figs. 2-5 may be explained as follows.

In general, *PCC* represents linear similarity between two time series data and depends strongly on the shape or the wave-form of the data. As seen in the case of AT in Fig. 3, *PCC* was reduced because the shapes of the two signals were not similar and were uncorrelated even if their R-waves were synchronized.

It is true that independence means decorrelation, but decorrelation does not always mean independence. Mutual information and the χ^2-statistic can quantify dependence including nonlinear relations [7], and thus it can be expected that *MI* and χ^2-statistic will be capable of more robust detection of fatal arrhythmia than the Pearson's correlation coefficient.

4.3. *Why χ^2-statistic was Excellent*

Theoretically, if the joint frequency distribution is nearly uniform, mutual information and the χ^2-statistic are proportional to each other [6]. The distribution of VF is nearly uniform because VF is caused by chaotic excitation of the ventricular muscle. In case of VT, the heart rate tends to be higher than in SR/AT and the behavior of the two ventricles are asynchronous, thus the number of bins whose frequency becomes zero is relatively lower than in SR/AT. Therefore, it can be interpreted that in case of VT/VF, mutual information and the χ^2-statistic were proportional to each other, and their values were reduced in comparison with SR/AT.

On the other hand, in case of SR/AT, the joint frequency is concentrated on the bin corresponding to the base line of the ECG in the two dimensional distribution, and the distribution becomes sparse. This characteristic indicates that the difference between mutual information and the χ^2-statistic of SR/AT becomes larger than that of VT/VF. Furthermore, it has also been found theoretically that in the more sparse distribution, the less the χ^2-statistic approximates to χ^2-distribution [6]. This fact implies that χ^2-statistic is more sensitive to the change in independence than mutual information.

Judging from the receiver operating characteristic (ROC) curves and the ROC areas shown in Fig. 6, the χ^2-statistic is the best index for differentiating supraventricular and ventricular arrhythmia. Moreover, the χ^2-statistic does not need the logarithm calculations required in mutual information which is advantageous for implementation in a single-chip computer used in an ICD.

5. Conclusion

In this study, a new algorithm based on a joint probability density function has been developed to detect fatal arrhythmias to apply to ICDs.

In the algorithm, sampled values of two ECG signals measured in the right and the left ventricles are classified to create a two-dimensional frequency distribution. On the basis of the distribution, three kinds of indices, *i.e.*, mutual information, Pearson's χ^2-statistic, and Pearson's correlation coefficient, are calculated to judge the independence of the two ECG signals. The independence has been expected to differentiate sinus rhythm (SR) / atrial tachycardia (AT) from ventricular tachycardia (VT) / ventricular fibrillation (VF) because the two ECG signals in VT/VF are independent of each other while those in SR/AT are dependent on each other due to the common source of those signals through the His-Purkinje bundle.

An experiment with a dog showed that the proposed method using the χ^2-statistic can accurately distinguish between supraventricular and ventricular arrhythmias without having to detect either R-wave or P-wave.

However, it can be predicted that the proposed method will not work well in the case of a cardiac conduction defect such as a bundle branch block which would always prevent synchronization of the two ventricular ECG signals.

Acknowledgments

The authors thank the support given by 2007 Tohoku University Global COE Program "Global Nano-Biomedical Engineering Education and Research Network Centre."

References

1. F. Eberhardt, W. Peters, F. Bode *et al.*, R wave undersensing caused by an algorithm intended to enhance sensing specifity in an implantable cardioverter defibrillator. *PACE* **26**, 1776–1777 (2003).
2. E. Aliot, R. Nitzsche and A. Ripart, Arrhythmia detection by dual-chamber implantable cardioverter defibrillators. A review of current algorithms. *Europace* **6**, 273–286 (2004).
3. J. Brugada, L. Mont, M. Figueiredo *et al.*, Enhanced detection criteria in implantable defibrillators. *J Cardiovasc Electrophysiol.* **9**, 261–268 (1998).
4. J. Neuzner, H. F. Pitschner and M. Schlepper, Programmable VT detection enhancements in implantable cardioverter defibrillator therapy. *PACE* **18**, 539–547 (1995).
5. D. Hoyer, U. Leder, H. Hoyer *et al.*, Mutual information and phase dependencies: measures of reduced nonlinear cardiorespiratory interactions after myocardial infarction. *Med Eng Phys* **24**, 33–43 (2002).
6. A. Hyvärinen, J. Karhunen and E. Oja, *Independent Component Analysis,* John Wiley & Sons (2001).
7. C. Omichi, S. T. Lamp, S. F. Lin *et al.*, Intra-cellular Ca dynamics in ventricular fib-rillation. *Am J Physiol Heart Circ Physiol* **286**, H1836–H1844 (2004).

QUANTITATIVE EVALUATION OF EFFECTS OF VISUALLY INDUCED MOTION SICKNESS USING PHOTOPLETHYSMOGRAPHY

MAKOTO ABE [1)*], MAKOTO YOSHIZAWA [2)†], NORIHIRO SUGITA [1)], AKIRA TANAKA [3)], SHIGERU CHIBA [4)], TOMOYUKI YAMBE [5)†], SHIN-ICHI NITTA [5)]

1) Department of Electrical and Communication Engineering, Graduate School of Engineering, Tohoku University, 6-6-05, Aoba, Aramaki, Aoba-ku, Sendai, Miyagi 980-8579, Japan

2) Research Division on Advanced Information Technology, Cyberscience Center, Tohoku University, 6-6-05, Aoba, Aramaki, Aoba-ku, Sendai, Miyagi 980-8579, Japan

3) Department of Human Support System, Faculty of Symbiotic Systems Science, Fukushima University, 1, Kanayagawa, Fukushima, Fukushima 960-1296, Japan

4) Sharp Corporation, 1-9-2, Nakase, Mihama-ku, Chiba, Chiba 261-8520, Japan

5) Institute of Development, Aging and Cancer, Tohoku University, 4-1, Seiryo-cho, Aoba-ku, Sendai, Miyagi 980-8575, Japan

Viewing of moving video images displayed on a wide-field display or a screen sometimes causes visually induced motion sickness (VIMS); it also causes symptoms related to the autonomic nervous system such as nausea, vomiting, and dizziness. A previous study indicated that the maximum cross-correlation coefficient (ρ_{max}) between blood pressure variability (BPV) and heart rate variability (HRV) whose frequency components are limited to approximately 0.1 Hz is a useful parameter for evaluating the effects of VIMS on humans. The present study proposes a new method for determining ρ_{max} by using finger photoplethysmography and without the measurement of either continuous blood pressure or ECG. The validity of the proposed method was determined by performing the following two experiments: (I) experiment involving the Valsalva maneuver and (II) experiment involving viewing of a moving video image. The experimental results showed that the proposed method was as successful as the conventional method.

Keywords: visually induced motion sickness, photoplethysmography, heart rate variability, blood pressure variability.

---

[*] Makoto Abe is a Tohoku University Global COE Research Assistant.

[†] Makoto Yoshizawa and Tomoyuki Yambe are Tohoku University Global COE Members.

1. Introduction

Recently, our exposure to various artificial visual images in video games, movies, web contents, virtual reality systems, *etc*., has increased dramatically. Some of these images are likely to cause undesirable effects. One such effect is visually induced motion sickness (VIMS) that causes symptoms related to the autonomic nervous system, such as nausea, vomiting, and dizziness. In previous studies, subjective methods with questionnaires were frequently used to determine the effects of VIMS [1,2]. However, these methods are not useful for estimating the effects of VIMS from time series data.

To quantify the effects of VIMS as time series, the authors have previously proposed a physiological index ρ_{max}, which represents the maximum cross-correlation between blood pressure variability and heart rate variability whose frequency components are limited to the Mayer waveband [3,4]. In order to determine the value of ρ_{max}, it is necessary to measure blood pressure continuously with a large and expensive device. The present study proposes an easier method for determining ρ_{max} using only a finger photoplethysmography (PPG) and without measuring either continuous blood pressure or ECG. Independent component analysis (ICA) of PPG was used to determine an event-related parameter. To establish the validity of the proposed method, two experiments wherein subjects performed the Valsalva maneuver and watched a moving video image were carried out.

2. Methods

The ICA used in our method is described as follows:

(1) Let $x_1(k)$, $x_2(k)$, ..., $x_m(k)$ be m feature variables extracted from a PPG signal at the k-th beat. Define a feature vector $\boldsymbol{x}(k)$ as $\boldsymbol{x}(k) = [x_1(k), x_2(k), ..., x_m(k)]^T$.

(2) Let $s_1(k)$, $s_2(k)$, ..., $s_n(k)$ be n unknown physiological parameters that are independent of one another at the k-th beat. Define a parameter vector $\boldsymbol{s}(k)$ as $\boldsymbol{s}(k) = [s_1(k), s_2(k), ..., s_n(k)]^T$.

(3) Assume that the feature vector $\boldsymbol{x}(k)$ is given by a linear combination of $s_1(k)$, $s_2(k)$, ..., $s_n(k)$ as follows:

$$\boldsymbol{x}(k) = \boldsymbol{A}\boldsymbol{s}(k) \tag{3}$$

where the $m \times n$ matrix $\boldsymbol{A}$ is an unknown constant mixing matrix consisting of coefficients of the linear combination. Let K be the number of beats observed in an experiment. Define an $m \times K$ matrix $\boldsymbol{X}$ and an $n \times K$ matrix $\boldsymbol{S}$ as $\boldsymbol{X} = [\boldsymbol{x}(1), \boldsymbol{x}(2), ..., \boldsymbol{x}(K)]$ and $\boldsymbol{S} = [\boldsymbol{s}(1), \boldsymbol{s}(2), ..., \boldsymbol{s}(K)]$, respectively. Thus, the matrix $\boldsymbol{X}$ is assumed to be given by $\boldsymbol{S}$ as follows:

$$X = AS \tag{4}$$

(4) The ICA is applied to estimate the A from X. The independent component S can be obtained from

$$S = A^{+}X \tag{5}$$

where A^{+} is the pseudoinverse matrix of A.

In this study, the fast fixed point algorithm (Fast-ICA) developed by Hyvärinen and Oja was used to linearly separate S from X [5,6]. In addition, m was set to 7.

Figure 1 shows an example of the PPG signal with seven feature variables used in the present study. These variables are defined at every beat as follows:

(1) *FFI*: foot-to-foot interval of the PPG signal
(2) t_d: interval between the time required to maximize the PPG signal to the time required to minimize it
(3) $t_{maxslope}$: time taken to maximize the slope of the PPG signal
(4) PW_{bias}: minimum value of the PPG signal
(5) PW_{max}: maximum value of the PPG signal
(6) DPW_{max}: value of the PPG signal at $t_{maxslope}$
(7) *NPWA*: area of the PPG signal normalized by *FFI*

These parameters provide information on various hemodynamic states such as blood pressure and vascular compliance. For example, *NPWA* represents the mean value of the pulsatile component of the arterial blood volume and is a candidate of substitution of *BP*.

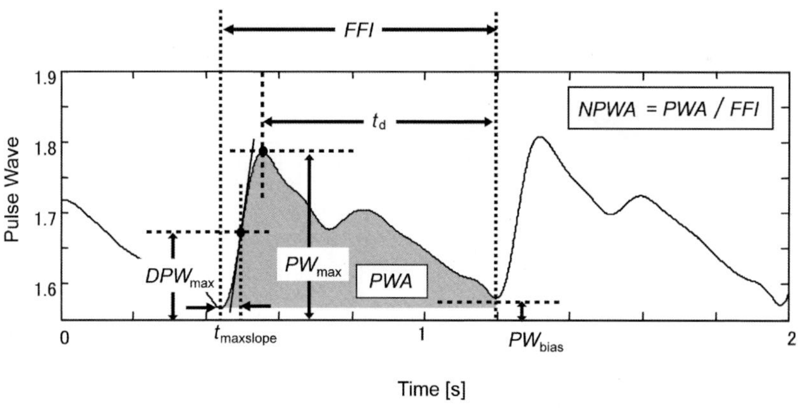

Figure 1. Example of PPG signal.

3. Experiments

In this study, 29 healthy subjects (21 males and 8 females, 23.2 ± 3.7 years) participated in two consecutive experiments. In the first experiment, the subjects were asked to perform the Valsalva maneuver, in order to obtain the constant mixing matrix $\boldsymbol{A}$ for each subject. In the second experiment, the subjects watched a moving video image. These experiments are described below in detail.

1) Experiment I (Valsalva maneuver): Each subject preformed the Valsalva maneuver by conducting a maximal forced expiration against a closed glottis and holding this condition for 1 min. It was found that the baroreflex sensitivity of the subject during the Valsalva maneuver decreases and hence $\rho_{\max}$ decreased. The subject's ECG, continuous blood pressure, and finger PPG were recorded during the experiment. The subjects performed two Valsalva maneuvers during this experiment. During the resting condition, the subjects were instructed to breathe normally.

2) Experiment II (viewing a moving video image): After Experiment I, the subjects watched a moving video image projected by an LCD projector, as shown in Fig. 2. The video was recorded during a continuous movement using a handheld camera; hence, the subjects were at an increased risk of experiencing VIMS. Before and after watching the video, the subjects observed a still picture of a landscape for 5 min as control. The physiological parameters measured in this experiment were the same as those measured in Experiment I. After the second experiment, the subjects were instructed to answer a simulator sickness questionnaire (SSQ) [1]. The total score (*TS*) of the SSQ was calculated in order to subjectively evaluate the intensity of VIMS.

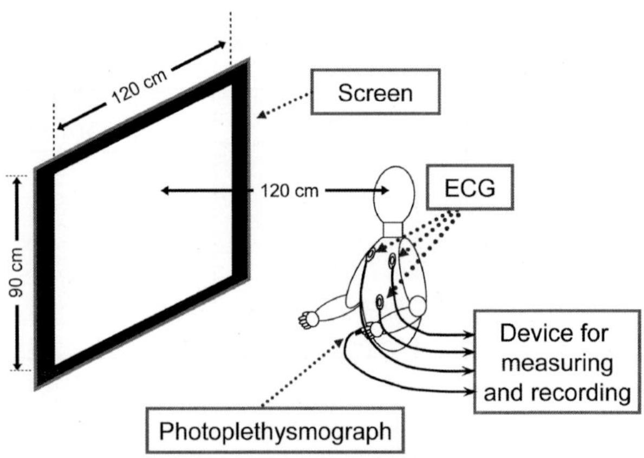

Figure 2. Setup of Experiment II.

4. Analyses

On the basis of matrix $\boldsymbol{X}$ obtained from Experiment I, the mixing matrix $\boldsymbol{A}$, its pseudoinverse $\boldsymbol{A}^+$, and the independent component matrix $\boldsymbol{S}$ were calculated using the ICA for each subject. n time series data of the independent component IC_l is defined as $l = 1, 2, \ldots, n$ as $IC_l = \{s_l(1), s_l(2), \ldots, s_l(K)\}$ with a data size of K corresponding to the duration of the experiment. Let the time series $\rho_{\max}(BP)$ and $\rho_{\max}(IC_l)$ represent $\rho_{\max}$ between HR and BP and $\rho_{\max}$ between HR and an independent component IC_l, respectively.

Let j^* denote the optimal number of l that is required to minimize the mean square error between $\rho_{\max}(ES)$ and $\rho_{\max}(IC_l)$ as follows:

$$j^* = \arg\min_{l=1,\cdots,n} E[\{\rho_{\max}(ES) - \rho_{\max}(IC_l)\}^2] \tag{6}$$

where $\rho_{\max}(ES)$ represents an event signal in the Valsalva maneuver. It is defined as follows:

$$\rho_{\max}(ES) = \begin{cases} 1 & \text{if } 0 \le t < 310, \\ & \phantom{\text{if }} 360 \le t < 550, \\ & \phantom{\text{if }} 600 \le t \le 900 \quad \text{in s} \\ 0.5 & \text{if } 310 \le t < 360, \\ & \phantom{\text{if }} 550 \le t < 600 \quad \text{in s} \end{cases} \tag{7}$$

where t represents the duration of the experiment.

Applying $\boldsymbol{A}^+$ obtained above to the data measured in Experiment II yields another parameter IC_l. If $l = j^*$ in Eq. (7), which is based on the data obtained from Experiment I, IC_{j^*} represents the independent component time series similar to that of the event signal $\rho_{\max}(ES)$. Let $\rho_{\max}(IC_{j^*})$ represent $\rho_{\max}$ between HR and IC_{j^*}. $\rho_{\max}(IC_{j^*})$ can be obtained without measuring BP. Similarly, $\rho_{\max}$ was also calculated from the data obtained from Experiment II.

5. Results and Discussion

5.1. *Experiment I (Valsalva maneuver)*

In Experiment I, data of all the 29 subjects could be successfully obtained and analyzed. The relationship between the number of independent components n and the root mean square error obtained from $\rho_{\max}(BP)$ and $\rho_{\max}(IC_{j^*})$ indicated that the optimal value of n was 4. Therefore, n was set to 4 for all experiments. Figure 3 shows the changes in $\rho_{\max}(BP)$ and $\rho_{\max}(IC_{j^*})$ with time. Each value of $\rho_{\max}$ is the average value for 29 subjects estimated every 30 s. The standard deviation (SD) of these values is shown in the lower part of the figure. These results indicate that the change in $\rho_{\max}(IC_{j*})$ with time is similar to that in $\rho_{\max}(BP)$. Note that $\rho_{\max}(IC_{j*})$, which was calculated without measuring BP, was similar to $\rho_{\max}(BP)$.

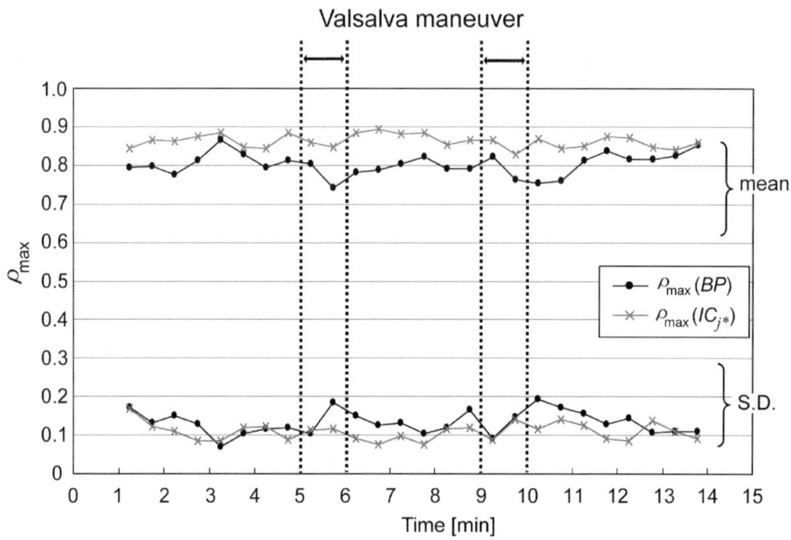

Figure 3. Changes in $\rho_{\max}$ in Experiment I (Valsalva maneuver). Each value of $\rho_{\max}$ was the average value for 29 subjects.

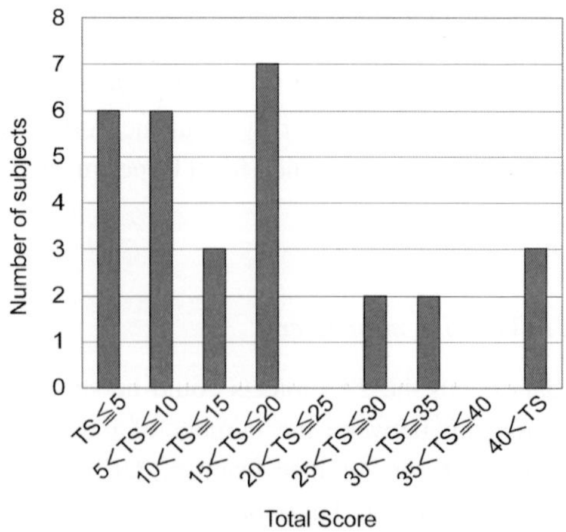

Figure 4. Results of subjective evaluation based on the total score (TS) of SSQ.

5.2. *Experiment II (Presentation of the Swaying Video Image)*

The data of all the 29 subjects could be successfully obtained in Experiment II. Figure 4 shows the result of subjective evaluation based on the SSQ. In general, VIMS causes symptoms such as simulator sickness. Therefore, it was estimated that the threshold of the *TS* for determining whether the subjects suffered from VIMS or not was 13.22; this represents the median of the *TS* over 29 subjects. The subjects were divided into two groups—sick group and healthy group. The sick group consisted of 15 subjects with a *TS* higher than 13.22, and the healthy group consisted of 14 subjects with a *TS* lower than 13.22.

Figures 5 and 6 show the changes in ρ_{max} (BP) and ρ_{max} (IC_{j*}) of the sick and healthy groups. In these figures, each value of ρ_{max} of the sick group is compared with that of the healthy group. The shaded area in these figures represents the time at which a significant difference in the values of ρ_{max} ($p < 0.05$) between the two groups was found by using Welch's *t*-test. ρ_{max} (BP) and ρ_{max} (IC_{j*}) of the sick group were significantly lower than those of the healthy group during the viewing of the video image. However, in some instances, ρ_{max} (BP) of the healthy group was significantly less than that of the sick group. This implies that the decrease in ρ_{max} (BP) reflects not only the effects of VIMS caused by watching a moving video image but also other effects such as a psychological effect. On the other hand, significant differences in ρ_{max} (IC_{j*}) between the sick and healthy groups shown in Fig. 6 indicates the effect of VIMS. This result implies that ρ_{max} (IC_{j*}) calculated using the constant mixing matrix obtained from the result of the Valsalva maneuver experiment was valid for estimating the effects of VIMS.

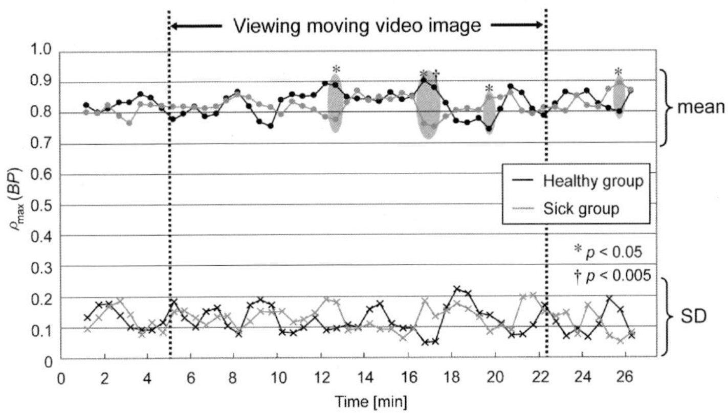

Figure 5. Comparison of ρ_{max} between sick and healthy groups on the basis of BP.

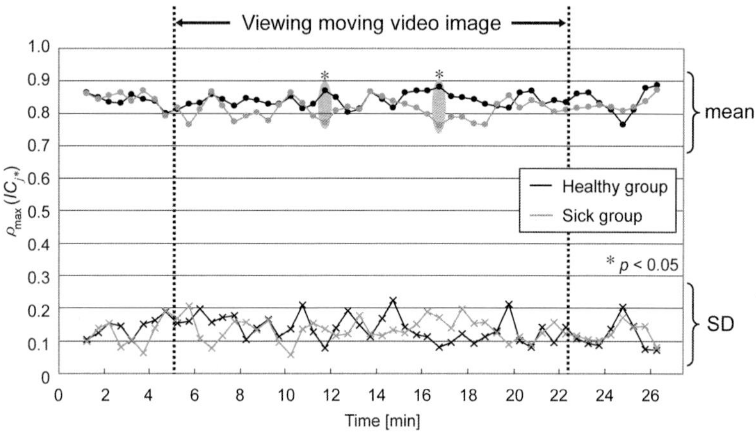

Figure 6. Comparison of $\rho_{\max}$ between the sick and healthy groups on the basis of *IC*.

6. Conclusion

In order to quantify the effects of VIMS, this study has proposed a new method for determining an event-related parameter by finger PPG using the ICA. From the experimental results, it was concluded that the proposed method could extract an independent component related to the event signal in order to maximize the cross-correlation between the heart rate and the independent component used for estimating the effects of VIMS. Furthermore, the cross-correlation coefficient $\rho_{\max}$ based on the independent component related to the event signal can be obtained without measuring the blood pressure.

However, it is important to determine the optimal value of *m* that is required to enhance the efficiency of the proposed method. In the future, we must develop a method to determine a constant mixing matrix using PPG and a mathematical theory or a calculation algorithm for ICA. In addition, it is important to understand physiological mechanisms that distinguish independent components.

Acknowledgment

Makoto Abe gratefully acknowledges the support of the Tohoku University Global COE Program: "Global Nano-Biomedical Engineering Network Centre."

References

1. R. S. Kennedy, N. E. Lane, K. S. Berbaum and M. G. Lilienthal, Simulator sickness questionnaire: An enhanced method for quantifying simulator sickness. *Int J Aviation Psychology* **3**, 203–220 (1993).
2. M. B. Jones, R. S. Kennedy and K. M. Stanney, Toward systematic control of cybersickness. *Presence, Teleoperators and Virtual Environments* **13**, 589–600 (2004).
3. N. Sugita, M. Yoshizawa, M. Abe, A. Tanaka, T. Watanabe, S. Chiba, T. Yambe and S. Nitta, Evaluation of adaptation to visually induced motion sickness based on the maximum cross-correlation between pulse transmission time and heart rate [Online]. *J Neuroeng Rehabil*, http://www.jneuroengrehab.com/content/4/1/35 (2007).
4. N. Sugita, M. Yoshizawa, A. Tanaka, K. Abe, S. Chiba, T. Yambe and S. Nitta, Quantitative evaluation of effects of visually-induced motion sickness based on causal coherence functions between blood pressure and heart rate. *Displays* **29**, 167–175 (2008).
5. A. Hyvärinen and E. Oja, A fast fixed-point algorithm for independent component analysis. *Neural Computation* **9**, 1483–1492 (1997).
6. A. Hyvärinen, Fast and robust fixed-point algorithms for independent component analysis. *IEEE Transactions on Neural Networks* **10**, 624–634 (1999).

SENTINEL LYMPH NODE BIOPSY AND MAPPING BY SILICA-COATED FLUORESCENT BEADS

LIMAN CONG *

1) Department of Surgical Oncology, Graduate School of Medicine, Tohoku University, 1-1 Seiryo-machi, Aoba-ku, Sendai 980-8574; Japan

MOTOHIRO TAKEDA

2) Division of Nano-Medical Science, Graduate School of Medicine, Tohoku University, 1-1 seiryo-machi, Aoba-ku, Sendai 980-8574, Japan

MIKA WATANABE

3) Department of Pathology, Tohoku University Hospital, 1-1 seiryo-machi, Aoba-ku, Sendai 980-8574, Japan

YOSHIO KOBAYASHI

4) Department of Chemical Engineering, Graduate School of Engineering, Ibaraki University., Hitachi316-8511, Japan

MASAKI KOBAYASHI

5) Department of Electronics, Tohoku Institute of Technology, 35-1kasumi-cho, Yagiyama, Taihaku-ku, Sendai 982-8577, Japan

NORIAKI OHUCHI

6) Department of Surgical Oncology, Graduate School of Medicine, Tohoku University, 1-1 Seiryo-machi, Aoba-ku, Sendai 980-8574, Japan

Sentinel lymph node biopsy (SLNB) has been developed as a new diagnostic and therapeutic modality in melanoma and breast cancer surgery. There are two major methods for detection of SN using blue dye and radioisotope. However, there are disadvantages for each method. In this study we used nano-sized fluorescent beads of extremely narrow diameter distribution. This paper describes silica-coating of fluorescent microspheres by means of a seeded polymerization technique based on the Stöber method with modifying TEOS concentration at 0.00038~0.2M. We have successfully performed silica coating of fluorescent microspheres. We then demonstrated an application of silica-coated fluorescent beads for sentinel lymph node mapping by an animal model. Taken together, the chemical, optical and *in vivo* data presented in this study demonstrate the potential role of silica-coated fluorescent beads for imaging in medical treatments.

Keywords: sentinel lymph node biopsy, silica-coated fluorescent nano-particle, Stöber method

---

* Liman CONG is a Tohoku University Global COE Research Assistant.

1. Introduction

The incidence and the mortality of cancer are increasing. Early and accurate diagnoses are the most important issues for improvement of mortality. Functional nano-particles and their detection techniques are the most expected issues for the improvement of cancer therapy.

A sentinel lymph node (SN) is a lymph node in a given lymphatic drainage group that is first to receive lymphatic flow from a primary tumor site. This node should be the first to become involved by metastasis from the tumor. Therefore, the histological status of the SN should be highly predictive of metastatic involvement of the lymph node group in which it is situated [1]. A tumor-negative SN virtually excludes involvement of a regional lymphatic system by surgical procedures. SN biopsy in breast cancer surgery has been developed to assess axillary nodal status accurately without removing most axillary contents, thus avoiding unnecessary axillary lymph node dissection Recently, support has increased for use of SN biopsy in staging patients with primary breast cancer [2,3].

Sentinel node navigation surgery is emerging as a “made-to-order surgery” for carcinomas in early stages. The sentinel node (SN) is the first lymph node on the lymphatic drainage pathway from the cancerous lesion. In cases where the SN has a metastasis, there is the possibility of another cancer positive lymph node. When the SN is negative for metastasis, we can consider that there will be no other cancer positive lymph node and lymph node dissection (LND) is necessary except the SN [4]. This leads to the avoidance of functional or organic complication after LND. There are two major methods for detection of SN using tracer molecules or particles. Dyes and/or radioisotopes have been employed for SN detection in standard methods, however each detection method has advantages and disadvantages. The dye method is not available for pre-operational detection of SNs. And there is limited number of hospitals in which radioisotope method is available for some regulations. Thus, a new non-invasive method that makes up for the disadvantages is expected.

In the present study, we tried to perform silica coating of 40nm fluorescent beads and to evaluate the utility of silica-coated fluorescence beads for SNB by a rat model.

2. Materials and Methods

2.1. *Chemicals*

The chemicals PVP (polyvinylpyrrolidone, K-30, Wako Pure Chemical Led, Osaka, Japan), TEOS (tetraethoxyorthosilicate, 99.5%, Wako Pure Chemical Led, Osaka, Japan), NH_4OH (25% aqueous solution, Wako Pure Chemicals Led, Osaka, Japan) and Ethanol (99.5% Wako Pure Chemicals Led Osaka, Japan), were used as received. All chemicals were used as received. Ultra pure deionized water (resistivity higher than 18 MΩ cm) was used in all the preparations.

2.2. *Silica-Coating Fluorescent Nano-Particles*

The Stöber method using TEOS was employed for silica-coating of the fluorescent nano-particles. Silica-coating of the fluorescent microspheres was carried out with ammonia-catalyzed reaction of TEOS in ethanol–water solution in a hermetically sealed reactor equipped with a magnetic stirrer at room temperature. Ethanol solution of TEOS was added to aqueous PVP solution under vigorous stirring after addition of the suspension of the fluorescent microspheres. Hydrolysis reaction of TEOS was initiated by the addition of the aqueous ammonia solution to form silica shell on the microspheres, which was added with reaction at room temperature for 12 hours or more.

The size of silica coated fluorescent nano-particles was estimated with a transmission electron microscopy (TEM).

2.3. *Electron Microscopy*

The silica-coated fluorescent nano-particles were observed by a transmission electron microscope (TEM). The TEM was (H-7600, Hitachi Science System Ltd., Hitachi, Japan) operated at 80~100kV accelerating voltage. Silica-coating fluorescent nano-particles trickled the particulate suspension into the collodion membrane attachment mesh (2634N, F-400, Nissin EM, Tokyo, Japan) directly and observed.

2.4. *Instrumentation*

We designed a laser scanning fluorescence imaging system, which consists of laser system, a resonant scanner (resonant frequency/200Hz), a cylindrical lens and a charge-coupled device (CCD) camera. We used a diode laser (wavelength 657nm 56ICS153/HS, CVL (Melles Griot), NM, USA) for excitation. The inguinal and femoral areas were continuously scanned with an area of 30×50mm. Illuminant recorder was 7.3×10^{-5} W/cm^2. The fluorescence image was observed using a CCD camera (ORCA II, Hamamatsu Photonics KK, Hamamatsu, Japan) with band-pass (filter confer wavelength/680nm, with fall width at half maximum /30mm).

3. Results and Discussion

We successfully made up silica-coated fluorescent nano-particles. The average size of silica coated fluorescent beads was 60nm (Fig. 1). We can clearly observe the silica-coated fluorescent nano-particles subcutaneously in the lymphatic pathway and inguinal lymph nodes (Fig. 2).

There are two advantages of the silica coated fluorescent nano-particles. The first is that it is physically and chemically stable and they can prevent adverse reaction when they are administered to a body. The second is that we can modify their size according to our purposes by changing thickness of silica shell. Based on the optical trait of the organism, we used the fluorescence of near infrared. It also needs an exploitation of the new measurement technique for detecting fluorescence in deep site of organs.

Safety of nano-particles is very important for medical applications. We have studied dynamics and distribution of silica coated nano-particles and we are studying distribution of silica-coated fluorescent nano-particles in the body by TEM and estimating LD_{50} [5].

Based on the optical trait of the organism, we used the fluorescence of near infrared, there also needs an exploitation of the new measurement technique for detecting fluorescence in deep site of organs [6,7].

In conclusion, novel silica-coated fluorescent nano-particles could be used for SNB. And silica-coating has advantages for modulation of the size and giving stability to nano-particles. Further precise studies should be performed to establish safety and utility with specific size for each purpose.

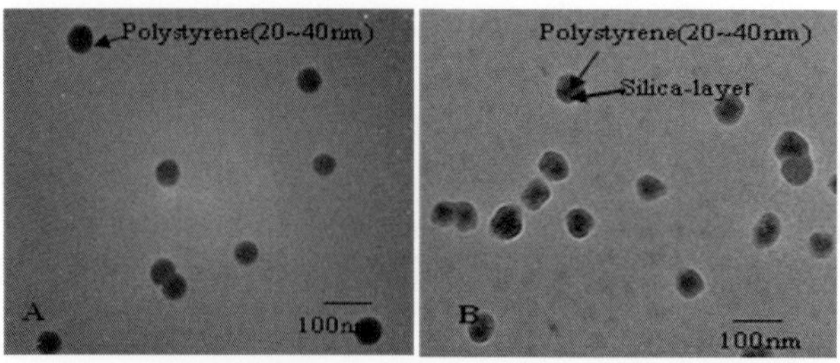

Figure 1. TEM images of FluoSpheres® (F-8789) and silica-coated fluorescent nano-particles.
A; FluoSpheres® (F-8789) beads at 40nm.
B; Silica-coated fluorescent nano-particles at 60nm.

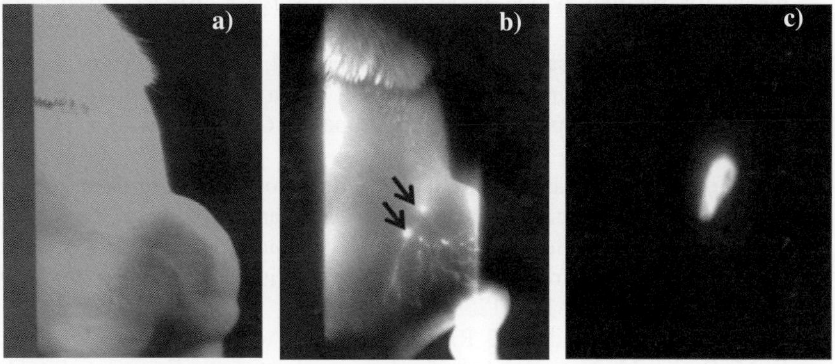

Figure 2. Sentinel lymph node images using silica-coated fluorescent nano-particles in the rat.
a) An image of inguinal region under weak light.
b) A fluorescent image of SNs 120min after detection (arrows).
c) A fluorescent image of an isolated SN.

Acknowledgments

This work was supported by Grants-in-aid for Research Projects, Promotion of Advanced Medical Technology (H18-Nano-001), from the Ministry of Health, Labor and Welfare, Japan and the Global Center of Excellence (GCOE) Program "Global Nano-Biomedical Engineering Education and Research Network Centre", the Japan Society for the Promotion of Science.

References

1. M. Nakajima, M. Takeda, M. Kobayashi, S. Suzuki and N. Ohuchi, Nano-sized fluorescent particles as new tracers for sentinel node detection: experimental model for decision of appropriate size and wavelength. *Cancer Sci.* **96**, 353–356 (2005).
2. Y. Kobayashi, K. Misawa, M. Takeda, M. Kobayashi, M. Satake, Y. Kawazoe, N. Ohuchi, A. Kasuya and M. Konno, Silica-coating of AgI semiconductor nanoparticles. *Colloids and Surfaces A: Physicochem. Eng. Aspects* **242**, 47–52 (2004).
3. Y. Kobayashi, K. Misawa, M. Kobayashi, M. Takeda, M. Konno, M. Satake, Y. Kawazoe, N. Ohuchi and A. Kasuya, Silica-coating of fluorescent polystyrene microspheres by a seeded polymerization technique and their photo-bleaching property. *Colloids and Surfaces A: Physicochem. Eng. Aspects* **251**, 197–201 (2004).
4. Y. Sakurai, M. Takeda, Y. Kawazoe, A. Kasuya, Y. Kobayashi, T. Kamei, M. Nakajima and N. Ohuchi, Nanosized silver iodide beads as new contrast

media for sentinel lymph node navigation surgery. *Breast Disease* **25**, 55–56 (2006).

5. Y. Kobayashi, K. Misawa, M. Kobayashi, and M. Takeda, Silica-coating of fluorescence polystyrene microspheres by a seeded polymerization technique and their photo-bleaching property. *Colloids and Surf A* **242**, 47–52 (2004).
6. J. Shen, M. Z. Gilcrease, G. V. Babiera, M. I. Ross, F. Meric-Bernstam, B. W. Feig, H. M. Kuerer, A. Francis, F. C. Ames and K. K. Hunt, Feasibility and Accuracy of Sentinel Lymph Node Biopsy After Preoperative Chemotherapy in Breast Cancer Patients With Documented Axillary Metastases. *CANCER* **109**, 1255–1263 (2007).
7. R. Sharma, J. A. Wendt, J. C. Rasmussen, K. E. Adams, M. V. Marshall and E. M. Sevick-M, New Horizons for Imaging Lymphatic Function. *Ann. N. Y. Acad. Sci.* **1131**, 13–36 (2008).

IN VIVO REAL-TIME TRACKING OF SINGLE PARTICLE IN TUMORS OF MICE

YOHEI HAMANAKA [1)*], MASAAKI KAWAI [1)], KOHSUKE GONDA [2)], MOTOHIRO TAKEDA [2)], NORIAKI OHUCHI [1)†]

1) Department of Surgical Oncology, Graduate School of Medicine, Tohoku University, 1-1 Seiryo-machi, Aoba-ku, Sendai 980-8574, Japan

2) Department of Nano-Medical Science, Graduate School of Medicine, Tohoku University, 1-1 Seiryo-machi, Aoba-ku, Sendai 980-8574, Japan

The real-time tracking of a single molecule is a very useful technique to demonstrate the dynamics of drugs *in vivo*. We have succeeded in capturing the specific delivery of trastuzumab conjugated with a Quantum dot and fluorescent substances of various sizes in animal models. These results revealed the particular movements of drugs or particles in the tumor tissues. Knowledge of the detailed movement of particles incorporated in drugs can lead to improvement of the design of drugs. We are applying this single molecular imaging technique to estimate the efficacy of a drug delivery system.

Keywords: molecular imaging, drug delivery system, Quantum dot, polymeric micelle

1. Introduction

Many kinds of drugs are used therapeutically for various diseases. Nevertheless, only a small number of drug molecules taken into the body reach the target tissues. Much of the administered drug may be useless because it is excreted from the body without reaching the target. It may also cause adverse effects on normal cells.

The ideal delivery of drugs sends the necessary quantities to the target cells based on the demand. A drug delivery system (DDS), which requires the integration of various specialties such as medicine, pharmacology, material engineering and bio-engineering, is a particular way to achieve this.

There are four basic concepts for an ideal DDS: (1)"controlled release of the drug", (2)"stabilization and life-time prolongation of the drug", (3)"absorption acceleration through an organic barrier" and (4)"targeting of the intended cells" [1]. Various modifications of drugs have been developed to attain them [2]. The

---

[*] Yohei Hamanaka is a Tohoku University Global COE Research Assistant.

[†] Noriaki Ohuchi is a Tohoku University Global COE Member.

major modification is capsuling of drugs by polymeric micelles, liposomes or metals. Some modified drugs have been already certificated as medicine. Therefore, designing a more efficient DDS remains a formidable challenge.

The dynamics of drugs *in vivo* have been hardly observed. It is important to know where and how long a drug stays in the tissue, where the drug is released from its carrier, and how the fragments of the carrier are excreted from the body.

The technique of *in vivo* real-time tracking of a single molecule is useful to know the dynamics of a drug. However, there have been some problems that have so far prevented us from accomplishing this.

Conjugating a fluorescent substance to the aiming molecule is the standard way to demonstrate the dynamics of the molecule. Fluorescent proteins such as GFP are often used. A crucial weak point of the florescent protein is that the illumination time is short. A quantum dot is the best substance to trace the dynamics of a single molecule because it has intense and stable illumination.

The movement caused by breathing or cardiac motion in animals is crucial to observe the dynamics of a single molecule *in vivo*. Furthermore, the resolution of the lens of a microscope cannot be applied for single molecule imaging. We developed a confocal microscope with nano-order resolution by tracing the gravity center of the molecules.

We succeeded in the real-time tracking of trastuzumab conjugated with a quantum dot, from the tumor vessels to the perinuclear region of tumor cells [3]. This result could contribute to our analysis of the dynamics of drugs in the tissue. As a next step of the former study, we observed the tracking of different sizes of nano-particles *in vivo* to apply this technique to clarify the dynamics of the nano-particles.

2. Materials and Methods

2.1. *Fluorescent Beads and Quantum Dots*

We used Fluorescent beads (Fluospheres; Invitrogen Molecular Probes, Eugene, Oregon, U.S.A.) and Quantum dots (Quantum dot 705 ITK kit; Quantum Dot Corp., Hayward, California, U.S.A.). The Fluospheres were high-quality ultraclean polystyrene microspheres; the sizes we selected for study were 40nm, 100nm and 200nm in diameter. The Quantum dot 705 was 20nm in diameter.

2.2. *Cell Line and Mouse Model*

The human breast cancer cell line KPL-4 was kindly provided by Dr.J.Kurebayashi (Kawasaki Medical School, Kurashiki, Japan) [4]. KPL-4 cells were cultured in DMEM supplemented with 5% fetal bovine serum (FBS).

A suspension of KPL-4 cells (1.0×10^7cells/100μl DMEM) was subcutaneously transplanted into the dorsal skin of female Balb/c nu/nu mice at 5-7weeks of age (Charles River Japan, Yokohama, Japan). Several weeks after the inoculation, mice with tumor volumes of 100 mm^3-200mm^3 were selected. The mice were anesthetized by an intraperitoneal injection of a mixture of ketamine and wylazine at doses of 95mg/kg and 5mg/kg, respectively. The Fluospheres and Quantum dots were injected into the tail vein of the mice.

The dorsal skin fold chambers were used to fix the mouse tumors over the objective lens of an inverted microscope. A polyvinyl chloride plate with a small window was mounted on the stage to fix the extended double layer of dorsal skin including the tumor. The skin was sutured around the window and fixed. The mouse was fixed on the stage designed to stabilize the chamber. The tumor was exposed by oval skin and subcutaneous incisions measuring up to 10mm in diameter and then it was placed on aneutral saline-mounted coverslip.

2.3. In vivo *Imaging Systems*

The optics system for the observations primarily consisted of an epifluorescence microscope (IX71; Olympus, Tokyo, Japan) with modifications, a Nipkow lens confocal unit (CSU10; Yokokawa, Tokyo, Japan) and an electron multiplier CCD camera (iXon887; Andor, Belfast, Northern Ireland).The object lens (×60, Numerical Aperture 1.45) was moved by a piezo actuator with a feedback loop (Nanocontrol, Tokyo, Japan) to stabilize the position of the focus. A computer controlled the piezo actuator in synchronization with the image acquisitions such that the object lens remained within the exposure time of the CCD camera. An area measuring approximately 30μm×30μm was illuminated by a green laser (532nm; CrystaLaser, Reno, Nevada, U.S.A.). This system can capture images of a single nanoparticle at a video rate of 33ms/frame.

The xy position of each fluorescent spot was calculated by fitting to a two-dimensional Gaussian curve. Single molecules could be identified by their fluorescence intensity. The resolution of the x and y directions of images taken at an exposure time of 33ms was 30nm, while also taking the standard deviation into consideration.

3. Results

3.1. *Trajectories of Single Nano-particles in Perivascular, Interstitial and Intercellular Areas*

Nano-particles appeared in the tumor vasculature as soon as they were injected into the tail vein in the mice. We tracked 20nm, 40nm and 100nm particles for *in vivo* imaging, because the 200nm particle could not be seen out of the vasculature. The trajectories of the particles were classified into two kinds of movement. One was random diffusion which is known as Brownian motion and the other was directed movement from the vasculature to the lymphatic system. The movement of the 20nm particles in the perivascular area was mainly directed, while that of the larger particles was mainly random diffusion. The main movement of the 20nm particles changed into random diffusion in the interstitial region after 2 hours. In addition, the particles were seen in the intercellular region after 3 hours. In addition, the proportion of random diffusion increased in the tumor cells.

3.2. *Mean Square Displacements (MSDs), Velocities and Diffusion Coefficients in Different Regions of Tumors*

The MSD was calculated by the positional data for each particle. The MSDs of individual particles were defined by the following equation:

$$\mathrm{MSD}(n\Delta t) = \frac{1}{N-n}\sum_{i=1}^{N-n}[(x_{i+n} - x_i)^2 + (y_{i+n} - y_i)^2]$$

$$\mathrm{MSD}\,(\Delta t) = 4\mathrm{D}\Delta t + v^2(\Delta t)^2$$

$$\lim_{t\to 0} \mathrm{MSD}(\Delta t) = 4\mathrm{D}\Delta t$$

The MSD plots of each molecule size formed convex curves. This means that the movements of each molecule consisted of directed movement with random diffusion. The steepness of the curves of the interstitial region was lower than that of the perivascular region. That of the intercellular region was lower than the others.

The velocities changed in relation to the particle position and the particle size. Although the velocities were weakly dependent on the size, they were fastest in the perivascular region and slowest in the intercellular region for all sizes.

The curves of the diffusion coefficients of the 3 sizes were convex, inversely related to the diameter.

4. Discussion

The molecular mechanism underlying the movement and its cessation during the delivery of drugs or particles in animal models is the fundamental basis of drug delivery. There have been many different approaches to tumor-targeting nano-carriers. Polymeric micelles and liposomes are good candidates for nano-carriers for anticancer drugs. There is still little understanding of the biological behavior of nano-carriers, including such crucial features as their transport in the blood circulation, interstitial movement, translocation into the cytoplasm, and the final fate in the tissue.

We have succeeded in capturing the specific delivery of trastuzumab conjugated with a Quantum dot and different sizes of fluorescent substances in animal models. This approach is thus considered to provide new insight into particle behavior in complex biological environments. Such new insight will allow for rational improvements in particle design to increase the therapeutic index of the tumor-targeting nano-carriers.

Among the various kinds of nano-carriers, polymeric micelles have attracted much attention owing to their advantages such as very small size in a range of 10nm-100 nm, and high structural stability. They are composed of block copolymers which consist of poly-ethylene glycol and poly-aspartic acid (or poly-glutamic acid). Drugs are incorporated into a micelle's inner core by both chemical conjugation and physical entrapment [5]. However, the dynamics of polymeric micelles in the body have not yet been fully elucidated. It is important to clarify such dynamics, especially the moment and the location of drug release from the carrier, in order to design more effective polymeric micelles.

The technique of *in vivo* real-time tracking of a single molecule has the potential to clarify the molecular dynamics of drugs and their carriers, which may thus lead to an improved estimation of the efficacy of the drug delivery system.

Acknowledgments

The authors acknowledge the support of 2007 Tohoku University Global COE Program "Global Nano-Biomedical Engineering Education and Research Network Centre".

References

1. Y. Tabata, Significance of release technology in tissue engineering. *Drug Discov Today* **10**, 1639–1646 (2005).
2. M. Ferrari, Cancer nanotechnology: Opportunities and challenges. *Nat Rev Cancer* **5**, 161–171 (2005).
3. H. Tada, H. Higuchi, T. M. Watanabe and N. Ohuchi, *In vivo* real-time tracking of single quantum dots conjugated with monoclonal anti-HER2 antibody in tumors of mice. *Cancer Res* **67**, 1138–1144 (2007).
4. J. Kurebayashi, T. Otsuki, C. K. Tang *et al.*, Isolation and characterization of a new human breast cancer cell line, KPL-4, expressing the Erb B family receptors and interleukin-6. *Br J Cancer* **79**, 707–717 (1999).
5. M. Yokoyama, Drug targeting with nano-sized carrier systems. *J Artif Organs*. **8**, 77–84 (2005).

ANALYSIS OF POWER SPECTRUM AND FRACTAL DIMENSION DURING UNDULATION PUMP VENTRICULAR ASSISTANCE

HONGJIAN LIU [1)*], YASUYUKI SHIRAISHI [1)], XIUMIN ZHANG [2)], YUN LUO [1)], TOMOYUKI YAMBE [1)†]

1) Department of Medical Engineering and Cardiology, Institute of Development, Aging and Cancer, Tohoku University, 4-1 Seiryo-machi, Aoba-ku, Sendai 980-8575, Japan

2) School of Public Health, Jilin University, Changchun, China

The aim of this study was to assess cardiac autonomic nervous activities during long-term left heart ventricular assist by analyzing fractal dimension and spectral power of heart rate variability. An undulation pump ventricular assist device (UPVAD) was installed in the thoracic cavity of a goat by left thoracotomy. The inflow and outflow cannulae were sutured to the apex of the left ventricle and to the descending aorta, respectively. The driving cable was wired percutaneously to an external controller. The electrocardiogram was recorded at a sampling frequency of 1 kHz. The low frequency (LF) and high frequency (HF) peaks of HR were evaluated by power spectral analysis. The fractal dimension (FD) of the time series data was analyzed by the box-counting method. During the UPVAD support, HF power increased from 3.8 to 4.8 and the ratio of LF to HF decreased from 1.3 to 0.9. FD was 1.3 and 1.52 with natural and artificial circulation (significantly increase), respectively. These results indicated that both sympathetic and parasympathetic activities of the cardiac autonomic nerves changed during long-term UPVAD. UPVAD support may trigger a response from sympatho-vagal interaction at the vagal nuclei in the brain stem, as well as decrease sympathetic activity via a secondary increase in parasympathetic activity. Artificial circulation with UPVAD alters the fractal properties of HRV and increases the complexity of HRV.

Keywords: fractal dimension; power spectrum; undulation pump ventricular assist device; autonomic nervous system; heart rate variability.

1. Introduction

Many studies have reported that spectral analysis of heart rate variability (HRV) is a powerful and noninvasive tool for quantifying autonomic nervous system activity. Spectral analysis of short-term HR time series reveals a high-frequency peak (HF: 0.15–0.5 Hz) that is related to respiratory sinus arrhythmia and

---

[*] Hongjian Liu is a Tohoku University Global COE Fellow.

[†] Tomoyuki Yambe is a Tohoku University Global COE Member.

parasympathetic activity, a low-frequency peak (LF: 0.04–0.15 Hz) that is related to the sympathetic, vagal and baroreceptor mechanisms, and the LF/HF ratio that is considered by some investigators to reflect the sympatho-vagal balance [1].

Fractal dimension (FD) is used to analyse HR time series, as these series are essentially non-linear in nature. There are several ways to determine FD, which measures the space-filling propensity and complexity of the time series. Altered fractal correlation properties have been observed in patients with heart disease, and an increase in short-term fractal properties in patients with non-complicated coronary artery disease. Altered complexity of RRI dynamics, measured by approximate entropy, has also been described in various cardiovascular diseases [1,2].

Patients with severe heart failure do not have a LF/HF ratio in the RR interval, despite high levels of sympathetic activation and this may be secondary to abnormalities in the central autonomic regulation [3]. Left ventricular assist devices (LVAD) are used to hemodynamically support heart failure patients awaiting transplantation, and by themselves may promote recovery of cardiac function. Mechanically unloading the heart with an LVAD can decrease heart size, improve cardiac function, decrease plasma neurohormones and cytokines, and decrease ventricular expression of atrial natriuretic peptide [4]. Mechanical unloading has been improved to the extent that cardiac transplantation has become unnecessary in some patients. This recovery has increased interest in expanding the role of LVAD in treating heart failure [5]. We therefore examined whether use of power spectral analysis and fractal analysis of heart rate variability improved circulatory homeostasis and autonomic nervous system activity after LVAD implantation.

2. Materials and Methods

2.1. *Animal*

A female goat was used in the animal experiment that was performed after being approved by the Research Animal Resource Committee Guidelines of the Institute of Experimental Animals, Tohoku University School of Medicine (Japan).

2.2. *Implantation of UPVAD*

UPVAD (The University of Tokyo) is composed of a pairs of undulation pumps, motors, and undulation shafts. Each undulation pump consists of a pump housing, a disk, and a pair of membranes. A jellyfish valve was inserted into the outflow cannula to prevent diastolic backflow. The motor was manufactured with DC brushless motor parts and is controlled by pulse wave modulation with a one-chip microcomputer system. The rotation of the motor is converted by the undulation shaft to the waving motion of the disk. As the disk undulates, fluid is transferred from the inlet port to the outlet port.

The goat was placed on the surgical table in the lateral position under halothane anaesthesia, and the UPVAD was implanted with left ventricle to aorta (LV-Ao) bypass. The UPVAD was implanted into the thoracic cavity of the female goat by a left thoracotomy. The inflow and outflow cannulae were sutured to the apex of the left ventricle and to the descending aorta. These cannulae were connected to a computer-controlled undulation pump. The driving cable was wired percutaneously to an external controller (Fig. 1).

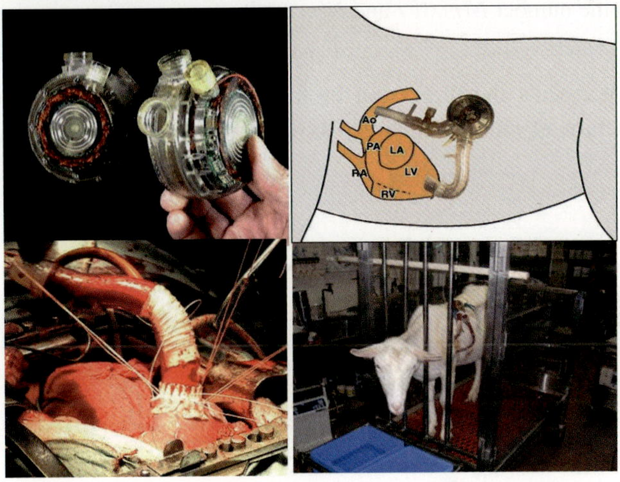

Figure 1. Undulation pump ventricular assist device (UPVAD).

2.3. *Experimental Protocol*

Testing was started one month after surgery. The experiment was performed daily in a quiet room in the morning between 10:00 and 12:00 hr every other day. While the animal was awake, during UPVAD support and without assistance, the electrocardiogram (ECG) was digitized by an A/D converter (TEAC, LX-10) at a sampling rate of 1 kHz and stored on a hard disk. The R-R interval was re-sampled by interpolation (using equidistant cubic splines) from the original signals.

2.4. *Analytical Method*

In the present study, we used a spectral analysis that is based on the fast Fourier transform (FFT) method. Two-minute periods of the HRV time series was computed based on the FFT.

FD analysis was used to evaluate the characteristics of fractals associated with HRV. We calculated FD by changing the coarse graining level, using a technique known as box-counting. Before calculating the fractal dimension, the time series data of the ECG were reconstructed in a phase plane. Through the use of a return map on a beat-to-beat basis, the heart rate was embedded into a two-dimensional phase plane. We first quantified the time-series data on a beat-to-beat basis, where the value of one beat was plotted on the *x*-axis and that of the next beat on the *y*-axis, we then divided the phase plane into several squares with side length *r*. The number of squares that contain at least one point is denoted as the number $N(r)$. If r and $N(r)$ are known,

$$r^{-D} \propto N(r)$$

where D represents the dimension of this structure of the return map. For a fractal structure, D value is expressed as a fraction (Fig. 2).

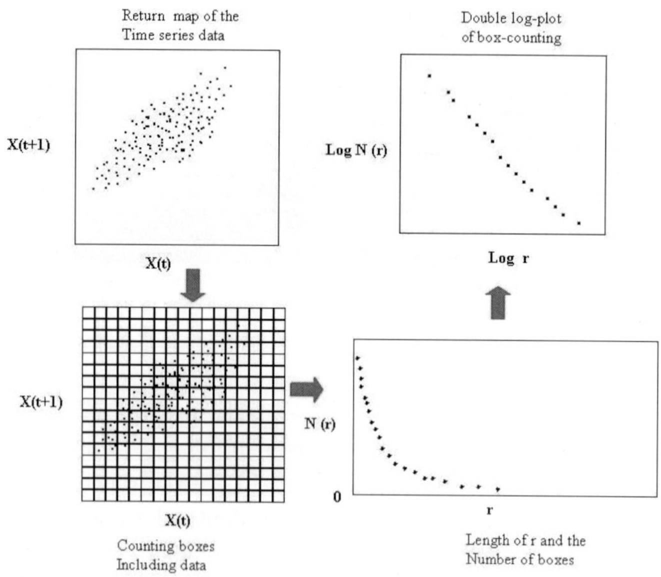

Figure 2. Box-counting of the return map.

2.5. *Statistical Analysis*

Data were expressed as mean ± S.D. A paired *t*-test was used to examine the statistical difference of HF, LF/HF and FD between with and without UPVAD. $p<0.05$ was considered to be statistically significant in all statistical analyses.

3. Results

AoP did not change significantly following the switch from without UPVAD support to with UPVAD support (98 ± 2 mmHg vs. 97 ± 9 mmHg), and HR tended to decrease, though not significantly (99 ± 5 beats/min vs. 92 ± 3 beats/min). Figure 3 shows an example of the time series and power spectral data of HRV between with and without UPVAD. A LF peak around 0.1 Hz and a HF peak around 0.3 Hz were clearly observed in the spectral data of both with and without UPVAD. HF was higher on UPVAD support than without UPVAD support (4.80 ± 0.13 vs. 3.80 ± 0.91). LF/HF tended to decrease under UPVAD support (0.90 ± 0.12 vs.1.30 ± 0.56). FD was significantly higher in the presence of UPVAD support than without UPVAD support (1.52 ± 0.12 vs. 1.30 ± 0.06).

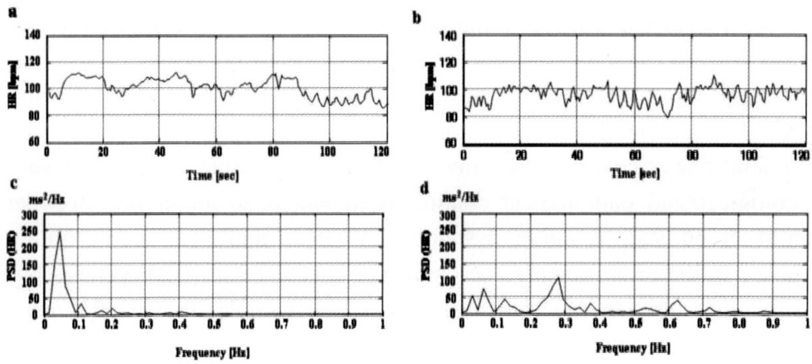

Figure 3. The examples of power spectral analysis with and without UPVAD assistance. (a) Time series data of HR without UPVAD. (b) Time series data of HR with UPVAD. (c) Power spectrum of HR without UPVAD. (d) Power spectrum of HR with UPVAD.

4. Discussion

In this study, we investigated the changes in cardiac autonomic nervous function during UPVAD support by analyzing heart rate variability. The spectral peaks were observed at the same frequency bands for HRV both with and without UPVAD. The power of the high frequency was higher during UPVAD support than without, suggesting that the increase in cardiac parasympathetic nervous activity was larger during the UPVAD support. The improvement seems to be due to an increase in cardiac function of the healthy goats. Activity of brain stem neurons that are involved in the regulation of cardiovascular function contains distinct LF and HF oscillatory components [3].

The primary intrinsic control of ventricular contractility by neurohormonal beta-adrenergic stimulation, preload and length-dependent activation, the HR under normal conditions, and the amplification of the force–frequency relation during beta-adrenergic receptor stimulation also play a role in ventricular contractility [6]. Dysfunctional β-adrenergic receptor (βAR) signaling in heart failure includes receptor downregulation and impaired signaling through the remaining receptors. Heart failure is associated with an increased level of circulating catecholamines. LVAD support has been shown to improve left ventricular βAR signaling after long-term LVAD support. These studies have shown that LVAD support can lead to upregulation of myocardial βAR density to near normal [7]. LF tended to be attenuated and the peripheral vascular resistance tended to decrease with LVAD assistance. LVAD assistance could also supply energy to the vascular system to decrease the workload (preload and afterload) on the heart and increase cardiac output [8]. In the future, we plan to carry out animal experiments using chronic heart failure goat models to clarify whether UPVAD support can improve automatic control in heart failure.

The fractal process in HRV has been investigated as a possible indicator of complex interactions that might reflect the number of inputs to HR controllers. The fractal component has been investigated as a possible indicator of various physiological and pathological conditions in which an autonomic disorder is often observed, such as aging, psychological stress, and electrical instability of the heart. In the present study, we confirmed that FD with UPVAD was bigger than that without UPVAD, suggesting that the complexity of HRV may be different from that with UPVAD.

Our study in a goat model of UPVAD assistance is consistent with the idea that UPVAD support increases parasympathetic outflow to the heart and reduces sympathetic vasomotor tone. UPVAD support may cause a response from sympatho-vagal interaction at the vagal nuclei in the brain stem, as well as decrease sympathetic activity via a secondary increase in parasympathetic activity.

Acknowledgments

This work was partly supported by Tohoku University Global COE Program "Global Nano-Biomedical Engineering Education and Research Network Centre".

References

1. V. K. Yeragani, E. Sobolewski, V. C. Jampala, J. Kay, S. Yeragani and G. Igel, Fractal dimension and approximate entropy of heart period and heart rate: awake versus sleep differences and methodological issues. *Clin Sci (Lond)* **95**, 295–301 (1998).
2. J. S. Perkiomaki, V. Jokinen, J. Tapanainen, K. E. Airaksinen and H. V. Huikuri, Autonomic markers as predictors of nonfatal acute coronary events after myocardial infarction. *Ann Noninvasive Electrocardiol* **13**, 120–129 (2008).
3. R. L. Cooley, N. Montano, C. Cogliati, P. van de Borne, W. Richenbacher, R. Oren and V. K. Somers, Evidence for a central origin of the low-frequency oscillation in RR-interval variability. *Circulation* **98**, 556–561 (1998).
4. M. L. Ogletree-Hughes, L. B. Stull, W. E. Sweet, N. G. Smedira, P. M. McCarthy and C. S. Moravec, Mechanical unloading restores beta-adrenergic responsiveness and reverses receptor downregulation in the failing human heart. *Circulation* **104**, 881–886 (2001).
5. S. Miyagawa, Y. Sawa, N. Fukushima, M. Nishimura, G. Matsumiya, S. Taketani, K. Horiguchi, S. Ohtake and H. Matsuda, Analysis of sympathetic nerve activity in end-stage cardiomyopathy patients receiving left ventricular support. *J Heart Lung Transplant* **20**, 1181–1187 (2001).
6. J. Ross, Jr., T. Miura, M. Kambayashi, G. P. Eising and K. H. Ryu, Adrenergic control of the force-frequency relation. *Circulation* **92**, 2327–2332 (1995).
7. P. K. Pandalai, C. F. Bulcao, W. H. Merrill and S. A. Akhter, Restoration of myocardial beta-adrenergic receptor signaling after left ventricular assist device support. *J Thorac Cardiovasc Surg* **131**, 975–980 (2006).
8. V. Jeevanandam, D. Jayakar, A. S. Anderson, S. Martin, W. Piccione, Jr., A. L. Heroux, J. Wynne, L. W. Stephenson, J. Hsu, P. S. Freed and A. Kantrowitz, Circulatory assistance with a permanent implantable IABP: initial human experience. *Circulation* **106** (Suppl 1), I183–I188 (2002).

FABRICATION OF TRANSPARENT ARTERIOLE MEMBRANE MODELS

TAKUMA NAKANO *

Department of Bioengineering and Robotics, Tohoku University
6-6-01 Aramaki-Aoba, Aoba-ku, Sendai, 980-8579, Japan

SEIICHI IKEDA

Department of Micro-Nano Systems Engineering, Nagoya University
Furo-cho, Chikusa-ku, Nagoya, 464-8603, Japan

TOSHIO FUKUDA

Department of Micro-Nano Systems Engineering, Nagoya University
Furo-cho, Chikusa-ku, Nagoya, 464-8603, Japan

TAKEHISA MATSUDA

Kanazawa Institute of Technology
7-1 Ohgigaoka, Nonoichi, 921 - 8501, Japan

MAKOTO NEGORO

Fujita Health University
1-98 Dengakugakubo, Kutsukake-cho, Toyoake, 470-1192, Japan

FUMIHITO ARAI †

Department of Bioengineering and Robotics, Tohoku University
6-6-01 Aramaki-Aoba, Aoba-ku, Sendai, 980-8579, Japan

We fabricated a transparent arteriole membrane model by grayscale lithography. We employed a wax and polyvinyl alcohol mixture as a novel molding material for fabricating a sacrificial model. Our objective was to complement previous surgical simulators for practice and rehearsal of medical treatments. Since block vessel models cannot recreate the moderate compliance that is generally seen in real blood vessels, we proposed a method for fabricating a transparent membranous arteriole model having a circular cross-section less than 500 μm in diameter. Here, we describe the fabrication of our model and the evaluation results of our new molding material.

* Takuma Nakano is a Tohoku University Global COE Research Assistant.
† Fumihito Arai is a Tohoku University Global COE Member.

1. Introduction

In the field of tissue engineering, many engineered approaches have been studied, including those employing artificial hearts, artificial joints, and synthetic vascular prostheses [1,2]. We have been developing tailor-made three-dimensional (3D) elastic membranous blood vessel models by employing 3D wax models [3,4]. We have also been developing a surgical simulator by connecting our elastic membranous models (Fig. 1).

Surgical simulators are used by physicians both to prepare for catheter treatments and develop new treatment methods. Thus, we constructed a pulsatile pump that reproduces patient-specific pulsative blood streaming, as evaluated by ultrasound or other measurement modalities. This device was connected to our aorta model and reproduced human-like pulsatile blood streaming inside the simulator. Such a vascular model can reproduce vascular pulsation along with streaming.

Because of the brittle nature of wax, the surgical simulator model needs to be greater than 500 μm in diameter. However, to better simulate an actual blood vessel environment, blood vessel models smaller than 500 μm in diameter are required. Furthermore, diseases such as arteria basilaris effect blood vessels under ϕ500 μm. Current surgical simulators, therefore, cannot be used for surgical rehearsal and training treatments of such diseases. Thus, a surgical simulator model with ϕ10–500 μm arterioles and capillaries is needed.

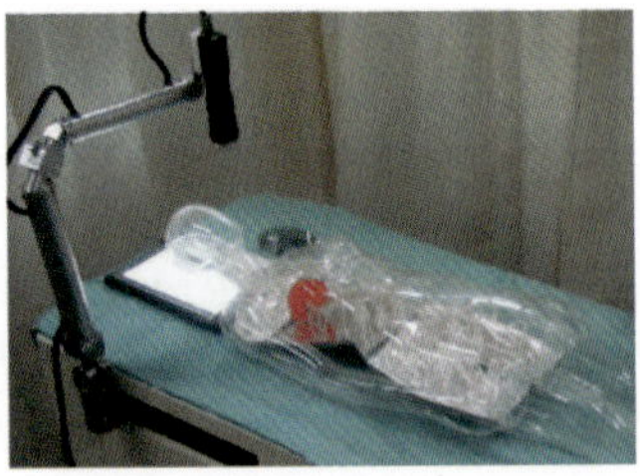

Figure 1. Surgical simulator for endovascular neurosugery fabricated by a photoelastic modality, allowing for the evaluation of stresses on the arterial structure applied during surgical procedures and blood streaming.

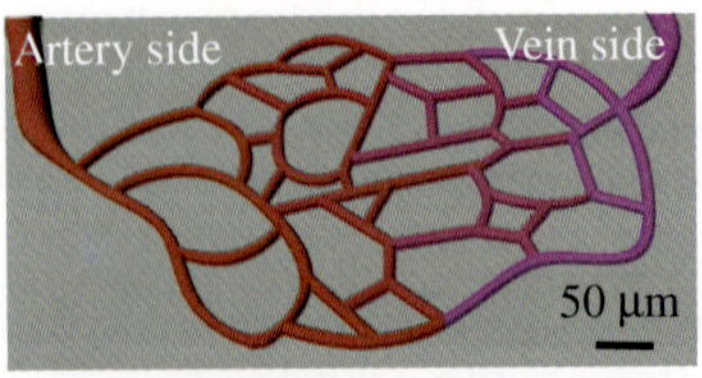

Figure 2. Designed arteriole and capillary vessel model.

To address this issue, we designed arteriole and capillary vessel models, as shown in Fig. 2. These models were fabricated by photolithography [5-7].

Figure 3 shows the multiscale fabrication methods used for different blood vessel models. We used an over exposure method, a reflow method, gray-scale lithography, and a layer stack molding machine. Among these, the over exposure method, the reflow method, and gray-scale lithography are used to generate fine structures. Due to limits of the fabrication accuracy, however, it is necessary to select the method appropriate for the fabrication of a model with the desired vessel diameter. After exposure, fabricated photoresist patterns are transcribed onto poly(dimethylsiloxane) (PDMS). Arteriole and capillary vessel block models with circular cross-sections are fabricated by bonding two patterned PDMS substrates using plasma treatment and heating (Fig. 4.).

Block models, however, cannot recreate the moderate compliance needed to mimic that of real blood vessels [8,9]. Thus, we believe that the use of arteriole and capillary vessel membrane models as surgical simulators can resolve this problem, and that sacrificial models are needed to improve membrane reproduction. However, because of the brittleness of wax, we were unable to fabricate sacrificial models by the previous fabrication method using a layer stack-molding machine.

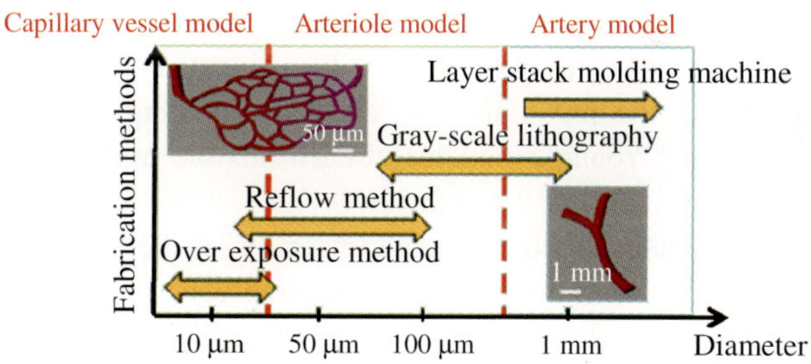

Figure 3. Fabrication methods for multiscale blood vessel models.

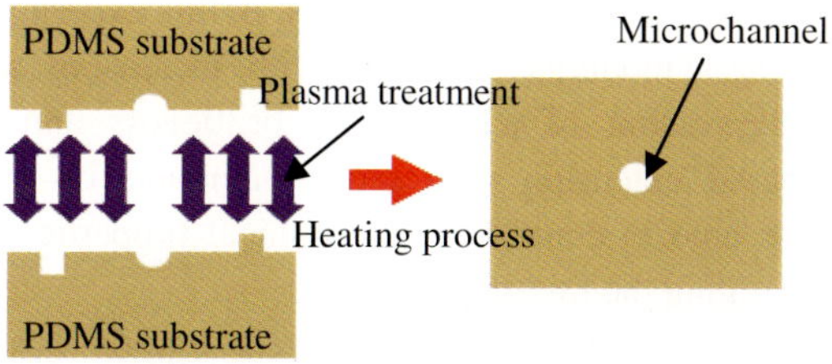

Figure 4. Concept behind the making of our capillary vessel model. After plasma treatment, we aligned two patterned PDMS substrates and bonded them to complete blood vessel model.

In this paper, we propose a novel method for fabricating transparent arteriole membranous models. We attempted to decrease the brittleness of wax by mixing it with PVA (polyvinyl alcohol). After fabricating sacrificial models using this mixture, arteriole membranous models were fabricated by dip coating.

The merits of making membranous models smaller than ϕ500 μm for surgical simulation include the following: (1) enables modeling of blood vessel diseases such as arteria basilaris that effect vessels with a diameter of 100 μm, and (2) allows fabrication of high-precision surgical simulators. In this paper, we also report our fabrication method, fabricated arteriole membranous model, and the evaluation results for the molding material (wax and PVA mixture).

2. Evaluation of Wax + PVA Mixture Material

For our fabrication process, a wax and PVA mixture was used to build the sacrificial models. These models were useful for fabricating membrane models smaller than 500 μm in diameter. When we used only wax for making the sacrificial models, a certain temperature had to be maintained in order to control the process. Furthermore, the fragility of wax proved unsuitable for our fabrication procedure. On the other hand, when PVA alone was used for making our sacrificial model, the model quickly deformed under its own weight when dip coated because PVA has a low bend strength. Therefore, we propose a mixture of wax and PVA for fabricating sacrificial models. We predict that the mixture has properties of both wax and PVA. We evaluated the properties of the mixture by changing the mix ratio. We evaluated melting time in different liquids (solubility) and Young's modulus of each mixture made using different mix ratios (wax:PVA = 1:0, 1:4, 2:3, 3:2, 4:1, 0:1).

For Young's modulus evaluation experiments, a tensile tester was used to measure Young's modulus. Samples fabricated using our wax and PVA mixture were 50 mm × 10 mm × 2 mm (long × wide × thick) in size. We plotted the measurement results as a load–displacement curve. We conducted the tensile experiment in triplicate for each wax:PVA mix ratio. We calculated Young's modulus using the curve data. Young's modulus for each mix ratio was calculated by averaging the experimental results. In this experimental results, we observed that Young's modulus increased with increasing amount of wax. Extension is an expression of a material's brittleness and ductibility. We observed an increased extension with greater amounts of PVA in the mix. Therefore, we were able to control the mechanical properties of the wax and PVA mixture by changing the mix ratio.

Melting time of each mix ratio in different liquids was determined by measuring the mass changes. Experiments were conducted to determine the solubility of each mix ratio. We used an ultrasound bath set at 50°C for our solubility experiments. We used DI water, acetone, and a mixture of DI water and acetone as solvents in the experiments. DI water easily melted PVA and acetone easily melted wax. We decided to use a 1:1 acetone to DI water mix to melt the wax and PVA combinations. Because of the physical characteristics of DI water, the mix samples melted easily. Although acetone can melt wax easily, it could only melt a few mixtures in our experiments. This was likely because wax was coated with PVA, so acetone could not reach the wax component of the mixtures. Thus, we assumed that the mixture of wax and PVA would melt easily by the DI water and acetone combination. This was verified by our experiment. Comparing these results, we found that all mixtures of wax and PVA could be melted in the DI water and acetone combination, but the solution rate was not constant. We observed that PVA absorbed DI water and increased in mass at the beginning of these experiments. Our findings indicate that a 1:1 combination can be used to melt wax and PVA mixed materials.

3. Fabrication of Arteriole Membrane Models

Since wax is fragile, we opted to use a mixture of wax and PVA as a material for our sacrificial model. During the fabrication process of the membrane models, we created model patterns by grayscale lithography and transcribed the resist patterns onto PDMS. We aligned two patterned PDMS substrates filled with the wax and PVA mixture on the patterned side, and dried them at room temperature. We fabricated our sacrificial models. This sacrificial model was coated with PVA to smooth the surface and was later coated with a transparent silicone resin by dip coating to achieve a membranous structure. After dip coating, the sacrificial model and PVA used for smoothing were dissolved by DI water and acetone, leaving a synthetic transparent arteriole membrane. Figure 5 shows a ϕ500 μm arteriole membranous model fabricated using this technique. The fabricated model has circular cross-section. The circularity of the inside of the

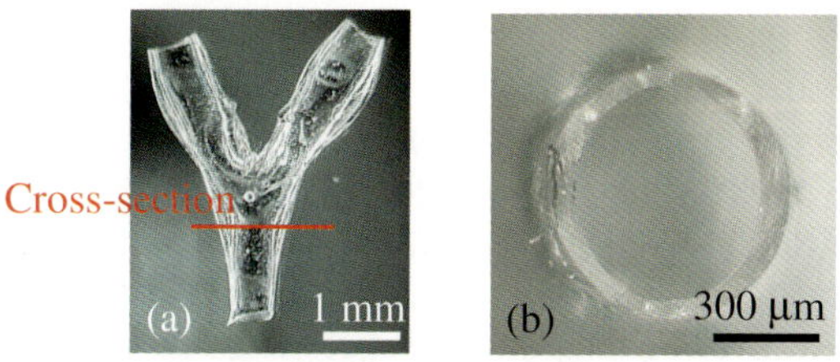

Figure 5. ϕ500 μm arteriole membrane model made by using a WAX and PVA mixture material and grayscale lithography. (a) Fabricated arteriole membrane model showing the cut line for (b), the cross-section of the fabricated arteriole membrane model.

membrane model was calculated by dividing the shortest axis by the longest axis; the calculated circularity of this model was 90%.

4. Conclusion and Discussion

In this paper, we proposed a fabrication method to obtain 100–500 μm diameter arteriole membrane models for surgical simulations. These blood vessel models were fabricated to complement with previously designed surgical simulators. These membrane models were designed to enable surgeons to practice and simulate treatments of diseases afflicting blood vessels of ϕ100–500 μm, and to provide a highly precise surgical simulator. We described the fabrication method used for our arteriole membranous model, as well as a detailed evaluation of the molding material. We fabricated our arteriole membrane models by grayscale lithography using a wax and PVA mixture. We determined that the use of a wax/PVA mix was essential and evaluated the mechanical and chemical properties of this material. Young's modulus of the mixed material increased with an increased amount of wax in our experiments. In addition, we observed that the extension of the mixture increased with the amount of PVA. We determined that a DI and acetone liquid was the most effective means for melting away the WAX and PVA mixture. Our new technique overcame the limits imposed by the brittleness of our previous sacrificial model.

Finally, we succeeded in making the membranous model with a hollow inner structure. This arteriole membrane model had circular inner cross-section with a calculated circularity of 90%.

Acknowledgments

This work was supported by a Grant-in-Aid for Scientific Research from the Ministry of Education, Culture, Sports, Science and Technology of Japan (No.17076015 and No.18206027), and Global COE Program, "Global Nano-Biomedical Engineering Education and Research Network Centre".

References

1. L. G. Griffith *et al.*, *Science* **295**, 1009–1014 (2002).
2. R. Langer and J. P. Vacanti, *Science* **260**, 920–926 (1993).
3. S. Ikeda, F. Arai, T. Fukuda *et al.*, *Journal of Robotics and Mechatronics* **17**, **3**, 327–334 (2005–2006).
4. S. Ikeda, F. Arai, T. Fukuda *et al.*, *Proceedings of MICCAI 2005, LNCS 3749*, USA, 925–932 (2005).
5. N. Futai, W. Gu and S. Takayama, *Adv. Mater* **16**, 1320–1323 (2004).

6. J. T. Borenstein, H. Terai, K. R. King *et al.*, *Biomedical Microdevices* **4**, 167–175 (2002).
7. M. Ikeuchi and K. Ikuta, *Proceedings of μTAS 2006*, 693–695 (2006).
8. T. Nakano, M. Tada, Y. C. Lin *et al.*, *Proceedings of MHS 2007*, 17 (1)–(6) (2007).
9. F. Arai, T. Nakano *et al.*, *Journal of Robotics and Mechatronics* **19**, 535–543 (2007).

ACHIEVEMENT OF MECHANICAL ASSISTANCE BY AN ARTIFICIAL MYOCARDIUM USING SHAPE MEMORY ALLOY FIBRE

YASUYUKI SHIRAISHI

Department of Medical Engineering and Cardiology, Institute of Development, Aging and Cancer, Tohoku University, 4-1 Seiryo-machi, Aoba-ku, Sendai 980-8575, Japan

TOMOYUKI YAMBE*

Department of Medical Engineering and Cardiology, Institute of Development, Aging and Cancer, Tohoku University, 4–1 Seiryo-machi, Aoba-ku, Sendai 980–8575, Japan

DAI HOMMA

Toki Corporation, 3–43–15 Ohmori-kita, Ohta-ku 143–0016, Japan

Recently, the ventricular assist devices are widely applied for a surgical treatment of the final stage of severe heart failure as the bridge to heart transplantation or the destination therapy. However, it was anticipated that the artificial components in the ventricular assist devices might cause the problems concerning thrombosis and infection. As heart failure involves the decrease in myocardial contractile function, the mechanical assistance by using an artificial myocardium might be effective. In this study, the authors developed a mechano-electric artificial myocardial assist system (artificial myocardium), which is capable of supporting natural contractile function from the outside of the ventricle.

1. Introduction

Chronic heart failure is functionally and structurally characterized by pathophysiological remodelling of the ventricle. In general, patients with severe heart failure, who present an increased ventricular filling pressure or insufficient amount of blood supply to the tissue, are given medical or surgical treatment. Recently, artificial circulatory assistance by using ventricular assist devices, such as artificial hearts, has been provide, which is followed by heart transplantation. However, the deficiency of donor hearts might be a serious problem in the

---

* Tomoyuki Yambe is a Tohoku University Global COE Member.

world. And the transplantation waiting period in this country extends to several years. Therefore, necessity of clinical application of artificial hearts with long-term durability has arisen [1-4]. As the size of the western ventricular assist devices, which are provided at present, is still big for the smaller body size Asian people, several artificial heart projects are being conducted in Japan, and one of these has started clinical trials.

There are also some devices or procedures suggested to be useful for the surgical treatment of severe heart failure, such as the ventricular CorCup, Myosprint, Coapsys or Dor's procedure. And recently, cell transplantation to repair or supplement the impaired heart tissue has been reported on as an alternative therapy for that [5-10]. There are many problems about the tissue reproduced *in vitro* or *in vivo* that are not yet solved. Moreover, any control of the implanted tissue might be impossible from the outside.

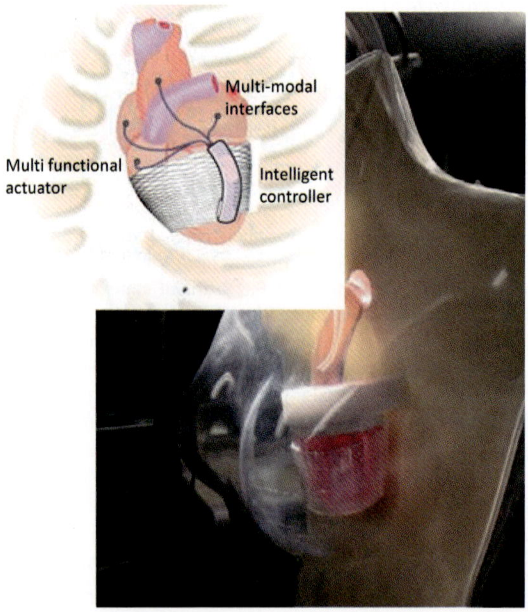

Figure 1. Schematic illustration and an installation in the mock of an artificial myocardium attached on the ventricular wall; the synchronous contraction can be achieved according to the natural physiological demand.

As the heart failure is caused by a decrease in the myocardial contractile function, the direct mechanical myocardial assistance in response to physiological deficit, *i.e.* a synchronous support of the contractile function from outside of the heart, might be effective. The purpose of this study was to develop an artificial myocardium, which would be capable of supporting the cardiac contraction directly by using the shape memory alloy fibre of a minute diameter based on nano-technology.

The authors have been developing a totally-implantable artificial myocardial assist device. The methodologies of the direct ventricular support systems were already reported on as direct mechanical ventricular assistance (DVMA) by Anstadt's or other groups, as well as the right ventricular assist device, which was invented and reported on at IDAC, Tohoku University [5-7, 11-16]. In this study, the authors developed a prototype system of the mechano-electric artificial myocardium by using a parallel-linked covalent shape memory alloy fibres, which was shown in Fig. 1, and its basic hemodynamic performance was examined in goat experiments.

2. Materials and Methods

2.1. *Basic Characteristics of the Fibre*

In general, Ti-Ni alloy is well-known as a material with the shape-memory effect [17-19]. The fibre material (Biometal, Toki Corporation), which was used in this study for the development of artificial myocardium, has the configuration of a covalent bond, and demonstrates a marked strain change as shown in Fig. 2, which is similar to the changes in the surface strain in natural ventricle [20]. The linearity of the recovery strain and the changes in electric resistance could be adjusted through the fabrication process, so that the strain of the fibre could be easily controlled by using the digital-servo system without potentiometers. Special features of the shape memory alloy fibre material (Biometal®) which was to be employed as the actuator of the artificial myocardium were as follows: a) composition of covalent and metallic bonding structure, b) 4–7 % shortening by Joule heating induced by the electric current input, c) linear characteristics of electric resistance against shortening, d) strong maximum contractile force of 10 N with 100-um-fibre, e) high durability of over one billion cycles, f) contractile frequent response by 1–3 Hz, g) selectable maltensitic temperature by fabrication processing from 45 to 70 Celsius, and h) elective diameter size smaller than 30 um. The contraction of the device can be controlled by an originally-designed microcomputer system.

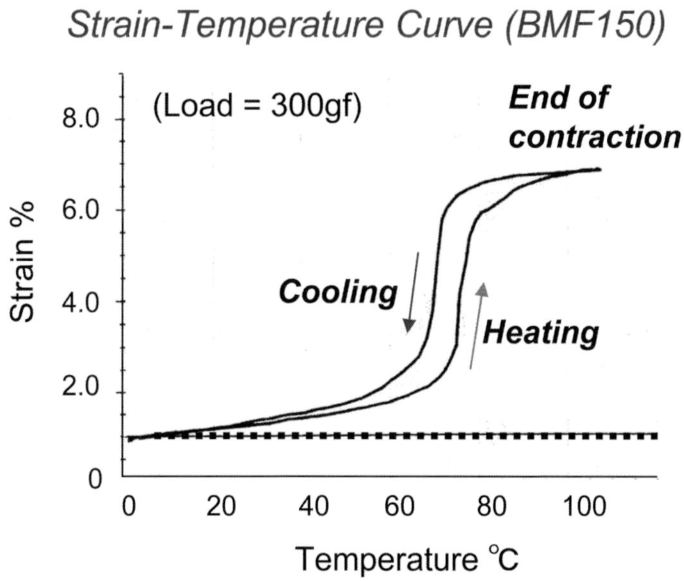

Figure 2. Typical relationship between strain and temperature obtained from the Biometal fibre (diameter: 150 μm). Because of the linearity between the strain and the electric resistance, the displacement can be controlled by the simple circuit and also the sense of force can be estimated.

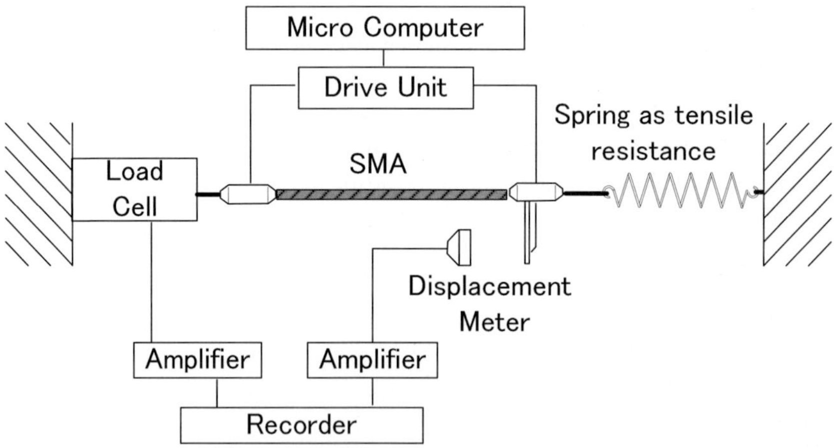

Figure 3. Schematic drawing of the measurement system for the stress-strain characteristics of the shape memory alloy fibre. The spring constant as the tensile resistance was selected from 1.7 to 4.1 N/mm.

The basic stress–strain characteristics were examined in a test circuit as shown in Fig. 3. Tensile force which was generated by the fibre and its displacement were measured simultaneously by a force transducer (Kyowa, LUR–A–50SA1) and a laser position sensor (Keyence, LB–01), respectively.

2.2. *Conceptual Structure of the Artificial Myocardium*

The newly-developed electro-mechanical artificial myocardium consists of the following components: a) an actuator which was made of parallel-linked shape memory alloy fibres, b) an originally-designed signal controller. The weight of each fibre was 14 mg, and the length was set to be 280 mm. The total weight of the whole actuator was around 15 g. The myocardial actuator shown in Fig. 4 was attached onto the heart, and it could support the natural contractile function from the outside of the ventricular wall. Its mechanical assistive motion was synchronized with the electrocardiogram so as not to obstruct the natural cardiac diastolic functions.

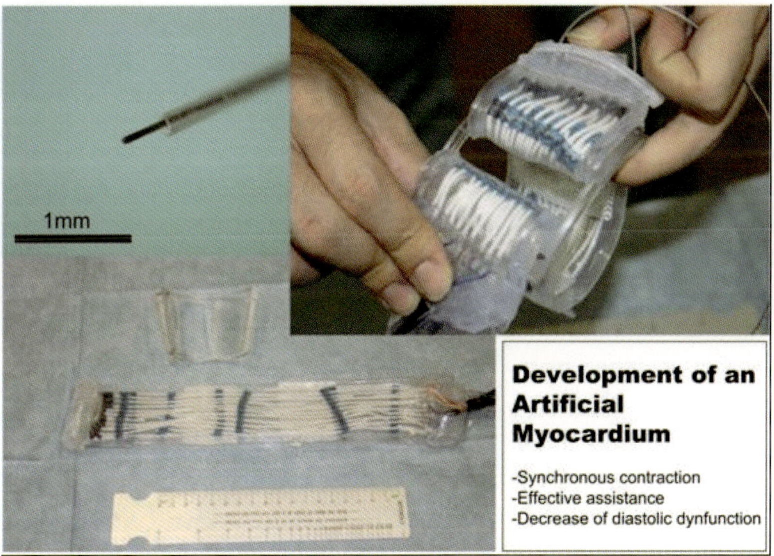

Figure 4. A shape memory alloy fibre covered by a silicone tubing (upper left), and a prototype of artificial myocardium developed in this study (bottom left); the artificial myocardial fibres were covered with waterproof polymers.

2.3. *Animal Experiments*

The animal experiments were performed in healthy female adult Japanese Saanen goats. Implantation of the device was performed as part of an open-chest cardiac procedure on a beating heart under the normal inhalation and anesthesia followed by endotoracheal intubation using 2.5 % halothane. The band-shaped myocardial assist device was installed into the thoracic cavity girdling from the apex to the base and one of the ends was parallel to the left ventricular long axis myocardial streamline on its free wall side [21]. Coronary vasculature was visually identified and avoided on the exterior of the heart during implantation. There was no direct suture on the tissue or muscles with the device. Each signal for the device contraction was regulated by the controller and synchronized with native electrocardiogram as shown in Fig. 5 and 6. The hemodynamic waveforms were obtained from healthy goats, the mean weight of which was 53 kg. Pulmonary and aortic blood pressures were measured by transducers and amplified with a polygraph (Fukuda Denshi, MCS5000). The aortic flow rate was also measured at the aortic root by an ultrasonic flowmeter (Transonic Systems, TS420). Each data was digitally recorded with a data recorder (TEAC, LX–10) by the sampling frequency over 1 kHz.

All animals were cared for in accordance with the rules and regulations set by "the Guideline for the Care and Use of Laboratory Animals" published by the National Institute of Health (NIH publication 85–23, revised 1985) as well as with "the Guidelines for Proper Conduct of Animal Experiments" formulated by Science Council of Japan (2006). The experimental protocols were approved by the Institutional Animal Care and Use Committee of Tohoku University.

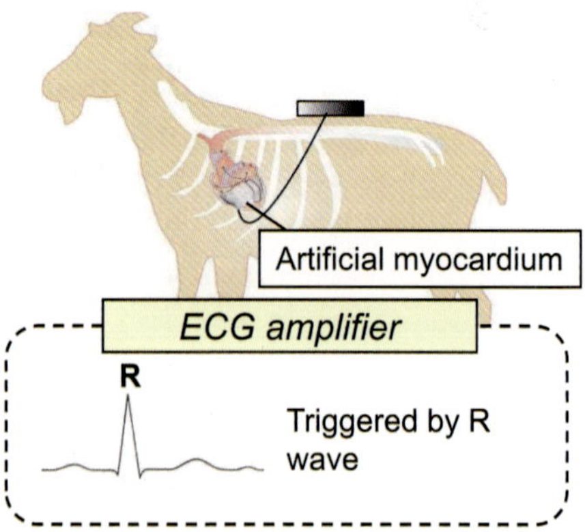

Figure 5. Schematic illustration of the signal connection for the control of mechanical myocardial contraction in the animal experiments.

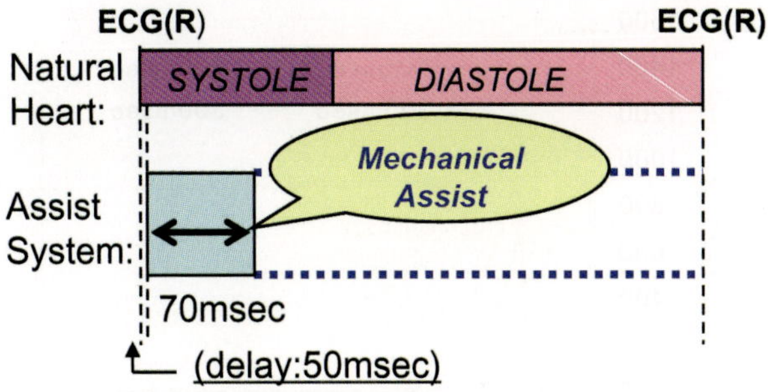

Figure 6. Schematic drawing of the contraction signal input for the artificial myocardium synchronized with electrocardiogram (R wave of ECG).

3. Results and Discussion

Figure 7 shows the transient response of the Biometal fibre under the different input conditions. The duty ratio was changed from 50 to 300 msec, and there was no discernible variation of the speed of the stress gain. As the actuation was conducted only by the cooling-and-heating process in the fibre, neither a sound nor an electric noise could be easily generated in each part.

The myocardial device developed was successfully installed into the goats' thoracic cavity. Prior to the installation of the device, it was covered with silicone rubber and PVC polymer. For the installation of the former electrohydraulic myocardial assist device, which was developed by the authors [14], it was necessary to remove at least the fifth costa to make enough room to fit it into the thoracic cavity. But in this study, by using shape memory fibres, the actuator itself was so small that it would require even a smaller volume in the thoracic cavity. Moreover, the procedure of the closed chest was found to be much simpler. However, any other complications which might have been associated with the operation were not confirmed in goats yet.

Hemodynamic waveforms were changed by the mechanical assistance as shown in Fig. 8. The aortic flow rate was increased by 23 % and the systolic left ventricular pressure was elevated by 6 % under the low cardiac output condition at 2.5 L/min by the mechanical assistance. Therefore, it was indicative of the point that the effective assistance might have been achieved by using those shape-memory alloy fibres.

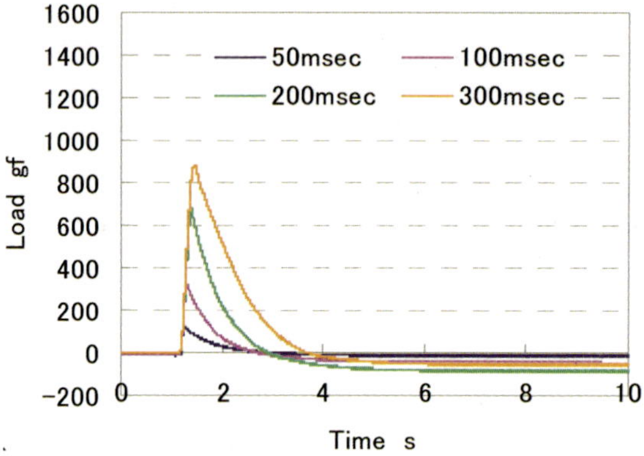

Figure 7. Basic characteristics of the transient response obtained from a single fibred module under the different pulse wave modulation input conditions; the duty of the input was set to be 50, 100, 200, 300 msec respectively at the room temperature (25°C).

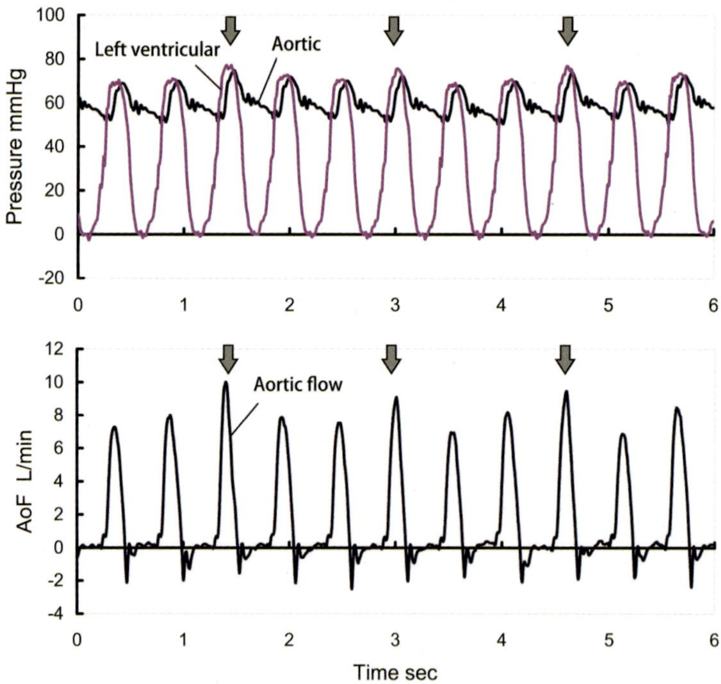

Figure 8. Changes in hemodynamic waveforms obtained in a goat; the arrows indicated the mechanical contractile assistance by the artificial myocardial developed. The assistance was carried out one third of the natural heart beat.

There were no significant changes in the left ventricular end diastolic pressure. Then there was no severe diastolic dysfunction caused by the mechanical assistance using the artificial myocardium.

4. Conclusion

The newly-designed mechano-electric artificial myocardium was developed by using a shape memory alloy fibre, which was capable to be totally installed into the thoracic cavity. It was easy to attach the device onto the ventricular wall. And also the preliminary examination of the performance of the device was conducted in goat experiments. The amelioration of the cardiac functions following the changes in the vascular hemodynamics were investigated by the mechanical assist. As our system could be a symbiotic autonomous system which is capable of assisting natural ventricular functions with physiological deficit, it might be useful for the application in patients with chronic heart failure, such as angina of effort, and also as the cardiac massage in life-saving emergency for the recovery from ventricular fibrillation, as an alternative circulatory support.

Acknowledgments

This work is supported by the Grant in Aid for Scientific Research of Ministry of Education, Culture, Sports, Science and Technology (17790938, 19689029, 20659213), and partially supported by Tohoku University Global COE Program Nano-Biomedical Engineering Education and Research Network Centre. We thank Prof H. Sasada, Mr. K. Kikuchi and Mr. T. Kumagai for kind support of animal cares as well as experimental preparations, and Mr. R. Sakata, Mr. Y. Sato, Mr. H. Lin, and Prof M. Umezu for kind corporation for device preparations.

References

1. J. D. Hosenpud *et al.*, *J Heart Lung Transplant.* **17**, 656–668 (1998).
2. E. A. Rose *et al.*, *New Eng J Med.* **345**, 1435 (2001).
3. J. W. Long *et al.*, *Congest Heart Fail.* **11**, 133–138 (2001).
4. S. G. Drakos *et al.*, *J Heart Lung Transplant.* **25**, 22–28 (2006).
5. G. L. Anstadt *et al.*, *Circulation.* **31**(Suppl. II), 43 (1965).
6. M. Anstadt *et al.*, *Resuscitation.* **21**, 7–23 (1991).
7. O. Kawaguchi *et al.*, *J Thorac Cardiovasc Surg.* **113**, 923–931 (1997).
8. K. Fukamachi and P. M. McCarthy, *J Card Surg.* **20**, S43–S47 (2005).
9. P. M. McCarthy *et al.*, *J Thorac Cardiovasc Surg.* **122**, 482–490 (2001).
10. T. Shimizu *et al.*, *Circ Res.* **90**, e40 (2002).

11. S. Nitta *et al.*, *MBE.* **83-49**, 45–51 (1983) (in Japanese).
12. T. Yambe *et al.*, *Biomed Pharmacother* **58** (Suppl. 1), S145–S149 (2004).
13. T. Yambe *et al.*, *Biomed Pharmacother* **57** (Suppl. 1), 122s–125s (2004).
14. Q. Wang *et al.*, *Artif Organs.* **28**, 853–857 (2004).
15. Y. Shiraishi *et al.*, *Conf Proc IEEE Eng Med Biol Soc.* **1**, 406–408 (2005).
16. Y. Shiraishi *et al.*, *Conf Proc IEEE Eng Med Biol Soc. 2007*, 3974–3977 (2007).
17. W. J. Buehler, J. Gilfrich and K. C. Wiley, *J Appl Phys.* **34**, 1465 (1963).
18. D. Homma, N. Iguchi *et al.*, *Proc 25th Japan Congress on Materials Research.* (1982).
19. P. N. Sawyer *et al.*, *Cardiovasc Diseases Bull Texas Heart Inst.* **3**, 65 (1976).
20. M. Uematsu *et al.*, *Conf Proc IEEE Eng Med Biol Soc.* **7**, 7640–7643 (2005).
21. G. Buckberg *et al.*, *J Thorac Cardiovasc Surg.* **136**, 578–589 (2008).

ANALYSIS OF BRCA1 ACCUMULATION AT DNA DOUBLE-STRAND BREAKS USING A MOLECULAR IMAGING TECHNIQUE

LEIZHEN WEI *, NATSUKO CHIBA †

Department of Molecular Immunology, Institute of Development, Aging and Cancer, Tohoku University, 4-1 Seiryo-machi Aoba-ku, Sendai, Miyagi, 980-8575, Japan

Breast cancer susceptibility gene 1 (BRCA1) is the first susceptibility gene to be linked to breast and ovarian cancer. Although mounting evidence has indicated that BRCA1 participates in the repair of DNA double-strand breaks (DSBs), its precise mechanism is still unclear. Here, we analyzed the *in situ* accumulation of BRCA1 at DSBs produced by laser micro-irradiation. The amino (N)- and the carboxy (C)-terminal fragments of BRCA1 accumulated independently at DSBs with distinct kinetics. The N-terminal BRCA1 fragment appeared immediately after laser-irradiation at DSBs. In contrast, the C-terminal fragment accumulated more slowly at DSBs. Interestingly, rapid accumulation of the BRCA1 N-terminus, but not the C-terminus, at DSBs depended on the Ku80 protein. Missense mutations found within the N-terminus of BRCA1 in cancers significantly reduced its accumulation at DSBs. The results suggest that the N-terminus of BRCA1 plays important roles in the early response to DSBs in a Ku80-dependent manner.

1. Introduction

Germline mutations in the breast cancer susceptibility gene BRCA1 predispose women to breast and ovarian cancers [1,2]. The human BRCA1 protein is composed of 1863 amino acid (a.a.) residues and contains a RING domain in its N-terminus. Two tandem BRCT domains, which are frequently found in DNA repair proteins and function as a binding module for phospho-serine peptides, are present in its C-terminus. The N-terminal region of BRCA1 directly interacts with BARD1, and its association with BARD1 enhances the ubiquitin polymerase activity of BRCA1 [3].

BRCA1 is involved in many cellular processes, including DNA repair, transcription, cell cycle regulation, chromatin remodeling, and apoptosis. The tumor suppressor activity of BRCA1 is primarily attributed to its involvement in DNA repair. For example, mouse and human cells deficient in BRCA1 are more sensitive to DNA damage, including ionizing irradiation and drugs that produce

---

* Leizhen Wei is a Tohoku University Global COE Fellow.

† Natsuko Chiba is a Tohoku University Global COE Member.

DSBs or inter-strand DNA cross-links. Clinically observed missense mutations often result in a non-functional BRCA1 protein that has lost the ability to repair DNA damage [4]. BRCA1 interacts with a number of DNA repair factors such as Rad51, the Mre11-Rad50-Nbs1 (MRN) complex, BLM, and the DNA helicase BACH1 (also called BRIP1or FANCJ) [5]. BRCA1 localizes to nuclear foci during S-phase of the cell cycle. Various types of DNA damage result in BRCA1 hyperphosphorylation and alterations in its localization to nuclear foci [6]. Normally, BRCA1 redistributes to nuclear foci where multiple DNA repair factors accumulate. BRCA1 co-localizes with phosphorylated H2AX (γH2AX), a protein that is rapidly phosphorylated at DSB sites after DNA damage. Furthermore, BRCA1 co-localizes with Rad51, which mediates homologous recombination (HR), and/or with an MRN complex, which is involved in both HR and non-homologous end-joining (NHEJ) [7]. These observations suggest that BRCA1 accumulates at sites of DNA damage, forms potentially distinct types of protein complexes, and functions in multiple aspects of DNA repair.

DNA DSBs can be produced directly by ionizing radiation and by some chemicals, or indirectly by the blockage of replication forks. Repairing DSBs correctly is critical for maintaining genome integrity. NHEJ and HR are the major cellular mechanisms that repair DSBs. In the NHEJ pathway, the DNA-dependent protein kinase catalytic subunit (DNA-PKcs) and a Ku heterodimer of Ku80 and Ku70 are recruited to the sites of DNA DSBs, followed by subsequent recruitment of XRCC4-ligase IV. In contrast, Rad51 and replication protein A (RPA) are essential factors in the HR pathway. MRN complexes are involved at early stages of both the HR and NHEJ pathways. Although BRCA1 is thought to be primarily involved in the HR pathway, it has been implicated in the NHEJ pathway as well. However, the precise mechanism by which BRCA1 accumulates at DSB sites and what role it plays in DSB repair pathways is not yet fully understood. Therefore, we collaborated with Dr. Akira Yasui of the Department of Molecular Genetics, Institute of Development, Aging, and Cancer, Tohoku University to address this issue. He has established a laser light micro-irradiation system to create various types of DNA damage in living cells, including DSBs, single-strand breaks, and base damage, which has enabled us to detect BRCA1 accumulation at the site of DSBs [8,9]. In this study, we examined the real-time accumulation of endogenous BRCA1 and green-fluorescent-protein (GFP)-tagged BRCA1 at DSBs induced by laser-irradiation. The behavior of various deletion and missense mutants of BRCA1 was also examined. We found that the immediate recruitment of BRCA1 at DSBs after laser-irradiation is mediated by an interaction of the N-terminal portion of BRCA1 with Ku80.

2. Results

2.1. *Endogenous BRCA1 Accumulates at Laser Micro-irradiated Sites*

After DSBs were introduced into the nuclei of human Saos-2 cells by laser micro-irradiation, the cells were fixed and processed for immunofluorescence. Endogenous BRCA1 clearly accumulated at the irradiated sites (Fig. 1). Accumulation of phosphorylated H2AX (γH2AX), a protein that is rapidly phosphorylated at DSB sites after DNA damage, was also observed. The time course of accumulation of fluorescence intensity for each protein showed that both the accumulation and clearance of BRCA1 were slower than that of γH2AX.

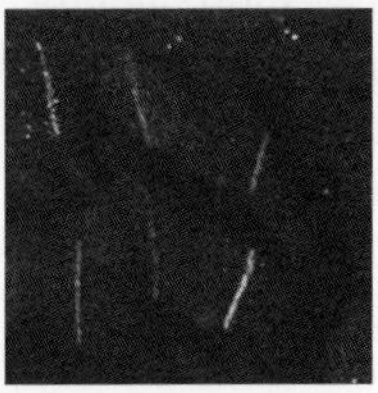

Figure 1. Accumulation of endogenous BRCA1 at laser micro-irradiated sites. Immunochemical detection of BRCA1 in Saos-2 cells after 500 scans of 405 nm laser irradiation. Cells were fixed after laser irradiation and stained with antibodies against BRCA1.

2.2. *BRCA1 Accumulation at Laser-irradiated Sites is a Response to DNA DSBs*

We next examined the real time localization of GFP-tagged BRCA1 in living cells after DSB induction. GFP-BRCA1 was transfected into Saos-2 cells and the cells were then laser-irradiated. As shown in Fig. 2, GFP-BRCA1 clearly accumulated at the irradiated sites. The mean intensity of GFP-BRCA1 at such sites was quantified, and the accumulation of BRCA1 at the irradiated sites was found to be gradual. These results indicate that the accumulation kinetics of overexpressed GFP-BRCA1 mimics that of endogenous BRCA1.

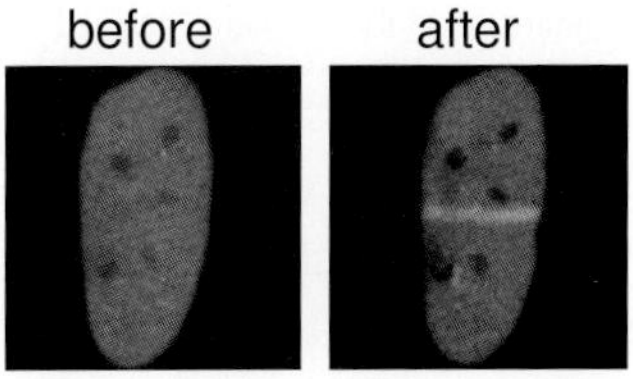

Figure 2. GFP-tagged BRCA1 accumulates at DSB sites in Saos-2 cells.

2.3. *BRCA1 Localizes at DSBs Via Its N- and C-terminal Regions*

We next identified the regions in BRCA1 that mediate this accumulation. Because the RING domain is located within a.a. 1-302 and the BRCT domains are located within a.a. 1528-1863, we hypothesized that deletion of either of these two regions would suppress BRCA1 accumulation at DSBs. The full-length BRCA1, as well as the deletion mutants Δ305-770, Δ775-1292, Δ1-302, and Δ1527-1863 were constructed and transfected into Saos-2 cells. However, each of the four deletion mutant BRCA1 proteins accumulated at DSBs. Therefore, a 303-1526 a.a. BRCA1 fragment was constructed, which contained only the central region of BRCA1 and lacked both the N- and C-terminal regions. This BRCA1 fragment did not accumulate at irradiated sites. These data suggest that either the N- or C-terminal region alone is sufficient for BRCA1 accumulation at DSBs. Subsequently, a.a. 1-304 and a.a. 1527-1863 fragments were constructed and each was shown to accumulate at laser-induced linear DSBs.

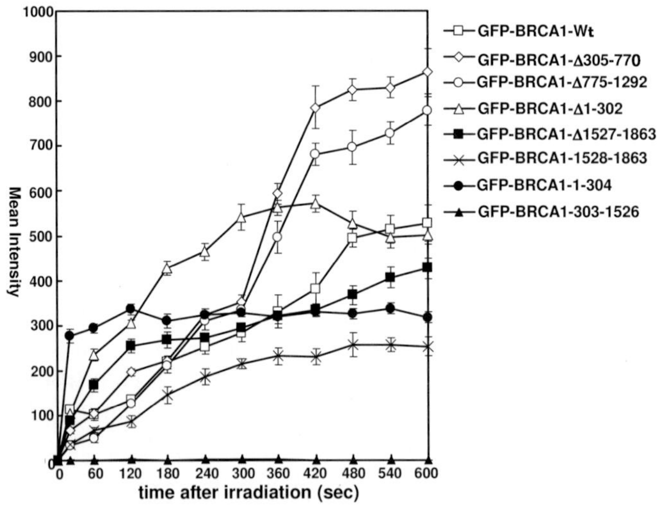

Figure 3. Kinetics of accumulation of full-length and various fragments of BRCA1 after laser irradiation in Saos-2 cells. Standard deviations were derived from at least three independent experiments.

The mean intensity of GFP-BRCA1 at accumulation sites was quantified and the kinetics of accumulation of the full-length protein and the BRCA1 fragments were examined for 600 s after irradiation (Fig. 3). Interestingly, the N-terminal (a.a. 1-304) fragment rapidly and maximally accumulated at DSBs within 20 s, whereas the C-terminal (a.a.1527-1863) fragment slowly and gradually accumulated, reaching a plateau at 360 s. Full-length BRCA1 exhibited a pattern of accumulation that reflected the combination of the N- and C-terminal fragment patterns. Taken together, the above kinetic analyses suggest that the N- and C-terminal regions of BRCA1 may accumulate at DSBs through different mechanisms.

2.4. *Accumulation of BRCA1 at DSBs Requires Another DNA Repair Factor*

Next, we used factor-deficient cells to examine whether BRCA1 accumulation at DSBs is dependent on specific DNA repair factors. The N- and C-terminal fragments accumulated at laser-induced DSBs in ATM-deficient AT1KY/T-n cells, H2AX-/- MEF cells, CHO-derived CHO9 cells, and XR-C1 (DNA-PKcs -/-) cells. However, in XRV15B (Ku80 -/-) cells, the N-terminal fragment of BRCA1 failed to accumulate at DSB sites, whereas the C-terminal fragment did accumulate. Therefore, accumulation of the N-terminal fragment, but not the C-terminal fragment, of BRCA1 at DSBs is Ku80 dependent.

2.5. *Missense Mutations of BRCA1 Reduce Its Accumulation*

We examined whether missense mutations affected the accumulation of BRCA1 at DSBs. Four missense mutations were selected, on the basis of their relatively frequent occurrence in the Breast Cancer Information Core database (www.nhgri.nih.gov). However, the pathological significance of these mutations has not been verified.

GFP-BRCA1 constructs, each harboring one of the missense mutations, were generated. Each of the four mutants accumulated at the irradiated sites, but the fluorescence intensity was markedly diminished relative to that of the wild-type BRCA1 fragments. All mutant constructs were expressed at levels similar to that of the wild-type BRCA1 fragments, as confirmed by Western-blot analysis. These accumulation kinetics suggest that these four missense mutations are likely to have a pathological significance.

3. Discussion

In this study, we laser micro-irradiated living cells and demonstrated that endogenous BRCA1 as well as GFP-BRCA1 accumulated at DSBs. Using various deletion mutants and fragments of BRCA1, we found that both the N- (a.a. 1-304) and C- (a.a. 1528-1863) terminal regions of BRCA1 accumulate at DSBs, and detailed kinetic analyses revealed distinct behaviors of the N- and C-termini of BRCA1. The N-terminal region of BRCA1 immediately accumulated at DSBs, whereas the C-terminal region slowly and gradually accumulated (Fig. 3). We propose that the N-terminal region may function to guide BRCA1 to DSBs in an early phase of the repair process, and that the C-terminal region may be responsible for accumulation in a later phase of repair.

The N-terminus of BRCA1 accumulated at DSBs in a Ku80-dependent manner. Accordingly to Kim *et al.*, other NHEJ factors, Ku70 and DNA-PKcs, only transiently localize to laser-irradiated sites [10]. In our analysis, the N-terminus of BRCA1 accumulated only briefly, suggesting that BRCA1 functions with NHEJ factors in the early phase of the repair process, and that it dissociates from DSB sites together with NHEJ factors.

The missense mutations that we analyzed significantly reduced BRCA1 accumulation at DSBs. We found for the first time that these four missense mutations could be pathological mutations for this reason. These data suggest that the rapid accumulation of BRCA1 at DSBs via its N-terminus is important for its tumor suppressor activity.

4. Conclusion

The N and C-terminal fragments of BRCA1 accumulated independently at DSBs with distinct mechanisms. The N-terminus of BRCA1 plays important roles in the early response to DSBs in a Ku80-dependent manner. The missense mutations that we analyzed could be pathological mutations.

Acknowledgments

This study was supported by Grants-in-Aid from the Ministry of Education, Science, Sports and Culture of Japan. We acknowledges the support of the Tohoku University Global COE Program "Global Nano-Biomedical Engineering Education and Research Network Centre".

References

1. M. C. King, J. H. Marks and J. B. Mandell, Breast and ovarian cancer risks due to inherited mutations in BRCA1 and BRCA2. *Science* **302**, 643–646 (2003).
2. Y. Miki, J. Swensen, D. Shattuck-Eidens, P. A. Futreal, K. Harshman, S. Tavtigian, Q. Liu, C. Cochran *et al.*, A strong candidate for the breast and ovarian cancer susceptibility gene BRCA1. *Science* **266**, 66–71(1994).
3. A. Chen, F. E. Kleiman, J. L. Manley, T. Ouchi and Z. Q. Pan, Autoubiquitination of the BRCA1*BARD1 RING ubiquitin ligase. *J Biol Chem* **277**, 22085–22092 (2002).

4. R. Scully, S. Ganesan, K. Vlasakova, J. Chen, M. Socolovsky and D. M. Livingston, Genetic analysis of BRCA1 function in a defined tumor cell line. *Mol Cell* **4**, 1093–1099 (1999).
5. Y. Wang, D. Cortez, P. Yazdi, N. Neff, S. J. Elledge and J. Qin, BASC, a super complex of BRCA1-associated proteins involved in the recognition and repair of aberrant DNA structures. *Genes Dev* **14**, 927–939 (2000).
6. R. Scully, J. Chen, R. L. Ochs, K. Keegan, M. Hoekstra, J. Feunteun and D. M. Livingston, Dynamic changes of BRCA1 subnuclear location and phosphorylation state are initiated by DNA damage. *Cell* **90**, 425–435 (1997).
7. Q. Zhong, C. F. Chen, S. Li, Y. Chen, C. C. Wang, J. Xiao, P. L. Chen, Z. D. Sharp and W. H. Lee, Association of BRCA1 with the hRad50-hMre11-p95 complex and the DNA damage response [see comments]. *Science* **285**, 747–750 (1999).
8. L. Lan, S. Nakajima, K. Komatsu, A. Nussenzweig, A. Shimamoto, J. Oshima and A. Yasui, Accumulation of Werner protein at DNA double-strand breaks in human cells. *J Cell Sci* **118**, 4153–4162 (2005).
9. L. Lan, S. Nakajima, Y. Oohata, M. Takao, S. Okano, M. Masutani, S. H. Wilson and A. Yasui, *In situ* analysis of repair processes for oxidative DNA damage in mammalian cells. *Proc Natl Acad Sci U S A* **101**,13738–13743 (2004).
10. J. S. Kim, T. B. Krasieva, H. Kurumizaka, D. J. Chen, A. M. Taylor and K. Yokomori, Independent and sequential recruitment of NHEJ and HR factors to DNA damage sites in mammalian cells. *J Cell Biol* **170**, 341–347 (2005).

AUTHOR INDEX